Android 开发与实战

赵书兰 编著

电子工业出版社
Publishing House of Electronics Industry
北京·BEIJING

内 容 简 介

计算机便携化是未来的发展趋势，而 Android 作为最受欢迎的手机和平板电脑的操作系统之一，从其诞生到现在的短短几年里，凭借其开源开放性、优异的用户体验和极为方便的开发方式，赢得了广大用户和开发者的青睐。

本书循序渐进地介绍 Android 应用开发的相关知识，内容覆盖了 Android 用户界面编程。首先介绍 Android 界面布局、控件、菜单、对话框等内容；然后进一步介绍 Android 视图与动画、Android 数据存储、Android 传递消息与联网、Android 多媒体等内容；最后综合介绍 Android 辅助工具。

本书可作为程序开发人员进行 Android 开发的参考书，也可供高校师生学习参考。

未经许可，不得以任何方式复制或抄袭本书之部分或全部内容。
版权所有，侵权必究。

图书在版编目（CIP）数据

Android 开发与实战 / 赵书兰编著. —北京：电子工业出版社，2013.6
ISBN 978-7-121-20612-2

Ⅰ. ①A… Ⅱ. ①赵… Ⅲ. ①移动终端－应用程序－程序设计 Ⅳ. ①TN929.53

中国版本图书馆 CIP 数据核字（2013）第 120156 号

策划编辑：陈韦凯
责任编辑：毕军志
印　　刷：北京七彩京通数码快印有限公司
装　　订：北京七彩京通数码快印有限公司
出版发行：电子工业出版社
　　　　　北京市海淀区万寿路 173 信箱　邮编 100036
开　　本：787×1 092　1/16　印张：34.75　字数：934.6 千字
版　　次：2013 年 6 月第 1 次印刷
印　　次：2023 年 9 月第 2 次印刷
定　　价：69.00 元

凡所购买电子工业出版社图书有缺损问题，请向购买书店调换。若书店售缺，请与本社发行部联系，联系及邮购电话：(010) 88254888。

质量投诉请发邮件至 zlts@phei.com.cn，盗版侵权举报请发邮件至 dbqq@phei.com.cn。
服务热线：(010) 88258888。

前　言

计算机便携化是未来的发展趋势。在人们的习惯里，很容易把计算机理解为主机、显示器、键盘的"组合"，即使后来出现了笔记本电脑，其实依然摆脱不了主机、显示器与键盘的组合。对于这种传统的计算机，用户必须"安静"地坐下来，打开它，然后才能使用计算机。但用户并不能完全满足通过这种方式使用计算机，有时用户需要在车上查看、管理公司的运营状况，有时用户需要在等飞机时查看、管理自己的证券交易情况，有时用户需要随时玩玩游戏松弛一下神经⋯⋯在这些需求场景下，用户需要更加便携化的计算机，这也是平板电脑大行其道的重要原因。

3G 通信是移动通信市场经历了第一代模拟技术的移动通信业务，在第二代数字移动通信市场的蓬勃发展中被引入的。在当今 Internet 数据业务的不断升温中，在固定接入速率（HDSL、ADSL、VDSL）不断提升的背景下，3G 移动通信系统也看到了市场的曙光，益发为电信运营商、通信设备制造商和普通用户所关注。而 3G 时代的到来，使得更多内容丰富的应用程序被布置在手机上成为可能，如视频通话、视频点播、移动互联网冲浪、在线看书/听歌、内容分享等。为了实现这些需求，需要有一个好的移动开发平台来支持。

2007 年 11 月推出的 Android 平台，是任何公司及个人都可免费获取到源代码及 SDK 的手机开发平台。由于其开放性和优异性能，Android 平台得到了包括各大手机厂商和著名的移动运营商在内的业界的广泛支持。

随着 Android 平台在市场占有率上的稳步上升，采用 Java 语言开发的 Android 应用会越来越多。不过需要指出的是，运行 Android 平台的硬件只是手机、平板电脑等便携式设备，这些设备的计算能力、数据存储能力都是有限的，不太可能在 Android 平台上部署大型企业级应用，因此 Android 应用可能以纯粹客户端应用的角色出现，然后通过网络与传统大型应用交互，充当大型企业应用的客户端。

目前最新的 Android 版本为 4.1，其具有一些新特点：更快、更流畅、更灵敏；特效动画的帧速率提高至 60fps，增加了三倍缓冲；增强了通知栏；全新搜索将会带来全新的 UI、智能语音搜索和 Google Now 三项新功能；桌面插件自动调整大小；加强无障碍操作；语言和输入法扩展；新的输入类型和功能；新的连接类型。

鉴于 Android 作为新的平台、新技术，为了帮助众多开发人员和爱好者进入 Android 开发领域并提高程序开发水平，特编写了本书。

本书共分 10 章。

第 1 章：给出了 Android 概述，主要包括智能手机操作系统现状、Android 发展史、特性、搭建 Android 开发环境及 Android 应用工程文件组成等内容。

第 2 章：介绍了 Android 界面布局，主要包括线性布局、相对布局、表格布局、绝对布局等内容。

第 3 章：介绍了 Widget 组件布局，主要包括 Button 控件、TextView 控件、EditText 控件、RadioButton 控件等内容。

第 4 章：介绍了 Android 菜单，主要包括 Menu 菜单、MenuItem 菜单、SubMenu 菜单及 ContextMenu 菜单等内容。

第 5 章：介绍了 Android 对话框，主要包括 AlertDialog 对话框、DatePickerDialog 与 TimePickerDialog 对话框、ProgressDailog 对话框及 Notification 通知等内容。

第 6 章：介绍了 Android 视图与动画，主要包括 Android 图像、Android 绘图、Android 图形特效处理与 Android 动画等内容。

第 7 章：介绍了 Android 数据存储，主要包括 SharedPreferences 存储、文件存储数据、SQLite 数据库存储、ContentProvider 存储等内容。

第 8 章：介绍了 Android 传递消息与联网，主要包括电话管理器、信息处理、联网等内容。

第 9 章：介绍了 Android 多媒体，主要包括音频/视频的播放、录制音频及照相机等内容。

第 10 章：介绍了 Android 辅助工具，主要包括 Map 地图、蓝牙等内容。

本书主要由赵书兰编写，此外参加编写的还有周品、赵书梅、赵新芬、栾颖、刘志为、丁伟雄、雷晓平、李娅、杨文茵、何正风、周灵、余智豪、崔如春和张德丰。

由于作者的水平有限，加之时间较紧，书中难免会存在不足之处，敬请广大读者批评指正。

编著者

2013 年 3 月

目 录

第1章 Android 概述 (1)
- 1.1 智能手机操作系统现状 (1)
  - 1.1.1 智能手机的定义 (1)
  - 1.1.2 智能手机的特点 (2)
  - 1.1.3 3G 智能手机的基本要求 (2)
  - 1.1.4 智能手机的操作系统 (3)
- 1.2 Android 简介 (5)
  - 1.2.1 Android 发展史 (5)
  - 1.2.2 Android 特性 (8)
  - 1.2.3 Android 系统架构 (10)
  - 1.2.4 Android 组件 (12)
- 1.3 搭建 Android 开发环境 (14)
  - 1.3.1 Android 开发前的准备工作 (14)
  - 1.3.2 安装 JDK (15)
  - 1.3.3 安装 Eclipse (17)
  - 1.3.4 安装 Android SDK (18)
  - 1.3.5 安装 ADT 插件 (20)
  - 1.3.6 设定 Android SDK 主目录 (23)
  - 1.3.7 创建 Android 虚拟设备 (24)
  - 1.3.8 运行 AVD (26)
- 1.4 Android 应用工程文件组成 (27)
- 1.5 应用程序的生命周期 (30)
  - 1.5.1 进行优先级 (31)
  - 1.5.2 Activity 的生命周期 (32)
  - 1.5.3 Activity 生命周期调用顺序 (35)
  - 1.5.4 Service 的生命周期 (36)
  - 1.5.5 Android 生命周期综合实例 (37)
- 1.6 Android 的活动 (39)
  - 1.6.1 应用活动的样本与主题 (41)
  - 1.6.2 显示对话框 (42)
  - 1.6.3 显示进度条对话框 (45)

第2章 Android 界面布局 (49)
- 2.1 控件类概述 (49)
  - 2.1.1 View 简介 (49)
  - 2.1.2 ViewGroup 简介 (50)
- 2.2 线性布局 (50)
  - 2.2.1 LinearLayout 简介 (50)
  - 2.2.2 线性布局实例介绍 (51)
- 2.3 表格布局 (52)
  - 2.3.1 TableLayout 简介 (52)
  - 2.3.2 表格布局实例介绍 (53)
- 2.4 相对布局 (56)
  - 2.4.1 RelativeLayout 简介 (57)
  - 2.4.2 相对布局实例介绍 (58)
- 2.5 绝对布局 (59)
  - 2.5.1 AbsoluteLayout 简介 (59)
  - 2.5.2 绝对布局实例介绍 (60)
- 2.6 帧布局 (61)
  - 2.6.1 FrameLayout 简介 (61)
  - 2.6.2 帧布局实例介绍 (62)
- 2.7 列表布局 (63)
  - 2.7.1 ViewGroup 简介 (63)
  - 2.7.2 列表布局实例介绍 (63)

第3章 Widget 组件布局 (66)
- 3.1 Button 控件 (67)
  - 3.1.1 setOnClickListener 方法 (67)
  - 3.1.2 setOnLongClickListener 方法 (68)
  - 3.1.3 setOnFocusChangeListener 方法 (69)
  - 3.1.4 setOnTouchListener 方法 (70)
  - 3.1.5 setPressed 方法 (71)
  - 3.1.6 setClickable 方法 (72)
  - 3.1.7 setLongClickable 方法 (73)
  - 3.1.8 Button 控件综合实例 (74)
- 3.2 TextView 控件 (75)

· V ·

- 3.2.1 append 方法 …………… (75)
- 3.2.2 addTextChangedListener 方法 …………… (76)
- 3.2.3 setText 方法 …………… (78)
- 3.2.4 setTextSize 方法 …………… (79)
- 3.2.5 setTypeface 方法 …………… (80)
- 3.2.6 setTextColor 方法 …………… (82)
- 3.2.7 setHeight 方法 …………… (82)
- 3.2.8 setBackgroundColor 方法 …… (83)
- 3.2.9 getHeight 方法 …………… (84)
- 3.2.10 getWidth 方法 …………… (86)
- 3.2.11 setPadding 方法 …………… (87)
- 3.2.12 getPaddingLeft 方法 …………… (88)
- 3.2.13 getPaddingTop 方法 …………… (88)
- 3.2.14 getPaddingrRight 方法 …… (88)
- 3.2.15 getPaddingBottonm 方法 … (88)
- 3.2.16 getCurrentTextColor 方法 … (89)
- 3.2.17 getText 方法 …………… (91)
- 3.2.18 TextView 控件的综合实例 …………… (91)
- 3.3 EditText 控件 …………… (98)
  - 3.3.1 setText 方法 …………… (98)
  - 3.3.2 getText 方法 …………… (100)
  - 3.3.3 setSelection 方法 …………… (101)
  - 3.3.4 setHint 方法 …………… (102)
  - 3.2.5 setOnKeyListener 方法 …… (103)
  - 3.3.6 EditText 控件的综合实例 … (104)
- 3.4 RadioButton 控件 …………… (106)
  - 3.4.1 setOnCheckedChangedListener 方法 …………… (106)
  - 3.4.2 check 方法 …………… (109)
  - 3.4.3 clearCheck 方法 …………… (109)
- 3.5 CheckBox 控件 …………… (110)
  - 3.5.1 isChecked 方法 …………… (111)
  - 3.5.2 setChecked 方法 …………… (113)
  - 3.5.3 toggle 方法 …………… (113)
  - 3.5.4 setOnCheckedChangeListener 方法 …………… (115)
- 3.6 Toast 控件 …………… (117)
  - 3.6.1 cancel 方法 …………… (117)
- 3.6.2 getDuration 方法 …………… (117)
- 3.6.3 getGravity 方法 …………… (118)
- 3.6.4 getHorizontalMargi 方法与 getVerticalMargin 方法 … (119)
- 3.6.5 makeText 方法 …………… (120)
- 3.6.6 setView 方法 …………… (121)
- 3.6.7 getView 方法 …………… (122)
- 3.6.8 setGravity 方法 …………… (123)
- 3.6.9 getXOffset 方法与 getYOffset 方法 …………… (124)
- 3.6.10 setDuration 方法 …………… (125)
- 3.6.11 setMargin 方法 …………… (126)
- 3.6.12 setText 方法 …………… (127)
- 3.6.13 show 方法 …………… (128)
- 3.6.14 Toast 控件的综合实例 … (129)
- 3.7 ImageView 控件 …………… (134)
  - 3.7.1 setAdjustViewBounds 方法 …………… (134)
  - 3.7.2 setScaleType 方法 …………… (135)
  - 3.7.3 setSelected 方法 …………… (135)
  - 3.7.4 setImageURI 方法 …………… (135)
  - 3.7.5 setAdjustViewBounds 方法 …………… (136)
  - 3.7.6 setAlpha 方法 …………… (136)
  - 3.7.7 setImageResource 方法 …… (136)
  - 3.7.8 ImageView 控件综合实例 …………… (136)
- 3.8 ProgressBar 控件 …………… (139)
  - 3.8.1 ProgressBar 相关方法 …… (139)
  - 3.8.2 ProgressBar 相关类型 …… (139)
- 3.9 Spinner 控件 …………… (147)
  - 3.9.1 setAdapter 方法 …………… (147)
  - 3.9.2 setPrompt 方法 …………… (147)
  - 3.9.3 setPromptId 方法 …………… (149)
  - 3.9.4 setOnltemSelectedListener 方法 …………… (149)
- 3.10 AutoCompleteTextView 控件 … (151)
  - 3.10.1 setAdapter 方法 …………… (151)
  - 3.10.2 setThreshold 方法 …………… (152)
  - 3.10.3 setCompletionHint 方法 … (152)

3.10.4　setDropDownBackgroundResource
　　　　　方法……………………（154）
　　3.10.5　setDropDownBackgroundDrawable
　　　　　方法……………………（154）
　　3.10.6　MultiAutoCompleteTextView
　　　　　类………………………（155）
3.11　AnalogClock 控件………………（157）
3.12　DatePicker 与 TimePicker 控件…（161）
　　3.12.1　DatePicker 控件…………（161）
　　3.12.2　TimerPicker 控件…………（161）
　　3.12.3　DatePicker 与 TimePicker
　　　　　控件综合实例……………（162）
3.13　SeekBar 控件……………………（164）
3.14　RatingBar 控件…………………（166）
　　3.14.1　RatingBar 类方法…………（167）
　　3.14.2　RatingBar 控件综合
　　　　　实例………………………（167）
3.15　Tab 控件…………………………（170）
3.16　Gallery 控件……………………（173）
　　3.16.1　Gallery 类方法……………（173）
　　3.16.2　Gallery 控件综合实例……（178）
3.17　ToggleButton 控件………………（182）
　　3.17.1　ToggleButton 类方法……（182）
　　3.17.2　ToggleButton 类实现……（183）

第 4 章　Android 菜单……………………（185）

4.1　Menu 菜单…………………………（185）
　　4.1.1　Menu 菜单方法………………（185）
　　4.1.2　Menu 菜单综合实例…………（192）
4.2　MenuItem 菜单……………………（196）
　　4.2.1　MenuItem 菜单方法…………（196）
　　4.2.2　MenuItem 菜单综合实例……（200）
4.3　SubMenu 菜单……………………（204）
　　4.3.1　SubMenu 菜单方法…………（204）
　　4.3.2　SubMenu 菜单综合
　　　　　实例………………………（208）
4.4　ContextMenu 菜单…………………（209）
　　4.4.1　ContextMenu 菜单方法……（210）
　　4.4.2　ContextMenu 菜单综合
　　　　　实例………………………（216）

第 5 章　Android 对话框…………………（219）

5.1　AlertDialog 对话框…………………（219）
　　5.1.1　创建提示对话框………………（219）
　　5.1.2　创建列表对话框………………（223）
　　5.1.3　单选列表对话框………………（225）
　　5.1.4　复选列表对话框………………（227）
　　5.1.5　AlertDialog 对话框综合
　　　　　实例………………………（229）
5.2　DatePickerDialog 与 TimePickerDialog
　　对话框………………………………（233）
　　5.2.1　DatePickerDialog 与 TimePickerDialog
　　　　　概述………………………（233）
　　5.2.2　DatePickerDialog 与 TimePickerDialog
　　　　　对话框综合实例…………（234）
5.3　ProgressDailog 对话框……………（237）
　　5.3.1　ProgressDailog 对话框
　　　　　方法………………………（237）
　　5.3.2　ProgressDailog 进度条对话框
　　　　　综合实例…………………（238）
5.4　Notification 通知…………………（242）
　　5.4.1　常用的 Notification…………（242）
　　5.4.2　带进度条的 Notification……（246）

第 6 章　Android 视图与动画……………（250）

6.1　Android 图像………………………（250）
　　6.1.1　ImageSwitcher 类……………（250）
　　6.1.2　ScrollView 类…………………（253）
　　6.1.3　GridView 类…………………（258）
　　6.1.4　WebView 类…………………（263）
6.2　Android 绘图………………………（264）
　　6.2.1　Paint 类………………………（265）
　　6.2.2　Canvas 类……………………（268）
　　6.2.3　Canvas 与 Paint 类综合
　　　　　实例………………………（274）
　　6.2.4　Path 类………………………（279）
6.3　Android 图形特效处理……………（281）
　　6.3.1　Matrix 控制变换………………（282）
　　6.3.2　drawBitmapMesh 扭曲
　　　　　图像………………………（290）
　　6.3.3　渲染效果………………………（294）
6.4　Android 动画………………………（299）

· Ⅶ ·

	6.4.1	Animation 类 ……………	（299）
	6.4.2	Tween 动画 ………………	（300）
	6.4.3	Frame 动画 ………………	（305）
	6.4.4	Frame 动画与 Tween 动画	
		综合实例 ………………	（308）
	6.4.5	SurfaceView 类 …………	（312）
	6.4.6	动画组件	
		（ViewAnimator）………	（316）

第 7 章 Android 数据存储………… （321）

7.1	SharedPreferences 存储 ……………	（321）
	7.1.1 SharedPreferences 存储类	
	效率分析 ………………	（322）
	7.1.2 SharedPreferences 类实例··	（323）
7.2	文件存储数据 ………………………	（325）
	7.2.1 java.io 包的方法 …………	（326）
	7.2.2 openFileInput 和	
	openFileOutput ………	（331）
	7.2.3 从 resource 中的 raw 文件夹中	
	读取文件 ………………	（333）
	7.2.4 从 asset 中读取文件………	（334）
7.3	SQLite 数据库存储 …………………	（335）
	7.3.1 SQLite 数据库存储概述 ···	（335）
	7.3.2 SQLite 数据库存储分析···	（336）
	7.3.3 SQLite 数据库存储应用	
	实例 ………………	（341）
7.4	ContentProvider 存储………………	（347）
	7.4.1 ContentProvider 存储	
	分析 …………………	（347）
	7.4.2 Content Provider 存储	
	创建步骤 ………………	（350）
	7.4.3 Content Provider 应用	
	实例 …………………	（350）
7.5	网络存储 ……………………………	（358）

第 8 章 Android 传递消息与联网 …… （360）

8.1	电话管理器 …………………………	（360）
	8.1.1 网络与 SIM 卡获取信息··	（360）
	8.1.2 拨打电话 …………………	（364）
	8.1.3 监听手机来电 ……………	（368）
8.2	信息处理 ……………………………	（369）
	8.2.1 发送短信 …………………	（369）

	8.2.2 群发短信 …………………	（372）
8.3	发送邮件 ……………………………	（377）
8.4	实现震动 ……………………………	（379）
8.5	闹钟 …………………………………	（385）
	8.5.1 AlarmManager 类概述 ···	（385）
	8.5.2 设定闹钟实例 ……………	（385）
	8.5.3 更换墙纸实例 ……………	（395）
8.6	自动显示电量 ………………………	（398）
8.7	Wi-Fi 使用 …………………………	（402）
8.8	联网 …………………………………	（411）
	8.8.1 下载二进制数据 …………	（413）
	8.8.2 下载文本文件 ……………	（414）
	8.8.3 在线播放音乐 ……………	（416）

第 9 章 Android 多媒体 ……………… （425）

9.1	音频/视频的播放 ……………………	（425）
	9.1.1 MediaPlay 类……………	（425）
	9.1.2 SoundPool 类……………	（437）
	9.1.3 VideoView 类……………	（439）
	9.1.4 Android 的多媒体播放器	
	综合实例 ………………	（447）
9.2	录制音频 ……………………………	（452）
9.3	照相机 ………………………………	（456）
	9.3.1 照相机常用方法 …………	（456）
	9.3.2 照相机实例分析 …………	（464）

第 10 章 Android 辅助工具…………… （474）

10.1	Map 地图 …………………………	（474）
	10.1.1 位置服务实例 …………	（474）
	10.1.2 定位实例 ………………	（481）
	10.1.3 地址查询实例 …………	（485）
	10.1.4 导航实例 ………………	（490）
10.2	蓝牙 ………………………………	（499）
	10.2.1 RFCOMM 协议 ………	（499）
	10.2.2 MAC 硬件地址 ………	（499）
	10.2.3 编程实现蓝牙综合	
	实例 ………………	（501）
10.3	中国象棋 …………………………	（508）

参考文献 …………………………………… （547）

第 1 章 Android 概述

Android 作为一种手机开发平台，其建立在 Java 基础上，能够为手机软件开发提供快捷有效的解决方案。Android 的功能十分强大，已经成为移动平台开发领域的新热点。

1.1 智能手机操作系统现状

现如今的手机可谓五花八门，各式各样。其中深受商务人士喜爱的无疑是智能手机，个中原因除了外观气派、各种商务娱乐及网络安全功能强大是智能手机广受欢迎的更深层次原因。而随着智能手机人性化操作能力的加强和功能的不断提升，以往对于智能家族易用性和实用性的质疑在慢慢消退，取而代之的则是对智能手机操作系统之间的优劣讨论，以及对价格实惠且功能强大的操作系统的期盼。

1.1.1 智能手机的定义

智能手机（Smartphone)，是指"像个人电脑一样，具有独立的操作系统，可以由用户自行安装软件、游戏等第三方服务商提供的程序，通过此类程序来不断对手机的功能进行扩充，并可以通过移动通信网络来实现无线网络接入的这样一类手机的总称"。简而言之，智能手机就是一部像计算机一样可以通过安装软件来拓展手机基本功能的手机。

从广义上说，智能手机除了具备手机的通话功能外，还具备了 PDA 的大部分功能，特别是个人信息管理以及基于无线数据通信的浏览器和电子邮件功能。智能手机为用户提供了足够的屏幕尺寸和带宽，既方便随身携带，又为软件运行和内容服务提供了广阔的舞台，很多增值业务可以就此展开，例如，股票、新闻、天气、交通、商品、应用程序下载、音乐图片下载，等等。结合 3G 通信网络的支持，智能手机势必将成为一个功能强大，集通话、短信、网络接入、影视娱乐为一体的综合性个人手持终端设备。

"智能手机"这个说法主要是针对"功能手机（Featurephone)"而来的，本身并不意味着这个手机有多"智能（Smart)"；从另一个角度来讲，所谓的智能手机就是一台可以随意安装/卸载应用软件的手机（就像计算机那样)。功能手机是不能随意安装/卸载软件的，Java 的出现使后来的功能手机具备了安装 Java 应用程序的功能，但是 Java 程序的操作友好性、运行效率及对系统资源的操作都比智能手机差很多。

需要注意的是，虽然复制/粘贴功能被认为是重要的功能，Windows Mobile Professional、Symbian 等智能手机系统早就支持复制/粘贴功能，但 IOS 和 Windows Phone 却迟迟未能实现类似功能，不久前 Windows Phone 才宣布支持复制/粘贴功能。

智能手机通常使用的操作系统有：Symbian、Windows Mobile、Windows Phone、iOS、Linux

（含 Android、Maemo、MeeGo 和 WebOS）、Palm OS 和 BlackBerry OS。

1.1.2 智能手机的特点

1．优点

（1）具备无线接入互联网的能力，即需要支持 GSM 网络下的 GPRS 或者 CDMA 网络的 CDMA 1X 或 3G（WCDMA、CDMA、TD-SCDMA）网络，甚至 4G（HSPA+、FDD-LTE、TDD-LTE）。

（2）具有 PDA 的功能，包括 PIM（个人信息管理）、日程记事、任务安排、多媒体应用、浏览网页。

（3）具有开放性的操作系统，拥有独立的核心处理器（CPU）和内存，可以安装更多的应用程序，使智能手机的功能可以得到无限扩展。

（4）人性化，可以根据个人需要扩展机器功能。

（5）功能强大，扩展性能强，第三方软件支持多。

2．不足

智能手机的不足之处有：价格普遍较高，易用性较差，新手需要慢慢适应。对于计算机和手机不是很熟悉的用户，想玩转一个智能手机，不花点时间好好钻研是不行的，毕竟如今的智能手机就好比是一台缩小版的 PC。

其实智能手机易用性较差主要还是在手机界面上。一般普通手机多以人性化非常到位的 9 宫格和 12 宫格界面，让用户轻松上手。而智能手机就差在这里。不过如今诺基亚手机的 S60 界面已经相当人性化，就连不易上手的 Windows Mobile 界面也在逐渐改善中。而真正制约用户消费的还是价格，一般智能手机的价格都要高出非智能手机一大截。

1.1.3 3G 智能手机的基本要求

（1）高速度处理芯片。3G 手机不仅要能打电话、发短信，它还要处理音频、视频，甚至要支持多任务处理，这需要一颗功能强大、低功耗、具有多媒体处理能力的芯片。这样的芯片才能让手机不经常死机，不发热，不会让系统慢得如蜗牛。

（2）大存储芯片和存储扩展能力。如果要实现 3G 的大量应用功能，没有大存储就完全没有价值，一个完整的 GPS 导航图，要超过 1GB 的存储空间，而大量的视频、音频和多种应用都需要存储。因此要保证足够的内存存储或扩展存储，才能真正满足 3G 的应用。

（3）面积大、标准化、可触摸的显示屏。只有面积大和标准化的显示屏，才能让用户充分享受 3G 的应用。分辨率一般不低于 320×240。而支持手机的触屏功能是中国用户必不可少的。

（4）支持播放式的手机电视。以现在的技术，如果手机电视完全采用电信网的点播模式，网络很难承受，而且为了保证网络质量，运营商一般对于点播视频的流量都有所控制，因此，广播式的手机电视是手机娱乐的一个重要组成部分。

（5）支持 GPS 导航。GPS 导航不但可以帮助你很方便地找到目的地，而且还可以帮助找到你周围的兴趣点。未来的很多服务也会和位置结合起来，这是手机所特有的特点。

（6）操作系统必须支持新应用的安装，使用户的手机可以安装和定制自己的应用。

（7）配备大容量电池，并支持电池更换。3G 无论采用哪种低功耗的技术，电量的消耗都是一个大问题，必须要配备高容量的电池，1500mA·h 是标准配备，随着 3G 的流行，很可能未来外接移动电源也会成为一个标准配置。

（8）良好的人机交互界面。

1.1.4 智能手机的操作系统

1. Symbian

智能手机从产生发展到现在，Symbian 操作系统一直是现今手机领域中应用范围最广的操作系统，占据了当前手机市场的半壁江山，拥有相当多针对不同用户的界面。其中，诺基亚手机是其代表。最近诺基亚将把微软的 Silverlight 网络视频技术添加到其手机平台上。在谷歌推出 Android 手机后，开发过程中遇到新的困难，Symbian 与 Android 系统合并，并为 Android 手机提供一个单一的操作系统。Symbian 前身其实是一种名为 EPOC 的操作系统，它是一个实时性、多任务的纯 32 位操作系统，具有功耗低、内存占用少等特点，非常适合手机等移动设备使用，经过不断完善，可以支持 GPRS、蓝牙、SyncML 及 3G 技术。最重要的是 Symbian 是一个标准化的开放式平台，任何人都可以用支持 Symbian 的设备开发软件。但是，也存在以下的缺点：各类机型采用的处理器主频较低，兼容性较差，细节不够注意。

2. iOS 系统

iOS 的智能手机操作系统的原名为 iPhoneOS，其核心与 Mac OS X 的核心同样都源自于 Apple Darwin。它主要是给 iPhone 和 iPod touch 使用。就像其基于的 Mac OS X 操作系统一样，它也是以 Apple Darwin 为基础的。iPhoneOS 的系统架构分为四个层次：核心操作系统层（the Core OSlayer）、核心服务层（the Core Serviceslayer）、媒体层（the Media layer）、可轻触层（the Cocoa Touchlayer）。系统操作大约占用 512MB 的存储空间。

iOS 由两部分组成：操作系统和能在 iPhone 和 iPod touch 设备上运行原生程序的技术。由于 iPhone 是为移动终端而开发的，所以要解决的用户需求就与 Mac OS X 有些不同，尽管在底层的实现上 iPhone 与 Mac OS X 共享了一些底层技术。如果你是一名 Mac 开发人员，会在 iPhone OS 发现很多熟悉的技术，同时也会注意到 iPhone OS 的独有之处，例如，多触点接口（Multi-Touch interface）和加速器（accelerometer）。

3. Linux 系统

Linux 系统的智能手机依赖于开源的 Linux 内核，加上手机厂商根据硬件的优化而得的。这种系统的智能手机由于不需要为 Linux 内核付费，因此成本比较低。摩托罗拉公司曾经是 Linux 系统智能手机的主要厂商，推出了一系列的经典机型，如 V8、U9、A1210、A3000 等。

但是，由于 Linux 是开源操作系统，所以各大手机制造商往往各自独立研发。这就造成手机 Linux 系统林立，版本混乱，并且互相不兼容，可靠性差。而且，手机平台上的软件也没有通用性。这些原因导致 Linux 系统的智能手机缺乏竞争力。目前已经很少有 Linux 系统的智能手机生产。摩托罗拉公司也宣布将不再生产 Linux 系统的智能手机，而全面转向生产 Android 系统的智能手机。

4. BlackBerry 系统

"黑莓"BlackBerry 是美国市场占有率第一的智能手机,这得益于它的制造商 RIM(Research in Motion)较早地进入移动市场并且开发出适应美国市场的邮件系统。大家都知道 BlackBerry 的经典设计就是宽大的屏幕和便于输入的 QWERTY 键盘,所以 BlackBerry 一直是移动电邮的巨无霸。正因为它是正统的商务机,所以在多媒体播放方面的功能较弱,也许它在未来应该着力改善这个弱点,因为手机功能的整合是大势所趋,人们不会只满足于单一的功能。

BlackBerry 开始于 1998 年,RIM 的品牌战略顾问认为,无线电子邮件接收器挤在一起的小小的标准英文黑色键盘,看起来像是草莓表面的一粒粒种子,就起了这么一个有趣的名字。应该说,BlackBerry 与桌面 PC 同步堪称完美,它可以自动把 Outlook 邮件转寄到 BlackBerry 中,不过在用 BlackBerry 发邮件时,它会自动在邮件结尾加上"此邮件由 BlackBerry 发出"字样。

BlackBerry 在美国之外的影响微乎其微,我国最近已经在广州开始与 RIM 合作进行移动电邮的推广试验,不过目前看来收效甚微。大家都知道,我国对于电子邮件的依赖并不像美国人那么强,他们在电子邮件里讨论工作、安排日程,而我们则更倾向于当面交谈。可以说 BlackBerry 除了它那经典的外形,在中国的影响几乎为零。

5. Windows Mobile 系统

Windows Mobile 并不算是一个操作系统,只是微软旗下的一个品牌而已。目前微软的 Windows Mobile 系统已广泛用于智能手机和掌上电脑,虽然手机市场份额尚不及 Symbian,但正在加速追赶。Windows Mobile 系列操作系统包括 Pocket PC、Smart-Phone 和 Pocket PC Phone 三大平台体系。Windows Mobile 系列操作系统是在微软计算机的 Windows 操作系统上变化而来的。它采用弹出式菜单、左右键功能,操作形式类似 PC,PPC 版本更接近 PC。触摸笔的功能类似鼠标,有别于传统手机的操作,步骤相对烦琐,但熟悉计算机操作的人会更容易适应。它有以下的缺点:第一,对于不同的平台采用统一的代码编写;第二,沿用了微软 Windows 操作系统的界面,界面和操作都和计算机上的 Windows 十分接近,对于使用者来说十分熟悉又容易上手。

6. Palm 系统

Palm OS 是 Palm 公司的一种 32 位的嵌入式操作系统,它的操作界面采用触控式,差不多所有的控制选项都排列在屏幕上,使用触控笔便可进行所有操作。它本身所占的内存极小,基于 Palm 操作系统编写的应用程序所占的空间也很小,但可以运行众多的应用程序。Palm 操作系统本身不具有录音、MP3 播放功能等,需要另外加入第三方软件或硬件设备方可实现。Palm 在今年推出了最新手机操作系统"Nova"以及基于该操作系统的新款智能手机"Palm Pre"。新的 Nova 系统将会拥有类似 BlackBerry OS 的出色移动商务功能,同时也将具备像 Mac OS X 一样丰富的多媒体娱乐功能。

7. Mac OS X 系统

苹果手机的操作系统(Mac OS X)近几年也是智能手机的一个新亮点,它已超过微软跃居手机行业第二。现在,苹果手机又推出了新的手机操作系统——雪豹。Mac OS X 使用基于 BSD Unix 的内核,并带有 Unix 风格的内存管理和抢占式多任务处理,大大改进了内存管理,允许同时运行更多软件,这实质上消除了一个程序崩溃导致其他程序崩溃的可能性。它还具有极度华丽的图形用户界面、极高的运行效率和安全稳定性。其不足在于 Mac OS X 是一套封闭的操作系统,不支持第三方软件。

1.2 Android 简介

Android 一词的本义是"机器人",同时也是谷歌于 2007 年 11 月 5 日宣布的基于 Linux 平台的开源手机操作系统的名称,该平台由操作系统、中间件、用户界面和应用软件组成。

1.2.1 Android 发展史

2003 年 10 月,Andy Rubin 等人创建 Android 公司,并组建 Android 团队。

2005 年 8 月 17 日,谷歌低调收购了成立仅 22 个月的高科技企业 Android 及其团队。安迪鲁宾成为谷歌公司工程部副总裁,继续负责 Android 项目。

2007 年 11 月 5 日,谷歌正式向外界展示了这款名为 Android 的操作系统,并且在当天谷歌宣布建立一个全球性的联盟组织,该组织由 34 家手机制造商、软件开发商、电信运营商及芯片制造商共同组成。这一联盟将支持谷歌发布的手机操作系统及应用软件,将共同开发 Android 系统的开放源代码。

2008 年

5 月 28 日,Patrick Brady 于谷歌 I/O 大会上提出 Android HAL 架构图。

8 月 18 日,Android 获得美国联邦通信委员会的批准。

9 月 22 日,谷歌正式对外发布第一款 Android 手机——HTC G1(HTC dream)。

9 月 23 日,谷歌发布 Android 1.0。

9 月 24 日,全球业界都表示不看好 Android 操作系统,并且声称最多 1 年,Android 就会被谷歌关闭。

2009 年

4 月 30 日,Android 1.5 正式发布。

5 月 10 日,HTC G1 和 HTC G2 市场大卖,成为仅次于 iPhone 的热门机型。

9 月 25 日,Android 1.6 正式发布。

9 月 29 日,HTC Hero G3 广受欢迎,成为全球最受欢迎的机型。

10 月 28 日,Android 2.0 智能手机操作系统正式发布。

11 月 10 日,由于 Android 销售火热,Android 平台出现第一个恶意间谍软件:Mobile Spy。

2010 年

1 月 7 日,谷歌发布了旗下第一款自主品牌手机:Nexus one(HTC G5)。

2 月 3 日,Linux 内核开发者 Greg Kroah-Hartman 将 Android 的驱动程序从 Linux 内核"状态树"上除去。

5 月 19 日,谷歌正式对外发布 Android 2.2 智能操作系统。

5 月 20 日,谷歌对外正式展示了搭载 Android 系统的智能电视——Google TV,该电视为全球首台智能电视。

7 月 1 日,谷歌宣布正式与雅虎、亚马逊合作,并且在 Android 上推出了多项 Kindle 服务和雅虎服务。

7 月 9 日,美国 NDP 集团调查显示,Android 系统已占据了美国手机市场 28%的份额,以及全球 17%的市场份额。

8月12日，Android平台出现第一个木马病毒：Trojan-SMS.AndroidOS.FakePlayer.a。9月，Android应用数量超过9万个。

9月21日，谷歌对外公布数据，每日销售的Android设备用户数达到20万台。

10月26日，谷歌宣布Android达到第一个里程碑——电子市场上Android应用数量达到10万个。

12月7日，谷歌正式发布Android 2.3操作系统。

2011年

1月，谷歌对外宣布Android Market上的应用数量超过20万个。

1月，谷歌对外公布数据，每日销售的Android设备数达到了30万台。

2月，美国移动用户中36%拥有智能手机。当中，48%的智能手机用户选择Android智能手机。约32.1%的智能手机用户选择了苹果iPhone。黑莓及其他智能手机占有11.6%的份额。

2月2日，Android 3.0正式发布。

2月3日，谷歌发布了专用于平板电脑的Android 3.0蜂巢系统。

6月，Android在日本的智能手机操作系统市场占有率达到57%。

7月，Android在欧洲的智能手机操作系统市场占有率达到了22.3%。

7月，谷歌对外公布数据，Android每天的新用户达到55万人，Android设备用户总数达1.35亿台。

8月，谷歌收购摩托罗拉移动。

8月，谷歌对外宣布Android Market上的应用数量超过30万个。

8月2日，Android手机已占据全球智能机市场48%的份额，并在亚太地区市场占据统治地位，终结了Symbian（塞班系统）的霸主地位，跃居全球第一。

8月，Android在韩国的智能手机操作系统市场占有率达到了95%。

8月，Android系统在35个国家市场占有率第一，平均市场占有率达到48%。

8月，Android系统成为亚太地区第一大系统，市场占有率为亚太地区第一。

9月，Android在美国的智能手机操作系统市场占有率达到43%。

10月19日，谷歌正式发布Android 4.0操作系统。

11月，Android Market上提交审核的应用程序数量达到50万个。

11月初，谷歌对Android Market上的应用程序进行了大清理，据统计，此次共清理了约18万个应用程序，包括流氓应用、病毒软件、侵犯版权、低质量和滥竽充数的各种程序，谷歌将这一系列应用删除后，使得Android市场中的优质应用程序总数到31.5万个。

11月15日，Android在中国大陆的智能手机操作系统市场占有率达到了58%。

11月20日，谷歌宣布启动Android Market应用审核、取缔、清扫行为，定期对电子市场上存在的不合格、低质量、违法恶意的应用程序进行清理。

11月18日，美国NPD数据显示，Android和iOS平台上的游戏占有率都首度超过任天堂的DS掌机和索尼的PSP掌机，手机游戏玩家也超过了掌机玩家，游戏开发商更倾向于在Android和iOS手机上开发游戏。

11月18日，谷歌报告显示，通过谷歌服务器激活的Android设备用户总数已经超过2亿台，每天激活的新设备用户数超过55万台，而这仅仅是通过谷歌服务器激活的用户设备数。12月9日，谷歌对外宣布，Android Market的累计下载量已经突破100亿次，平均每月的下载量为10亿次。

12月18日，谷歌移动事业部副总裁Andy Rubin表示，每天激活的Android设备已达到70万台。

12月26日，Andy Rubin通过Twitter宣布，圣诞节的前两天24日和25日，共有370万台Android设备被激活。

2012年

1月4日，数据显示，Android Market上的应用程序数量突破40万个，每4个月增加10万个应用程序。

1月5日，数据显示，在Android Market上登记的Android开发者已经达到了10万名。1月20日，谷歌报告显示，通过谷歌服务器激活的Android设备用户总数已经超过2.5亿台，距离去年11月，在短短2个月时间内全球共卖出5000万台Andriod设备，平均每秒钟就卖出10台Andriod设备。

2月4日，comScore等数据市场研究机构的数据显示，Andriod在美国手机市场的占有率达到了47.3%。

2月5日，美国联邦政府宣布成立专门的Android实验室进行适用于军方的Android第三方ROM的定制工作。

2月15日，美国联邦政府总务署宣布已经采购了2万台Android手机供政府人员使用。2月22日，Android在中国大陆的智能手机操作系统市场占有率达到了68.4%。

2月28日，谷歌正式宣布，Android设备每天激活量达到85万台，通过谷歌服务器激活的Android设备用户总数突破3亿台。

2月28日，谷歌官方数据显示，Android Market上的应用程序数量已经突破45万个，一年内增加了30万个。

3月1日，谷歌对外宣布，Android Market的累计下载量已经突破130亿次。

3月4日，Android在印度的智能手机操作系统市场占有率达到了34%。

4月4日，Android在美国市场的占有率增长到50.1%。

5月1日，Android在东南亚各国的平均份额达到了49%。

5月31日，Android 4.04更新。

6月2日，Android 4.0系统目前已经有7.1%的（Android系统）用户。

6月13日，Android在意大利的市场占有率增至47.4%。西班牙增为78.4%，在英国的市场份额增为52.5%，在德国的市场份额增为68.6%，法国增为56%。

6月14日，过去一年，Android手机在欧洲国家的销量暴增1580%，在欧洲的市场份额达到60%。

6月15日，Android在全球的市场份额达到59%，领先于竞争对手苹果Mac OSX的23%和微软Windows Phone的2.2%。

6月28日，北京时间0:30在谷歌I/O大会上发布了Android 4.1 Jellybean系统。7月10日，Android开源项目技术负责人Jean-Baptiste M. Queru在谷歌论坛宣布谷歌发布Android 4.1（果冻豆）系统的源代码。

7月20日，国际刑事警察组织（国际刑警）宣布推出假货监察器程序，该程序能帮助国际刑警通过摄像头获取货物编号来监察出入境的货物是否为假货，该程序只支持Android系统。

9月12日消息，据国外媒体报道，市场研究公司IHS今天发表报告称，2013年Android手机累积销量将达到11亿部。

10月30日，Android 4.2沿用"果冻豆"这一名称，以反映这种最新操作系统与Android 4.1的相似性。

Android 5.0 将是下一代 Android 操作系统，2012 谷歌 I/O 大会在旧金山召开，谷歌宣布推出 Android 4.1 操作系统，代号为 Jelly Bean（果冻豆），并推出了全球首款搭载 Android 4.1 的 Nexus 7 平板。而在此之前，Jelly Bean（果冻豆）曾一度被认为是 Android 5.0 的代号。目前传言的 Android 5.0 的代号为 Key Lime Pie（酸橙派）。

2013 年

谷歌执行董事长埃里克·施密特在一场科技大会上预测：在 2013 年底前，全球安卓手机使用量将突破 10 亿台。他介绍说，目前每天激活的 Android 设备已达 150 万台。按当前的发展速度计算，预计 6～9 个月内 Android 智能手机全球使用量就会突破 10 亿台，在一两年内达到近 20 亿台。

1.2.2 Android 特性

Android 号称是首个为移动终端打造的真正开放和完整的移动平台，是安全开源免费的操作系统，任何人都可以获得和使用 Android 系统。谷歌公司还提供了 Android SDK，包括了进行 Android 应用开发所必需的工具和 API 接口。

1. Android 的特性

- 灵活的应用程序框架，可以随意重复使用或者替换手机的组件。
- 提供了专为移动设备优化的虚拟机——Dalvik 虚拟机。
- 拥有内部集成的浏览器——基于开源的 WebKit 引擎。
- 提供针对手机优化的图形库，包括定制的 2D 图形库和基于 OpenGL ES 1.0 的 3D 图形库。
- 使用集成了轻量级数据库管理系统 SQLite 作为结构化的数据存储。
- 娱乐功能丰富，支持多种媒体格式。
- 支持多种移动电话技术，如 GSM、WCDM 等。
- 支持 USB、蓝牙、Wi-Fi 等多种数据传输。
- 支持摄像头、GPS、光线传感器、加速传感器、温度传感器等多种传感器。
- 提供了丰富的开发工具，其中包括设备模拟器、调试工具、内存及性能分析图表和 Eclipse 集成开发环境插件等。

目前 Android 系统不但应用于智能手机，也在平板电脑市场急速扩张。2011 年年初数据显示，正式上市仅两年多的操作系统 Android 已经超越称霸 10 年的 Symbian 系统，并跃居全球最受欢迎的智能手机平台。

2. Android 的优势

1）开放性

在优势方面，首先就是 Android 平台的开发性，其开发平台允许任何移动终端厂商加入到 Android 联盟中来。显著的开放性可以使其拥有更多的开发者，随着用户和应用的日益丰富，一个崭新的平台也将很快走向成熟。

开发性对于 Android 的发展而言，有利于积累人气，这里的人气包括消费者和厂商，而对于消费者来讲，最大的受益正是丰富的软件资源。开放的平台也会带来更多竞争，如此一来，消费者将可以用更低的价位购得心仪的手机。

2）挣脱运营商的束缚

在过去很长的一段时间，特别是在欧美地区，手机应用往往受到运营商的制约，使用什么功能接入什么网络，几乎都受到运营商的控制。iPhone 上市后，用户可以更加方便地连接网络，运营商的制约相应减少。随着 EDGE、HSDPA 这些 2G 至 3G 移动网络的逐步过渡和提升，手机随意接入网络已不是运营商口中的笑谈，当你可以通过手机 IM 软件方便地进行即时聊天时，再回想不久前天价的彩信和图铃下载业务，是不是像噩梦一样？

互联网巨头谷歌推动的 Android 终端天生就有网络特色，将让用户离互联网更近。

3）丰富的硬件选择

丰富的硬件选择与 Android 平台的开放性相关，由于 Android 的开放性，众多的厂商会推出千奇百怪、功能各异的多种产品。功能上的差异和特色，却不会影响到数据同步、甚至软件的兼容，好比你从诺基亚 Symbian 风格手机一下改用苹果 iPhone，同时还可将 Symbian 中优秀的软件带到 iPhone 上使用，联系人等资料更是可以瞬间转移，是不是非常方便呢？

4）不受任何限制的开发商

Android 平台提供给第三方开发商一个十分宽泛、自由的环境，不会受到各种条条框框的阻扰，可想而知，会有多少新颖别致的软件诞生。但也有其两面性，血腥、暴力、情色方面的程序和游戏如何控制正是留给 Android 的难题之一。

5）无缝结合的谷歌应用

如今叱咤互联网的谷歌已经走过 10 年，从搜索巨人到全面的互联网渗透，谷歌服务如地图、邮件、搜索等已经成为连接用户和互联网的重要纽带，而 Android 平台手机将无缝结合这些优秀的谷歌服务。

3. Android 的不足

当然，"金无足赤"，相对于其他一些智能手机操作系统而言，由于进入市场时间不长，作为后起之秀的 Android 在现阶段也存在着以下一些不足。

1）安全和隐私

由于手机与互联网的紧密联系，个人隐私很难得到保护。除了上网过程留下的个人足迹，谷歌这个巨人也时刻站在你的身后，洞穿一切，因此，互联网的深入将会带来新一轮的隐私危机。

2）首先开卖 Android 手机的不是最大运营商

众所周知，T-Mobile 在美国纽约发布了 Android 首款手机 G1。但是在北美市场，最大的两家运营商是 AT&T 和 Verizon，而目前所知取得 Android 手机销售权的仅有 T-Mobile 和 Sprint，其中 T-Mobile 的 3G 网络相对于其他三家也要逊色不少，因此，用户可以购买 G1，但能否体验到最佳的 3G 网络服务则要另当别论。

3）运营商仍然能够影响 Android 手机

在国内市场，不少用户对移动定制机不满，感觉所购的手机像被人涂画了广告一般。这样的情况在国外市场同样出现。Android 手机的另一发售运营商 Sprint 就在定制机型中内置了其手机商店程序。

4）同类机型用户减少

不少手机论坛都会有针对某一型号的子论坛，交流某款手机的使用心得，并分享软件资源。而对于 Android 平台手机，由于厂商丰富，产品类型多样，这样使用同一款机型的用户越来越少，缺少统一机型的程序强化。

5）过分依赖开发商，缺少标准配置

在使用 PC 端的 Windows XP 系统的时候，都会内置微软 Windows Media Player 这样一个浏览器程序，用户可以选择更多样的播放器，如 Realplay 或暴风影音等。但入手开始使用默认的程序同样可以应付多样的需要。在 Android 平台中，由于其开放性，软件更多依赖第三方厂商，例如，Android 系统的 SDK 中就没有内置音乐播放器，全部依赖第三方开发，缺少了产品的统一性。

1.2.3 Android 系统架构

Android 系统是以 Linux 系统为基础的，谷歌将其按照功能特性划分为 4 层，自下而上分别是 Linux 内核、中间件、应用程序框架和应用程序，如图 1-1 所示。

图 1-1　Android 系统框架图

1. 应用程序

Android 系统内置了一些常用的应用程序，包括 Home 视图、联系人、电话、浏览器等。这些应用程序和用户自己编写的应用程序是完全并列的，同样都是采用 Java 语言编写的。而且，用户可以根据需要增加自己的应用程序，或者替换系统自带的应用程序。

2. 应用程序框架

应用程序框架提供了程序开发人员的接口，这是与 Android 程序员直接相关的部分。开发者可以用它开发应用，其中包括以下几方面。

- 丰富而又可扩展的视图（Views）：可以用来构建应用程序，它包括列表（lists）、网格（grids）、文本框（text boxes）、按钮（buttons），甚至可嵌入的 Web 浏览器。
- 内容提供器（Content Providers）：使得应用程序可以访问另一个应用程序的数据（如联系人数据库），或者共享它们自己的数据。
- 资源管理器（Resource Manager）：提供非代码资源的访问，如本地字符串、图形、布局文件（layoutfiles）。
- 通知管理器（Notification Manager）：使得应用程序可以在状态栏中显示自定义的提示信息。
- 活动管理器（Activity Manager）：用来管理应用程序生命周期并提供常用的导航回退功能。

3．中间件

中间件包括两部分：核心库（libraries）和 Android 运行时环境（Android runtime）。

1）核心库

核心库中主要包括一些 C/C++核心库，方便开发者进行应用的开发。

- 系统 C 库（libc）：专门为基于 embedded Linux 的设备定制的。
- 媒体库：支持多种常用的音频、视频格式回放和录制，同时支持静态图像文件。编码格式包括 MPEG4、H.264、MP3、AAC、AMR、JPG、PNG。
- SurfaceManager：对显示子系统的管理，并且为多个应用程序提供了 2D 和 3D 图层的无缝融合。
- WebKit/LibWebCore：Web 浏览引擎，支持 Android 浏览器和一个可嵌入的 Web 视图。
- SGL：底层的 2D 图形引擎。
- 3D libraries：基于 OpenGL ES 1.0 APIs 实现的 3D 引擎。
- FreeType：位图（bitmap）和矢量（vector）字体显示。
- SQLite：轻型关系型数据库引擎。

2）Android 运行时环境

- 运行时核心库：提供了 Java 库的大多数功能。
- Dalvik 虚拟机：依赖于 Linux 内核的一些功能，如线程机制和底层内存管理机制。同时虚拟机是基于寄存器的，Dalvik 采用简练、高效的 byte code 格式运行，它能够在低资耗和没有应用相互干扰的情况下并行执行多个应用，每一个 Android 应用程序都在它自己的进程中运行，都拥有一个独立的 Dalvik 虚拟机实例。Dalvik 虚拟机中可执行文件为.dex 文件，该格式文件针对小内存使用做了优化。所有的类都经由 Java 编译器编译，然后通过 SDK 中的"dx"工具转化成.dex 格式由虚拟机执行。

4．Linux 内核

Android 平台运行在 Linux 2.6 之上，其 Linux 内核部分相当于手机硬件层和软件层之间的一个抽象层。Android 的内核提供了显示驱动、摄像头驱动、闪存驱动、键盘驱动、Wi-Fi 驱动、音频驱动和电源管理等多项功能。此外，Android 为了让 Android 程序可以用于商业目的，将 Linux 系统中受 GNU 协议约束的部分进行了取代。

1.2.4 Android 组件

Android 开发有四大组件。
- 活动（Activity）：用于表现功能。
- 服务（Service）：后台运行服务，不提供界面呈现。
- 广播接收器（Broadcast Receiver）：用于接收广播。
- 内容提供商（Content Provider）：支持在多个应用中存储和读取数据，相当于数据库。

1. Activity

Activity 活动程序是 Android 系统中最常用，也是最基本的工作组件。直观上来说，Activity 活动程序相当于用户看到的手机界面。用户程序的每一个界面即为一个 Activity，用户通过 Activity 来与应用程序进行交互。Activity 通过视图 View 显示用户，视图 View 将用户的操作信息反馈给 Activity 来执行，如图 1-2 所示。

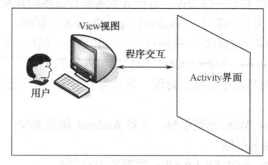

图 1-2　Activity 活动程序

一个 Android 应用程序可包含一个或多个 Activity，每个 Activity 都是独立的。一般来说，Android 程序在运行时会将其中一个 Activity 定为第一个显示的 Activity。在程序运行过程中，Activity 之间的跳转是通过 Intent 激活组件来完成的。Intent 激活组件负责完成 Activity 之间的切换和数据传递。

Android 四种 Activity 加载模流程图如图 1-3 所示。

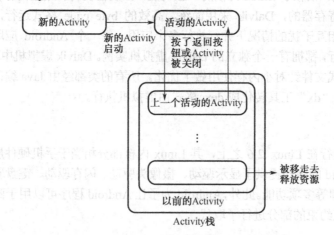

图 1-3　Android 四种 Activity 加载模流程图

Activity 活动程序在 Android 系统中具有不同的状态,这些状态构成了 Activity 的生命周期。

2. Service

Service 是 Android 系统中的一种组件,与 Activity 的级别差不多,但是它不能自己运行,只能在后台运行,并且可以和其他组件进行交互。Service 是没有界面的长生命周期的代码。Service 是一种程序,它可以运行很长时间,但是它却没有用户界面。例如,打开一个音乐播放器的程序,这个时候若想上网,那么打开 Android 浏览器,虽然已经进入了浏览器程序,但歌曲播放并没有停止,而是在后台继续一首接一首地播放。其实这个播放就是由播放音乐的 Service 控制的。当然这个播放音乐的 Service 也可以停止,例如,当播放列表里的歌曲播放完毕,或者用户按下了停止音乐播放的快捷键等。Service 可以在很多场合的应用中使用,例如,播放多媒体的时候用户启动了其他 Activity,此时程序要在后台继续播放;再如,检测 SD 卡上文件的变化,或者在后台记录用户地理信息位置的改变,等等。总之,服务是隐藏在后台的。

开启 Service 有两种方式。

- Context.startService():Service 会经历 onCreate→onStart(如果 Service 还没有运行,则 Android 先调用 onCreate()然后调用 onStart();如果 Service 已经运行,则只调用 onStart(),所以一个 Service 的 onStart 方法可能会重复调用多次);若使用 stopService 则直接调用 onDestroy,如果是调用者自己直接退出而没有调用 stopService,则 Service 会一直在后台运行。该 Service 的调用者再启动起来后可以通过 stopService 关闭 Service。

注意:多次调用 Context.startservice()不会嵌套(即使会有相应的 onStart()方法被调用),所以无论同一个服务被启动了多少次,一旦调用 Context.stopService()或者 stopSelf(),它都会被停止。

说明:传递给 startService()的 Intent 对象会传递给 onStart()方法。调用顺序为:onCreate → onStart(可多次调用)→onDestroy。

- Context.bindService():Service 会经历 onCreate()→onBind(),onBind 将返回给客户端一个 IBind 接口实例,IBind 允许客户端实施回调服务,比如得到 Service 运行的状态或其他操作。这个时候调用者(Context,如 Activity)会和 Service 绑定在一起,Context 退出了,Srevice 就会调用 onUnbind→onDestroyed 相应退出。

3. Broadcast Receiver

Broadcast Receiver 广播接收器负责接收 Android 系统中的广播通知信息,并做出相应的处理。当 Android 操作系统或某个应用程序发送广播时,另外的应用程序可以使用 Broadcast Receiver 广播接收器接收这个广播,并做出相应的处理。从此可看出,广播可以源自于 Android 系统,也可以来自于应用程序。

源自于 Android 系统的广播包括时区改变、电池电量低、更改语言选项等。来自应用程序的广播则可以是指定的功能状态,如数据下载完成后的通知。Android 应用程序可以拥有任意数量的广播接收器。所有的接收器均继承自 Broadcast Receiver 基类。

广播接收器没有像 Activity 那样的可视化用户界面,但可以采用很多种方式来吸引用户的注意,如闪烁指示灯、震动、播放声音等。

Broadcast Receiver 广播接收器需要注册和注销。Android 提供了两种方式来注册广播接收器,一种是在 Android Manifest.xml 中定义,另一种是直接在程序代码中动态设置。注册好的广播接收器并不一直在后台运行,只有相应的 Intent 才可调用。

在使用时，应用程序首先将需要广播的信息封装在 Intent 中，然后将 Intent 广播出去，另一个应用程序则通过 IntentFilter 对象来过滤所有的 Intent。最后，Broadcast Receiver 广播接收器用另一个 onReceive() 的回调方式来处理接收到的通知。

4．Content Provider

Content Provider 是 Android 提供的第三方应用数据的访问方案。

在 Android 中，对数据的保护是很严密的，除了放在 SD 卡中的数据，一个应用所持有的数据库、文件等内容，都是不允许直接访问的。Andorid 当然不会真地把每个应用都做成一座孤岛，它为所有应用都准备了一扇窗，这就是 Content Provider。应用想对外提供的数据，可以通过派生 Content Provider 类，封装成一枚 Content Provider，每个 Content Provider 都用一个 uri 作为独立的标识，形如：content://com.xxxxx。所有东西看着像 REST 的样子，但实际上，它比 REST 更为灵活。和 REST 类似，uri 也可以有两种类型，一种是带 id 的，另一种是列表的。但实现者不需要按照这个模式来做，给你 id 的 uri 也可以返回列表类型的数据，只要调用者明白，就无妨，不用苛求所谓的 REST。

另外，Content Provider 不和 REST 一样只有 uri 可用，还可以接受 Projection、Selection、OrderBy 等参数，这样，就可以像数据库那样进行投影、选择和排序。查询到的结果以 Cursor 的形式返回，调用者可以移动 Cursor 来访问各列的数据。

Content Provider 屏蔽了内部数据的存储细节，向外提供了上述统一的接口模型，这样的抽象层次大大简化了上层应用的书写，也对数据的整合提供了更方便的途径。Content Provider 内部常用数据库来实现，Android 提供了强大的 Sqlite 支持，但很多时候，你也可以封装文件或其他混合的数据。

在 Android 中，Content Resolver 是用来发起 Content Provider 的定位和访问的。不过它仅提供了同步访问的 Content Provider 的接口。但通常，Content Provider 需要访问的可能是数据库等大数据源，速度上不能足够快，否则会导致调用线程的拥塞。因此 Android 提供了一个 AsyncQueryHandler，帮助进行异步访问 Content Provider。

在各大组件中，Service 和 Content Provider 都需要持续访问。Service 如果是一个耗时的场景，往往会提供异步访问的接口，而 Content Provider 不论效率如何，提供的都是约定的同步访问接口。

1.3 搭建 Android 开发环境

在搭建环境前，需要了解安装开发工具所需要的硬件和软件配置条件。

1.3.1 Android 开发前的准备工作

1．系统基本要求

开发基于 Android 的应用软件所需要的开发环境如表 1-1 所示。

表 1-1　开发系统所需要参数

项　　目	版本要求	描　　述	备　　注
操作系统	Windows XP 以上版本、Mac OS、Linux	根据自己的计算机自行选择	选择最熟悉的操作系统
软件开发包	Android SDK	选择最新版本的 SDK	
IDE	Eclipse IDE+ADT	Eclipse3.3(Europa)、3.4 (Ganymede) 和 ADT(Android Development Tools)开发插件	选择"for Java Developer"
其他	JDK Apache Ant	Java SE Development Kit5 或 6 Linux 和 Mac 上使用 Apache Ant 1.6.5+、Windows 上使用 1.7+版本	（仅有 JRE 不行，必须有 JDK)，不兼容 Gnu Java 编译器(gcj)

2．软件开发工具

Anodroid 软件开发需要以下工具。

- JDK：Java 核心开发包。
- Eclipse：Java 集成开发环境。
- Android SDK：谷歌提供的 Android 开发包。
- ADT：Android 的 Eclipse 开发插件。

谷歌为 Android 提供了不同操作系统下的开发包，包括 Windows、Mac OS 和 Linux 操作系统。

- Windows XP（32 位）、Vista（32 位或 64 位）、Windows 7（32 位或 64 位）．
- Mac OS X 10.5.8 及其后面的版本（x86）。
- Linux Ubuntu（64 位的操作环境要求可以执行 32 位应用程序）。

另外，为了更好地运行 Android 开发环境，计算机硬件方面需要满足以下要求：

- 内存：512MB 以上。
- 硬盘：剩余 2GB 空间以上。
- CPU：P4 2.0GHz 以上。
- 显示器分辨率：1024×768 以上。

1.3.2　安装 JDK

在 Windows 平台上，搭建 Android 开发环境，首先下载并安装与开发环境相关的软件资源，这些资源主要包括 JDK、Eclipse、Android SDK 和 ADT 插件。

1．安装 JDK

在 Android 平台上，所有应用程序都是使用 Java 语言来编写的，所以要安装 Java 开发包 JDK（Java SE Development Kit），JDK 是 Java 开发时所必需的软件开发包。

安装 JDK 的过程比较简单，运行该程序后，根据安装提示选择安装路径，将 JDK 安装到指定的文件夹即可。默认安装目标为"C:\Program Files\Java\jdk1.6.0_10（jdk-6u10-rc2-bin-b32-windows-i586-p-12_sep_2008)"。

JDK 安装完毕后，进一步要设置 Java 的环境变量，即设置 bin 和 lib 文件夹的路径。其操作步骤如下（假定计算机操作系统为 Windows 7）：

（1）右击"计算机"，在弹出的快捷菜单中选择"属性"选项，在弹出的"系统"对话框中单击"高级系统设置"按钮，弹出"系统属性"对话框，如图 1-4 所示。

（2）在"系统属性"对话框的"高级"选项卡中，单击"环境变量"按钮，弹出"环境变量"对话框，如图 1-5 所示。

图 1-4 "系统属性"对话框　　　　　　　　图 1-5 "环境变量"对话框

（3）选中"系统变量"区域的"Path"变量，单击"编辑"按钮，弹出"编辑系统变量"对话框，如图 1-6 所示。

图 1-6 环境变量 Path 设置

（4）在该对话框的"变量值"文本框中添加"C:\Program Files\Java\jdk1.6.0_10\bin"，然后单击"确定"按钮即可完成设置。这样即设置了 bin 文件夹的路径。

（5）在"环境变量"对话框的"系统变量"区域，单击"新建"按钮，弹出"新建系统变量"对话框，如图 1-7 所示。

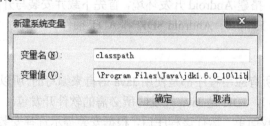

图 1-7 新建环境变量 classpath

（6）在图 1-7 中的"变量名(N)"右侧文本框中输入"classpath"，在"变量值(V)"右侧文本框中输入"C:\Program Files\Java\jdk1.6.0_10\lib"，即可设置好 lib 文件夹的路径。

完成以上操作后，一个典型的 Java 开发环境便设置好了。在正式开始下一步前先验证 Java 开发环境的设置是否成功。

在 Windows 7 系统中单击"开始"按钮，在弹出的窗口中选择"运行"，在运行框中输入"cmd"并按回车键，即可打开 CMD 窗口，在窗口中输入 java –version，则可显示所安装的 Java 版本信息，如图 1-8 所示。

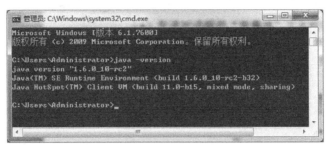

图 1-8 JDK 安装成功页面

1.3.3 安装 Eclipse

安装并设置好 JDK 后，即可接着安装 Eclipse 了。Eclipse 是一个非常强大的集成开发环境，可以支持 Java、C、C++等多种语言。由于 Android 开发使用 Java，因此需要下载 Java 版本的 Eclipse 集成开发环境。

将下载并保存的"eclipse-SDK-3.7（中文版）"解压到硬盘上的某个目录即可。Eclipse 集成开发环境无须安装，进入解压后的目录，双击可执行文件"eclipse.exe"，Eclipse 能自动找到用户前面安装的 JDK 路径。Eclipse 启动界面如图 1-9 所示。

图 1-9 Eclipse 启动界面

启动 Eclipse 开发环境桌面，将会看到选择工作空间的提示，如图 1-10 所示。

单击图 1-10 中的"OK"按钮，即可完成 Eclipse 的安装，系统进入 Eclipse 初始欢迎界面，如图 1-11 所示。接着单击图 1-11 左上角的"欢迎"按钮，即可进入 Eclipse 的开发环境界面，如图 1-12 所示。

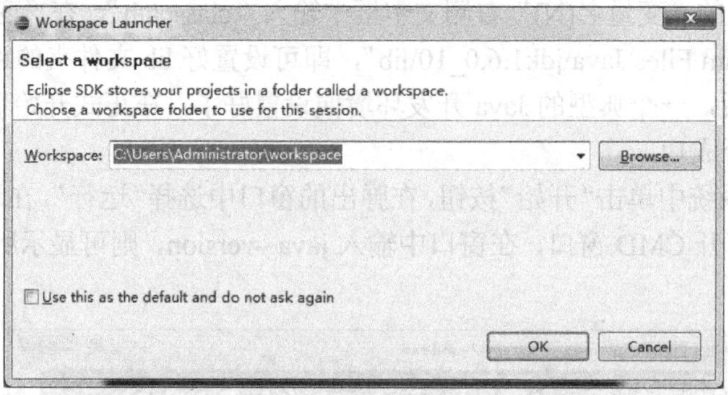

图 1-10　选择工作空间

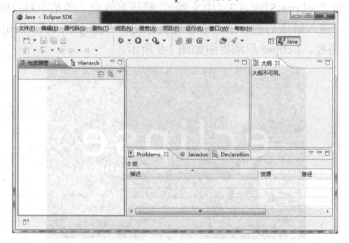

图 1-11　Eclipse 欢迎界面

图 1-12　Eclipse 开发界面

1.3.4　安装 Android SDK

将下载的 Android SDK 开发包解压到硬盘上的某个目录，该目录在后面配置 Android 开发工具 ADT 和使用 SDK 工具时都会用到。

解压后的文件夹有以下几个重要的文件：
- "add-ons"目录用来保存插件工具，目前为空。
- "platforms"目录用来保存不同版本的SDK数据包，目前为空。
- "tools"目录包含了Android的SDK工具。
- "SDK Manager.exe"为SDK管理工具，可以用来更新SDK数据包、管理Android模拟器等。
- "SDK Readme.txt"为Android SDK的说明文件。

Android SDK与Eclipse集成开发环境一样，不需要经过真正的安装过程，相当于解压之后就可以运行。读者在第一次运行SDK Manager时，需要下载Android各个版本的SDK数据包。

【操作步骤】

（1）双击"SDK Manager.exe"执行文件，程序将自动检测是否有更新的SDK数据包可下载，如图1-13所示。

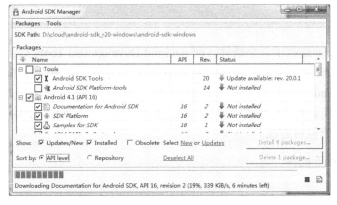

图1-13 运行SDK Manager.exe执行文件

（2）对于所要更新的内容，如果只要尝试一下Android 4.1，那么只选择"Android 4.1（API 16）"然后单击"Install X packages"按钮安装就可以了。如果你要在此SDK上开发应用程序和游戏应用，那么需要接受并遵守所有许可内容（Accept All），并单击"Install"按钮。

（3）接着将SDK tools目录的完整路径设置到系统变量Path中，这样便于在后面调用Android命令时，无须输入全部的绝对路径。设置系统变量Path的方法与JDK的环境变量值操作一致，在Path环境变量的"变量值（V）"文本框中添加";D:\cloud\android-sdk_r14-windows\android-sdk-windows\tools"即可，如图1-14所示。

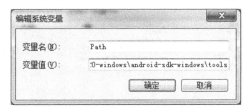

图1-14 设置Android SDK环境变量

最后检查Android SDK是否安装成功并能够正常运行。在Windows 7系统中单击"开始"按钮，在弹出的窗口中选择"运行"，在运行框中输入"cmd"并按回车键，即可打开CMD窗口，在窗口中输入"android –h"，则可显示所安装的Android SDK信息，如图1-15所示。

图 1-15 Android SDK 安装成功信息

1.3.5 安装 ADT 插件

Android 为 Eclipse 定制了一个插件，即 Android Development Tools（ADT），这个插件为用户提供了一个强大的综合环境，用于开发 Android 应用程序。ADT 扩展了 Eclipse 的功能，可以让用户快速地建立 Android 项目，创建应用程序界面，在基于 Android 框架 API 的基础上添加组件，以及用 SDK 工具集调试应用程序，甚至导出签名（或未签名）的 APKs 以便发行应用程序。

安装 ADT 插件有两种方法。

1. 手动安装 ADT 插件

【操作步骤】

（1）首先从 Android 官方开发网站下载 ADT 插件，找到目前最新版本 ADT-20.02，直接下载名为"ADT-20.02.zip"文件。

（2）将"ADT-20.02.zip"文件解压，将解压后的 plugins 包和 features 包中内容，分别复制到 Eclipse 对应的 plugins 和 features 文件夹中。

（3）重新启动 Eclipse，选择"窗口"菜单下的"首选项"命令，在弹出的"首选项"对话框左边多了"Android"项，如图 1-16 所示。

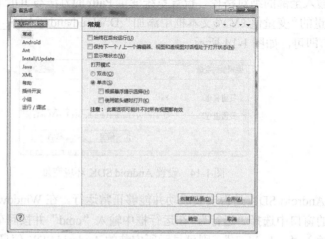

图 1-16 "首选项"对话框

（4）单击"Android"项，在该对话框右边的"SDK Location"文本框中，设置 Android SDK 的安装路径。此处设置为"C:\Users\Administrator\android-sdks"，对话框会列出当前可用的 SDK 版本和 Google API 版本。至此即完成了 Eclipse 开发环境下 ADT 插件的安装。

2．Eclipse 在线安装 ADT

除了手动设置 ADT 插件外，还可以采用更简单的在线更新 ADT 插件方法。

【操作步骤】

（1）打开 Eclipse 后，单击菜单栏中的"帮助"菜单下的"Install New Software…"选项，如图 1-17 所示。

图 1-17　添加插件

（2）在弹出的对话框中单击"Add…"按钮，如图 1-18 所示。

图 1-18　添加插件

（3）在弹出的"Add Site"对话框中分别输入名字和地址，名字可自己命名，如"abc"，但是在 Location 中必须输入插件的网络地址"http://dl-ssl.google.com/Android/eclipse"，单击"确定"按钮，如图 1-19 所示。

图1-19 插件网络地址

（4）单击图1-19中的"确定"按钮，此时在"Intall"界面会显示系统中可用的插件，如图1-20所示。

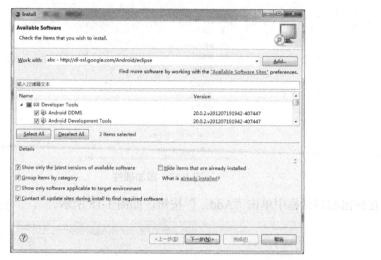

图1-20 插件列表

（5）选中图1-20中的"Android DDMS"和"Android Development Tools"选项，然后单击"下一步"按钮进入安装界面，如图1-21所示。

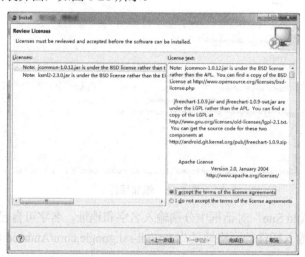

图1-21 插件安装界面

（6）选择"I accept the terms of the license agreements"项，单击"完成"按钮，即开始进行安装，如图1-22所示。

第 1 章　Android 概述

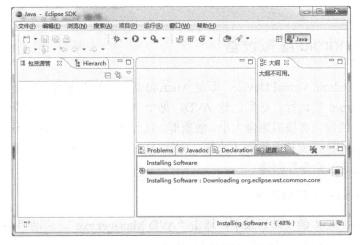

图 1-22　插件开始安装

注意：在第（6）步中，可能会发生计算插件占用资源的情况，过程有点慢。完成后会提示重启 Eclipse 来加载插件，等重启后就可以用了。不同版本的 Eclipse 安装插件的方法和步骤略有不同。

1.3.6　设定 Android SDK 主目录

安装好插件后，还需要做如下配置才可以使用 Eclipse 创建 Android 项目，需要设置 Android SDK 的主目录。

在图 1-16 中，选中左侧的"Android"项，在右侧设定 Android SDK 所在目录为 SDK Location，然后单击"确定"按钮完成设置，如图 1-23 所示。

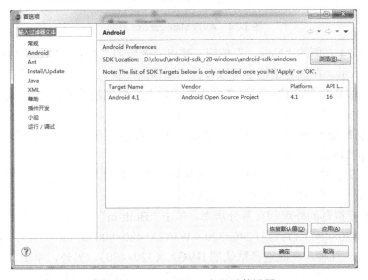

图 1-23　Android SDK 主目录的设置

1.3.7 创建 Android 虚拟设备

AVD 全称为 Android Virtual Device，即是 Android 运行的虚拟设备，它是 Android 的模拟器识别。建立的 Android 要运行，必须创建 AVD，每个 AVD 上可以配置很多的运行项目。创建 AVD 时，可以配置的选项有模拟影像大小、触摸屏、轨迹球、摄像头、屏幕分辨率、键盘、GSM、GPS、Audio 录放、SD 卡支持、缓存大小等。创建 AVD 的方法有两种：一种是通过 Eclipse 开发环境；另一种是通过命令行创建。

1. 通过 Eclipse 开发环境创建

（1）在 Android SDK 的安装目录下，双击"AVD Manager.exe"，启动"Android AVD"，弹出如图 1-24 所示的"Android Virtual Device Manager"窗口。

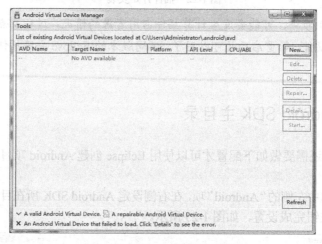

图 1-24　AVD Manager.exe 界面

（2）单击图 1-24 右侧的"New…"按钮，弹出一个新的"Android Virtual Device (AVD)"对话框，如图 1-25 所示。在该对话框中可以设置模拟器的配置，包括如下几项。

- Name：创建 AVD 的名称。可以在文本框中输入所要创建的 AVD 的名称，注意名称中不能有空格。
- Target：选择 Android 版本和 API 的等级。单击右边的下拉按钮，选择相应的 Android 版本和 API 的等级。
- SD Card：设置 SD 卡。在"Size"文本中指定 SD 卡的大小。另外，也可以在"File"文本框中设置已有的 SD 卡镜像文件的路径。
- Skin：设置模拟器的外观和屏幕分辨率。单击"Built-in"右边的下拉按钮，可以选择默认的 HVGA（320×480）、QVGA（240×320）、WVGA（480×800 或 480×854）、WQVGA（240×400 或 240×320）几种，在此选择默认的 HVGA（320×480）。另外，单击"Resolution"项，还可以自定义分辨率。不同版本的 Android 所设置的 Skin 参数有所不同。
- Hardware：设置模拟器支持的硬件设备的属性，包括影像大小、触摸屏、轨迹球、摄像头、屏幕分辨率、键盘、GSM、GPS、Audio 录放、SD 卡支持、缓存区大小等。单击该区域右边的"New…"按钮，在弹出的对话框中可以设置各项的属性。

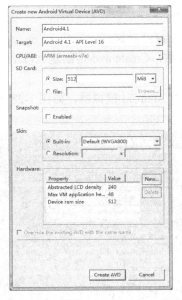

图 1-25　新建 AVD 时的 emulate 设置

（3）设置好模拟器的参数后，单击图 1-25 下边的"Create AVD"按钮即可创建一个 AVD。创建好的 AVD 将会显示在如图 1-26 所示的"Android Virtual Device Manager"窗口的文件列表中。

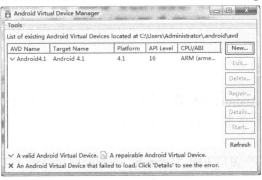

图 1-26　创建新的 AVD 界面

至此，已经创建了一个 Android 模拟器，使用同样的操作可以根据需要创建多个 AVD 模拟器。这样做的好处是，可以模拟程序在不同的 Android 版本上运行的兼容性。

2．命令行方式创建

（1）在 CMD 下输入"android list targets"，查看可用的 Android 平台，如图 1-27 所示。

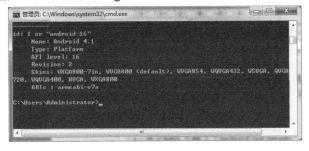

图 1-27　CMD 界面

（2）按照如下格式创建 AVD：

```
android create avd --target 2 --name my_avd
```

其中，android 是命令，后面是参数，Create avd 是创建 AVD，target 2 是等级，name 是 AVD 的名称。

以上代码创建名为 my_avd 的 Android 模拟器，如图 1-28 所示。

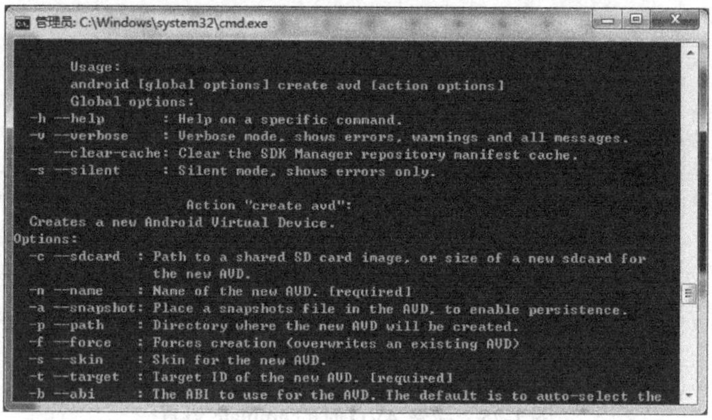

图 1-28　命令行创建 AVD

模拟器可以运行大部分的应用程序，但是实际操作中大部分时间是在真正机器上运行，那样效果更好且效率更高。

1.3.8　运行 AVD

创建好 AVD 后，运行 Android 模拟器有两种方式：一种是在"Android Virtual Device Manager"窗口中选中已创建的 AVD，单击右侧的"Start"按钮，即可弹出如图 1-29 所示的"Launch Options"窗口。

单击"Launch Options"窗口下的"Launch"按钮即成功启动 AVD，效果如图 1-30 所示。

图 1-29　"Launch Options"窗口

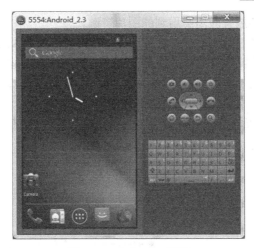

图 1-30　启动 AVD 界面

图 1-30 的各个控制按钮名称及其功能如表 1-2 所示。

表 1-2　AVD 的控制按钮功能

模拟器 AVD 的模拟按键	相应的图标	对应键盘上组合键	功　　能
相机按钮		Ctrl+V	控制摄像头
音量渐小按钮		Ctrl+S	控制音量大小
电源按钮		F7	设置电话模式，AVD 开关
音量增加按钮		Ctrl+F5	控制音量大小
挂断电话按钮		F4	模拟挂断电话功能
拨打电话按钮		F3	模拟拨打/接听电话功能
上/下/左/右按钮		键盘 5	确定按钮
中心按钮		键盘 4/8/2	上/下/左/右移动焦点
Home 按钮		Home	返回主界面
Menu 按钮		F2 或 Pageup	打开应用程序菜单
查询按钮		F5	在手机内部或上网查询
返回按钮		Esc	返回上一级界面

1.4　Android 应用工程文件组成

Android 的应用工程文件主要由以下几部分组成。
- src 目录：项目源文件都保存在这个目录中。
- R.java 文件：这个文件是 Eclipse 自动生成的，应用开发者不需要去修改里面的内容。
- Android Library：这是应用运行的 Android 库。
- assets 目录：里面主要放置多媒体等一些文件。
- res 目录：主要放置应用会用到的资源文件。

- drawable 目录：主要放置应用会用到的图片资源。
- layout 目录：主要放置用到的布局文件。这些布局文件都是 XML 文件。
- values 目录：主要放置字符串（strings.xml）、颜色（colors.xml）、数组（arrays.xml）。
- Androidmanifest.xml：相当于应用的配置文件。在这个文件中，必须声明应用的名称，应用所用到的 Activity、Service 以及 receiver 等。

在 Eclipse 中，一个基本的 Android 项目的目录结构如图 1-31 所示。

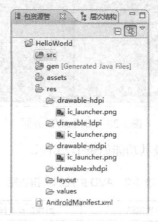

图 1-31　Android 应用工程文件组成

1．src 目录

与一般的 Java 项目一样，"src"目录下保存的是项目的所有包及源文件（.java），"res"目录下包含了项目中的所有资源。例如，程序图标（drawable）、布局文件（layout）和常量（value）等。不同的是，在 Java 项目中没有"gen"目录，也没有每个 Android 项目都必须有的 AndroidManfest.xml 文件。

".java"格式文件是在建立项目时自动生成的，这个文件是只读模式。不能手动添加或删除 R.java 文件，R.java 文件是定义该项目所有资源的索引文件。先来看看 HelloWorld 项目的 R.java 文件，代码如下：

```
package net.learn2develop.HelloWorld;
public final class R
{
    public static final class attr
    {
    }
    public static final class drawable
    {
        public static final int icon=0x7f020000;
    }
    public static final class layout
    {
        public static final int main=0x7f030000;
```

```
    }
    public static final class string
    {
        public static final int app_name=0x7f040001;
        public static final int hello=0x7f040000;
    }
}
```

从上述代码中，可以看到文件定义了很多常量，并且会发现这些常量的名字都与 res 文件夹中的文件名相同，这再次证明 .java 文件中所存储的是该项目所有资源的索引。有了这个文件，在程序中使用资源将变得更加方便，可以很快地找到要使用的资源，由于这个文件不能手动编辑，所以当用户在项目中加入了新的资源时，只需要刷新一下该项目，.java 文件便自动生成了所有资源的索引。

2．res 目录

在 res 目录下包含了该项目所使用到的资源文件，这里面的每一个文件或者资源都将在 R.java 文件中进行索引定义。文件类型主要有以下几类。

- 图片文件：分别提供了高分辨率（drawable-hdpi）、中分辨率（drawable-mdpi）和低分辨率（drawable-ldpi）的图片文件。
- 布局文件：在 layout 目录下，默认只有一个 main.xml，用户也可以添加更多的布局文件。
- 字符串：在 values 目录下的 strings.xml 文件中。

打开 main.xml 布局文件，代码为：

```xml
<?xml version="1.0" encoding="utf-8"?>
<LinearLayout xmlns:android="http://schemas.android.com/apk/res/android"   //线性布局
        android:orientation="vertical"
        android:layout_width="fill_parent"
        android:layout_height="fill_parent">
<TextView
    android:layout_width="fill_parent"
    android:layout_height="wrap_content"
    android:text="@string/hello"/>
</LinearLayout>
```

在该布局文件中，首先定义了采用线性布局，内部只有一个文本框控件。这个控件显示内容引用了 string 文件中 hello 变量。

其中，

- <LinearLayout>：线性版面配置，在这个标签中，所有元件都是按由上到下排队排成的。
- android:orientation：表示这个介质的版面配置方式是从上到下垂直地排列其内部的视图。
- android:layout_width：定义当前视图在屏幕上所占的宽度，fill_parent 即填充整个屏幕。
- android:layout_height：随着文字栏位的不同而改变这个视图的宽度或高度。

在上述布局代码中，使用了一个 TextView 来配置文件标签 Widget（构件），其中设置的属性 android:layout_width 为整个屏幕的宽度，android:layout_height 可以根据文字来改变高度，而

android:text 则设置了这个 TextView 要显示的文字内容，这里引用了@string 中的 hello 字符串，即 String.xml 文件中的 hello 所代表的字符串资源。Hello 字符串的内容"HelloWorld、HelloAndroid"就是用户在 HelloAndroid 项目运行时看到的字符串。

Strings.xml 文件的代码为：

```xml
<?xml version="1.0" encoding="utf-8"?>
<resources>
    <string name="hello">HelloWorld,HelloAndroid</string>
    <string name="app_name">HelloAndroid</string>
</resources>
```

3. AndroidManfest.xml 文件

在文件 AndroidManfest.xml 中包含了该项目中所使用的 Activity、Service、Receiver，以下代码为"HelloWorld"项目中的 AndroidManfest.xml 文件。

```xml
<?xml version="1.0" encoding="utf-8"?>
<manifest xmlns:android="http://schemas.android.com/apk/res/android"    //根节点
    package="PACKAGE"                                                    //包名
    android:versionCode="1"
    android:versionName="1.0" >
    <uses-sdk android:minSdkVersion="16" />                              //SDK 版本
    <instrumentation
        android:name="android.test.InstrumentationTestRunner"
        android:targetPackage="TEST_TARGET_PCKG" />
    <application                                                         //图标和应用程序名称
        android:icon="@drawable/ic_launcher"
        android:label="@string/app_name" >
        <activity android:name=".firstActivity"                          //默认启动的 Activity
            Android:label="@string/app_name" >                           //Activity 名称
            <intent-filter>
                <action android:name="android.intent.action.MAIN"/>
                <category android:name="android.intent.category.LAUNCHER"/>
            <uses-library android:name="android.test.runner" />
    </application>
</manifest>
```

1.5 应用程序的生命周期

应用程序的生命周期是指进程在 Android 系统中从启动到终止的所有阶段，也就是 Android 程序启动到停止的全过程。程序的生命周期由 Android 系统进行调试和控制。了解 Android 的生命周期是学习 Android 应用程序开发的基础。

在 Android 系统中，当某个 Activity 调用 startActivity(myIntent) 时，系统会在所有已经安

装的程序中寻找其 intent filter 和 myIntent 最匹配的一个 Activity，启动这个进程，并把这个 intent 通知给这个 Activity。这就是一个程序的"生"。例如，在 Home application 中选择"Web browser"，系统会根据这个 intent 找到并启动 Web browser 程序，显示 Web browser 的一个 Activity 用于浏览网页（这个启动过程有点类似在计算机上双击桌面上的一个图标，启动某个应用程序）。在 Android 中，所有的应用程序"生来就是平等的"，所以不光 Android 的核心程序甚至第三方程序也可以发出一个 intent 来启动另外一个程序中的一个 Activity。Android 的这种设计非常有利于"程序部件"的重用。

1.5.1 进行优先级

一个 Android 程序的进程是何时被系统结束的呢？通俗地说，一个即将被系统关闭的程序是系统在内存不足（low memory）时，根据"重要性层次"选出来的"牺牲品"。一个进程的重要性是根据其中运行的部件和部件的状态决定的。各种进程按照重要性从高到低排列，如图 1-32 所示。

1. 前台进程

前台进程是与用户正在交互的进程，也是 Android 系统中最重要的进程。处于前台进程一般包含以下四种情况。

- 进行中的 Activity 正在与用户进行交互。
- 进程服务被 Activity 调用，而且这个 Activity 正在与用户进行交互。

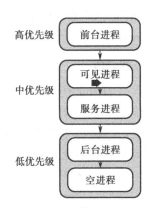

图 1-32　Android 进程优先级效果图

- 进程服务正在执行生命周期中的回调函数，如 onCreate()、onStart()或 onDestroy()。
- 进程的 BroadcastReceiver 正在执行 onReceive()函数。

Android 系统为多任务操作系统，当系统中的多个前台进程同时运行时，如果出现资源不足的情况，此时 Android 内核将自动清除部分前台进程，保证最主要的用户界面能够及时响应操作。

2. 可见进程

可见进程是指在屏幕上显示，但不在前台的程序。例如，一个前台进程以对话框的形式显示在该进程前面。这样的进程也很重要，它们只有在系统没有足够内存运行所有前台进程时，才会被结束。

3. 服务进程

服务进程在后台持续运行，如后台音乐播放、后台数据上传/下载等。这样的进程对用户来说一般很有用，所以只有当系统没有足够内存来维持所有的前台进程和可见进程时，才会被结束。

4. 后台进程

后台进程程序拥有一个用户不可见的 Activity。这样的程序在系统内存不足时，按照 LRU 的顺序被结束。

5. 空进程

空进程不包含任何活动的程序部件，系统可能随时关闭这类进程。

从某种意义上讲，垃圾收集机制把程序员从"内存管理噩梦"中解放出来，而 Android 的进程生命周期管理机制把用户从"任务管理噩梦"中解放出来。笔者见过一些 Nokia S60 用户和 Windows Mobile 用户，要么因为长期不关闭多余的应用程序而导致系统变慢，要么因为不时查看应用程序列表而影响使用体验。Android 使用 Java 作为应用程序 API，并且结合其独特的生命周期管理机制，同时为开发者和使用者提供最大程度的便利。

1.5.2 Activity 的生命周期

Activity 的生命周期是指应用程序的 Activity 从启动到销毁的全过程。Activity 的生命周期是在 Android 应用程序设计中最重要的内容，直接关系到用户程序的界面和功能。

1. 活动栈

每一个活动的状态是由它在活动栈中所处的位置所决定的，活动栈是当前所有正在运行的进程的后进先出的集合。当一个新的活动启动时，当前的前台屏幕就会移动到栈顶。如果用户使用 Back（返回）按钮返回到了刚才的活动，或者前台活动被关闭了，那么栈中的下一个活动就会移动到栈顶，变为活动状态。图 1-33 说明了这个过程。

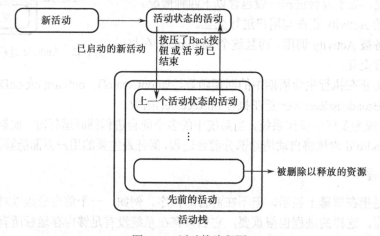

图 1-33 活动栈流程图

2. Activity 的状态

随着活动的创建和销毁，它们会按照图 1-33 所示的那样，从栈中移进移出。在这个过程中，它们也经历了活动、暂停、停止和非活动 4 种可能的状态，如图 1-34 所示。

（1）活动状态：当一个活动位于栈顶的时候，它是可见的、被关注的前台活动，这时它可以接收用户输入。Android 将会不惜一切代价来保持它处于活动状态，并根据需要来销毁栈下面部分的活动，以保证这个活动拥有它所需要的资源。当另一个活动变为活动状态时，这个活动就将被暂停。

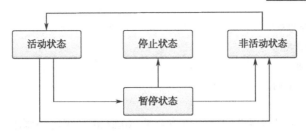

图 1-34　Activity 的状态

（2）暂停状态：在某些情况下，活动是可见的，但是没有被关注，此时它就处于暂停状态。当一个透明的或者非全屏的活动位于某个处于活动状态的活动之前时，这个透明的或者非全屏的活动就会达到这个状态。当活动被暂停的时候，它仍然会被当做近似于活动状态的状态，但是它不能接收用户的输入事件。在极端情况下，当一个活动变得完全不可见的时候，它就会变为停止状态。

（3）停止状态：当一个活动不可见的时候，它就处于停止状态。此时，活动仍然会停留在内存中，保存所有的状态和成员信息，然而当系统的其他地方要求使用内存的时候，它们就会成为被清除的首要候选对象。在一个活动停止的时候，保存数据和当前的 UI 状态是很重要的。一旦一个活动被退出或者关闭，它就会变为非活动状态。

（4）非活动状态：当一个活动被销毁之后，在它启动之前就处于非活动状态。处于非活动状态的活动已经从活动栈中移除了，因此，在它们可以被显示和使用之前，需要被重新启动。

在 Android 系统中，采用"栈"结构来管理 Activity，这是一种"后进先出"的原则，如图 1-35 所示。当一个 Activity 被启用时，将执行入栈操作。位于栈顶的 Activity 处于活动状态，其他 Activity 则处于暂停状态或者停止状态。当 Activity 关闭时，将执行出栈操作，从而变成非活动状态。当 Android 系统资源紧张时，Android 内核也会终止部分长久没有响应的 Activity，使之为非活动状态，从而释放系统资源。

图 1-35　Activity 的栈结构

3．管理 Activity 的生命周期

在 Activity 系统中，一般通过 Activity 的事件回调函数来管理 Activity 的生命周期。这些事件回调函数如下：

```
public class MyActivity extends Activity
{
    void onCreate(Bundle savedInstanceState);
    void onStart();      void onRestart();
```

```
    void onResume();
    void onPause();
    void onStop();
    void onDestroy();
}
```

这些 Activity 生命周期的事件回调函数将会被 Android 系统自动调用。用户也可以重载这些事件回调函数来完成自己的操作。Activity 生命周期的事件回调函数的说明如表 1-3 所示。

表 1-3　Activity 生命周期的事件回调函数

函　　数	是否可终止	描　　述
onCreate()	否	Activity 启动后第一个被调用的函数，常用来进行 Activity 的初始化，例如，创建 View、绑定数据或恢复信息等
onStrt()	否	当 Activity 显示在屏幕上时，该函数被调用
onRestart()	否	当 Activity 从停止状态进入活动状态前，调用该函数
onResume()	否	当 Activity 能够与用户交互，接受用户输入时，该函数被调用。此时的 Activity 位于栈的栈顶
onPause()	是	当 Activity 进入暂停状态时，该函数被调用，一般用来保存持久的数据或释放占用的资源
onStop()	是	当 Activity 进入停止状态时，该函数被调用
onDestroy()	是	在 Activity 被终止时，即进入非活动状态前，该函数被调用

在 Android 系统中，Activity 的生命周期，以及各个事件回调函数之间的跳转，如图 1-36 所示。

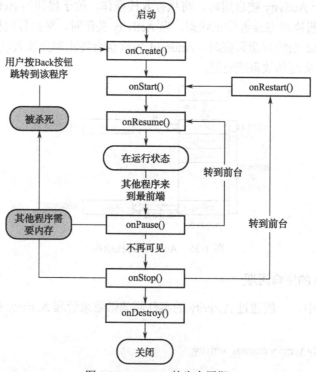

图 1-36　Activity 的生命周期

另外，当由于系统资源紧张时，Activity 可能会被终止。此时，Android 提供了相应的状态保存/恢复的事件回调函数，如表 1-4 所示。

表 1-4　Activity 状态保存/恢复的事件回调函数

函　数	是否可终止	描　述
onSaveInstanceState()	否	Android 系统因资源不足终止 Activity 前调用该函数,用以保存 Activity 的状态,供 onRestoreInstanceState()或 onCreate()恢复用
onRestoreInstanceState()	否	恢复 onSaveInstanceState()保存的 Activity 状态信息,在 onStart()和 onResume()之间被调用

1.5.3　Activity 生命周期调用顺序

一般来说,Activity 的生命周期可分为全生命周期、可视生命周期和活动生命周期三类。每种生命周期中包含不同的事件回调函数,各个事件回调函数的调用顺序也不同,如图 1-37 所示。

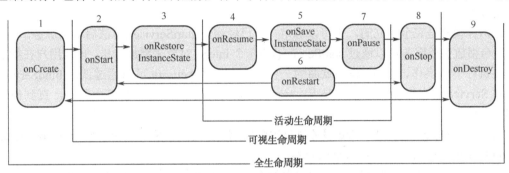

图 1-37　Activity 生命周期的调用顺序图

对于 Activity 全生命周期,事件回调函数的调用顺序为 onCreate()→onStart()→onResume()→onPause()→onStop()→onDestroy()。每一步调用的含义如下:

- 调用 onCreate()函数分配资源。
- 调用 onStart()将 Activity 显示在屏幕上。
- 调用 onResume()获取屏幕焦点。
- 调用 onPause()、onStop 和 onDestroy(),释放资源并销毁进程。

对于 Activity 可视生命周期,事件回调函数的调用顺序为 onSaveInstanceState()→onPause()→onStop()→onRestart()→onStart()→onResume()。每一步调用的含义如下:

- 调用 onSaveInstanceState()函数保存 Activity 状态。
- 调用 onPause()和 onStop(),停止对不可见 Activity 的更新。
- 调用 onRestart()恢复界面上需要更新的信息。
- 调用 onStart()和 onResume()重新显示 Activity,并接受用户交互。

对于活动生命周期,事件回调函数的调用顺序为 onSaveInstanceState()→onPause()→onResume()。每一步调用的含义如下:

- 调用 onSaveInstanceState()保存 Activity 的状态。
- 调用 onPause()停止与用户交互。
- 调用 onResume()恢复与用户的交互。

1.5.4　Service 的生命周期

Service 生命周期一般有两种使用方式。

Service 可以被启动或者允许被启动直到有人停止了它或者它自己停止了。在这种模式下，它通过 Context.startService()方法开始，通过 Context.stopService()方法停止。它可以通过 Service.stopSelf()方法或者 Service.stopSelfResult()方法来停止自己。只要调用一次 stopService()方法便可以停止服务，无论调用了多少次的启动服务方法。

Service 可以通过定义好的接口来编程，客户端建立一个与 Service 的链接，并使用此链接与 Service 进行通话。通过 Context.bindService()方法来绑定服务，通过 Context.unbindService()方法来关闭服务。多个客户端可以绑定同一个服务。如果 Service 还未被启动，bindService()方法可以启动服务。

这两种模式是完全独立的。你可以绑定一个已经通过 startService()方法启动的服务。例如，一个后台播放音乐服务可以通过 startService()和一个 intend 对象来播放音乐。可能用户在播放过程中要执行一些操作，例如，获取歌曲的一些信息，此时 Activity 可以通过调用 bindServices()方法与 Service 建立连接。这种情况下，stopServices()方法实际上不会停止服务，直到最后一次绑定关闭。

像一个 Activity 那样，Service 服务也有生命周期，也提供了事件回调函数：void onCreate()、void onStart(Intent intent)、void onDestroy()。

通过实现这三个生命周期方法，可以监听 Service 的两个嵌套循环的生命周期。

1．Service 整个生命周期

Service 整个生命周期是在 onCreate()和 onDestroy()方法之间。和 Activity 一样，在 onCreate()方法里初始化，在 onDestroy()方法里释放资源。例如，一个背景音乐播放服务可以在 onCreate()方法里播放，在 onDestroy()方法里停止。

2．Service 活动生命周期

Service 活动生命周期是在 onStart()之后，这个方法会处理通过 startServices()方法传递来的 intent 对象。音乐 Service 可以通过 intent 对象来找到要播放的音乐，然后开始后台播放。

Service 停止时没有相应的回调方法，即没有 onStop()方法。onCreate()方法和 onDestroy()方法针对的是所有的 Service，无论它们是否启动。然而，只有通过 startService()方法启动的 Service 才会被调用 onStart()方法。如果一个 Service 允许其他命令绑定，那么需要实现以下额外的方法：

　　　　IBinder onBind(Intent intent)
　　　　boolean onUnbind(Intent intent)
　　　　void onRebind(Intent intent)

onBind()回调方法会继续传递通过 bindService()传递来的 intent 对象。onUnbind()会处理传递给 unbindService()的 intent 对象。如果 Service 允许绑定，onBind()会返回客户端与服务互相联系的通信频道。如果建立了一个新的客户端与服务的链接，onUnbind()方法可以请求调用 onRebind()方法。

1.5.5　Android 生命周期综合实例

下面通过用 Java 代码来完成 Android 生命周期，其完整代码如下：

```java
import android.app.Activity;
import android.os.Bundle;

public class MyActivity extends Activity
{
    // 在完整生存期开始的时候调用
    @Override
    public void onCreate(Bundle icicle)
    {
        super.onCreate(icicle);
        // 初始化一个活动
    }
    // 在 onCreate 方法完成之后调用，用来恢复 UI 状态
    @Override
    public void onRestoreInstanceState(Bundle savedInstanceState)
    {
        super.onRestoreInstanceState(savedInstanceState);
        //从 savedInstanceState 中恢复 UI 状态
        // 这个 bundle 也被传递给 onCreate
    }
    // 在一个活动进程的后续的可见生存期之前调用
    @Override
    public void onRestart()
    {
        super.onRestart();
        // 当知道这个进程中的活动已经可见之后，载入改变
    }
    // 在可见生存期开始时调用
    @Override
    public void onStart()
    {
        super.onStart();
        // 既然活动可见，就应用任何要求的 UI 改变
    }
    // 在活动状态生存期开始时调用
    @Override
```

```java
public void onResume()
{
    super.onResume();
    //恢复活动所需要的任何暂停的 UI 更新、线程或者进程，但是当不活动
    //的时候，就挂起它们
}
// 在活动状态生存期结束的时候调用，用来保存 UI 状态改变
@Override
public void onSaveInstanceState(Bundle savedInstanceState)
{
    //把 UI 状态改变保存到 savedInstanceState
    //如果进程被销毁或者重启，那么这个 bundle 将被传递给
    //onCreate super.onSaveInstanceState(savedInstanceState)
}
// 在活动状态生存期结束时调用
@Override
public void onPause()
{
    //当活动不是前台的活动状态的活动时，挂起不需要更新的 UI 更新、
    //线程或者 CPU 密集的进程
    super.onPause();
}
// 在可见生存期结束时调用
@Override
public void onStop()
{
    // 当进程不可见的时候，挂起不需要的剩下的 UI 更新、线程或者处理，
    // 保存所有的编辑或者状态改变，因为这个进程可能会被销毁（从而为
    // 其他进程释放资源）
    super.onStop();
}
// 在完整生存期结束时调用
@Override
public void onDestroy()
{
    // 清空所有的资源，包括结束线程、关闭数据库链接等
    super.onDestroy();
}
}
```

1.6 Android 的活动

要了解一个活动所经历的各个阶段,最好的办法是创建一个新项目,实现各种事件,然后使活动经受各种用户交互的考验。

【例 1-1】理解一个活动的生命周期。

【实现步骤】

(1) 启动 Eclipse,创建一个名为 Activities 的工程。

(2) 在 MainActivity.java 文件中添加下列加框的语句:

```java
package com.android.tools.sdkcontroller.activities;
……
public class MainActivity extends BaseBindingActivity {
    String tag="Events";
    ……
    @Override
    public void onCreate(Bundle savedInstanceState) {
        super.onCreate(savedInstanceState);
        setContentView(R.layout.main);

        log.d(tag,"In the onCreate() event");

        mTextError = (TextView) findViewById(R.id.textError);
        mTextStatus = (TextView) findViewById(R.id.textStatus);

        WebView wv = (WebView) findViewById(R.id.webIntro);
        wv.loadUrl("file:///android_asset/intro_help.html");
        setupButtons();
    }
    @Override
    protected void onResume() {
        super.onResume();
        Log.d(tag,"In the onResume() event");
    }
    @Override
    protected void onPause() {
        super.onPause();
        Log.d(tag,"In the onPause() event");
    }
    @Override
```

```java
protected void onStop(){
    super.onStop();
    Log.d(tag,"In the onStop() event");
}

@Override
protected void onDestroy(){
    super.onDestroy();
    Log.d(tag,"In the onDestroy() event");
}           super.onStart();
    Log.d(tag, "In the onStart() event");
}
public void onRestart()
{
    super.onRestart();
    Log.d(tag, "In the onRestart() event");
}
public void onResume()
{
    super.onResume();
    Log.d(tag, "In the onResume() event");
}
public void onPause()
{
    super.onPause();
    Log.d(tag, "In the onPause() event");
}
public void onStop()
{
    super.onStop();
    Log.d(tag, "In the onStop() event");
}
public void onDestroy()
{
    super.onDestroy();
    Log.d(tag, "In the onDestroy() event");
}
}
```
……

（3）按 F11 键在 Android 模拟器上调试应用程序。
（4）当活动第一次被加载时，应该可以在 LogCat 窗口中看到以下内容（单击 Debug 透视图）：

08-07 03:41:27.779: DEBUG/Events(378):In the onCreate() event
08-07 03:41:27.779: DEBUG/Events(378):In the onStart() event
08-07 03:41:27.779: DEBUG/Events(378):In the onResume() event

（5）如果在 Android 模拟器上按 Back 按钮，可观察以下显示内容：

08-07 03:55:27.779: DEBUG/Events(378):In the onPause() event
08-07 03:55:27.779: DEBUG/Events(378):In the onStop() event
08-07 03:55:27.779: DEBUG/Events(378):In the onDestroy() event

（6）按住 Home 按钮不放，同时单击 Activities 图标可以看到以下内容：

08-07 04:05:27.779: DEBUG/Events(378):In the onCreate () event
08-07 04:05:27.779: DEBUG/Events(378):In the onStart () event
08-07 04:05:27.779: DEBUG/Events(378):In the onResume () event

（7）按下 Android 模拟器上的 Phone 按钮，当前活动就会被推到后台，观察 LogCat 窗口中的输出：

08-07 04:15:27.779: DEBUG/Events(378):In the onPause() event
08-07 04:15:27.779: DEBUG/Events(378):In the onStop() event

（8）注意，onDestroy()事件并没有被调用，表明这个活动仍旧在内存中。按 Back 按钮退出电话拨号程序，活动又再次显示了。观察 LogCat 窗口中的输出：

08-07 04:18:27.779: DEBUG/Events(378):In the onCreate() event
08-07 04:18:27.779: DEBUG/Events(378):In the onStart () event
08-07 04:18:27.779: DEBUG/Events(378):In the onResume () event

1.6.1 应用活动的样本与主题

默认情况下，一个活动占据整个屏幕。然而，也可以对活动应用一个对话框主题，使其显示为一个浮动对话框。例如，打算定制一个活动，以弹出窗口的形式显示它，用来提醒用户执行的一些操作。在这种情况下，以对话框形式显示活动来引起用户的注意是个很好的方法。

要对活动应用对话框主题，只要修改 AndroidManifest.xml 文件中 Activity 元素，添加 android:theme 属性即可：

```
<?xml version="1.0" encoding="utf-8"?>
<manifest xmlns:android="http://schemas.android.com/apk/res/android"
    package="net.learn2develop.Activities"
    android:versionCode="1"
    android:versionName="1.0">
    <application android:icon="@drawable/icon" android:label="@string/app_name">
        <activity android:name=".MainActivity"
            android:label="@string/app_name"
            android:theme="@android:style/Theme.Dialog" >
            <!-- android:theme="@android:style/Theme.Dialog" -->
```

```
            <intent-filter>
                <action android:name="android.intent.action.MAIN" />
                <category android:name="android.intent.category.LAUNCHER" />
            </intent-filter>
        </activity>
    </application>
    <uses-sdk android:minSdkVersion="9" />
</manifest>
```

这样可以使用活动显示一个对话框，如图 1-38 所示。

图 1-38　显示对话框

1.6.2　显示对话框

用户经常需要显示一个对话框窗口，以便从用户那里得到确认。这时，可以重写在 Activity 基类中定义的受保护的 onCreateDialog()方法来显示一个对话框。

【实现步骤】

（1）在 Eclipse 环境下，创建一个名为 ExampeDialog 的工程。

（2）打开 res/values 目录下的 strings.xml，在其中输入如下代码。

```
<resources>
    <string name="app_name">ExampleDialog</string>
    <string name="hello_world">ExampleDialog</string>
    <string name="menu_settings">Settings</string>
    <string name="title_activity_main">ExampleDialog</string>
</resources>
```

（3）打开 res/layout 目录下的 main.xml 文件，将其中已有的代码替换为如下代码。

```
<?xml version="1.0" encoding="utf-8"?>
<LinearLayout xmlns:android="http://schemas.android.com/apk/res/android"
    android:orientation="vertical"
    android:layout_width="fill_parent"
    android:layout_height="fill_parent" >
    <TextView
```

```
        android:layout_width="fill_parent"
        android:layout_height="wrap_content"
        android:text="@string/hello_world" />
    <Button
        android:id="@+id/btn_dialog"
        android:layout_width="fill_parent"
        android:layout_height="wrap_content"
        android:text="Click to display a dialog" />
</LinearLayout>
```

（4）打开 src/com.example.exampledialog 目录下的 MainActivity.java 文件，将其中代码代换为如下代码。

```java
package com.example.exampledialog;

import android.os.Bundle;
import android.app.Activity;
import android.app.AlertDialog;
import android.app.Dialog;
import android.content.DialogInterface;
import android.view.View;
import android.widget.Button;
import android.widget.Toast;

import android.app.ProgressDialog;
import android.os.Handler;
import android.os.Message;

public class MainActivity extends Activity {
    CharSequence[] items = { "Google", "IBM", "Microsoft" };
    boolean[] itemsChecked = new boolean [items.length];

    /** 当活动第一次被创建时调用 */
    @Override
    public void onCreate(Bundle savedInstanceState) {
        super.onCreate(savedInstanceState);
        setContentView(R.layout.main);

        Button btn = (Button) findViewById(R.id.btn_dialog);
        btn.setOnClickListener(new View.OnClickListener() {
            public void onClick(View v) {
                showDialog(0);
            }
```

```java
            });
        }
        @Override
        protected Dialog onCreateDialog(int id) {
            switch (id) {
            case 0:
                return new AlertDialog.Builder(this)
                    .setIcon(R.drawable.ic_action_search)
                    .setTitle("这是一个带简单文本的对话框")
                    setPositiveButton("确定", new DialogInterface.OnClickListener(){
                            public void onClick(DialogInterface dialog,
                            int whichButton)
                            {
                                Toast.makeText(getBaseContext(),
                                    "确定 clicked!", Toast.LENGTH_SHORT).show();
                            }
                    });
                    .setNegativeButton("取消", new
                        DialogInterface.OnClickListener() {
                        public void onClick(DialogInterface dialog,
                            int whichButton)
                        {
                            Toast.makeText(getBaseContext(),
                                "取消 clicked!", Toast.LENGTH_SHORT).show();
                        }
                    });
                    .setMultiChoiceItems(items, itemsChecked, new
                        DialogInterface.OnMultiChoiceClickListener() {
                            @Override
                        public void onClick(DialogInterface dialog, int which, boolean isChecked) {
                                Toast.makeText(getBaseContext(),
                                items[which] + (isChecked ? " 选中!": "未选中!"),
                                    Toast.LENGTH_SHORT).show();
                            }
                        }
                    )
                    .create();
            }
            return null;
        }
    }
```

按 F11 键在 Android 模拟器上调试应用程序。单击"Click to display a dialog"按钮得到如图 1-39 所示的效果。

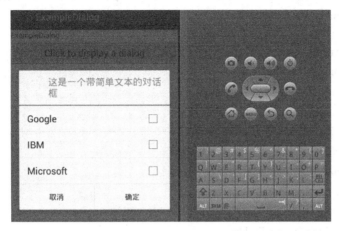

图 1-39　单击按钮显示对话框

以上代码使用 setPositionButton()和 setNegativeButton()方法分别设置了两个按钮："确定"和"取消"。还可以通过 setMultiChoiceItems()方法设置一个复选框列表供用户选择。对于 setMultiChoiceItems()方法，需要传入两个数组：一个是要显示的项列表，另一个包含了表明每个项是否被选中的值。当选中一个项时，使用 Toast 类来显示一条信息，如图 1-40 所示。

图 1-40　使用 Toast 类来显示信息效果

1.6.3　显示进度条对话框

除了前面介绍的普通对话框外，还可以创建进度条对话框。进度条对话框对于显示一些活动的进度有用，如下载操作的状态。

【实现步骤】
（1）使用 1.6.2 节创建的同一个工程，在 MainActivity.java 文件中添加以下加框代码。

```java
package com.example.exampledialog;

import android.os.Bundle;
import android.app.Activity;
import android.app.AlertDialog;
import android.app.Dialog;
import android.content.DialogInterface;
import android.view.View;
import android.widget.Button;
import android.widget.Toast;

import android.app.ProgressDialog;
import android.os.Handler;
import android.os.Message;

public class MainActivity extends Activity {
    CharSequence[] items = { "Google", "IBM", "Microsoft" };
    boolean[] itemsChecked = new boolean [items.length];

    private ProgressDialog _progressDialog;
    private int _progress = 0;
    private Handler _progressHandler;

    /** 当活动第一次被创建时调用*/
    @Override
    public void onCreate(Bundle savedInstanceState) {
        super.onCreate(savedInstanceState);
        setContentView(R.layout.main);

        Button btn = (Button) findViewById(R.id.btn_dialog);
        btn.setOnClickListener(new View.OnClickListener() {
            public void onClick(View v) {
                //使用一个 Handler 对象运行一个后台线程
                showDialog(1);
                _progress = 0;
                _progressDialog.setProgress(0);
                _progressHandler.sendEmptyMessage(0);
            }
        }
```

```java
        });
        //后台线程计数到100,每一次计数延迟100ms
        _progressHandler = new Handler() {
            public void handleMessage(Message msg) {
                super.handleMessage(msg);
                if (_progress >= 100) {
                    _progressDialog.dismiss();
                } else {
                    _progress++;
                    _progressDialog.incrementProgressBy(1);
                    _progressHandler.sendEmptyMessageDelayed(0, 100);
                }
            }
        };
    }

    @Override
    protected Dialog onCreateDialog(int id) {
        switch (id) {
        case 0:
            return new AlertDialog.Builder(this)
            ......
            ......
            .create();

            case 1:
            //创建一个 PropgressDialog 类的实例并设置其不同的属性
                _progressDialog = new ProgressDialog(this);
                _progressDialog.setIcon(R.drawable.ic_action_search);
                _progressDialog.setTitle("下载文件中...");

//设置在进度条对话框中显示的两个按钮
_progressDialog.setProgressStyle(ProgressDialog.STYLE_HORIZONTAL);
    _progressDialog.setButton(DialogInterface.BUTTON_POSITIVE, "隐藏", new
                DialogInterface.OnClickListener() {
                    public void onClick(DialogInterface dialog,
                        int whichButton)
                    {
                        Toast.makeText(getBaseContext(),"隐藏 clicked!", Toast.LENGTH_SHORT).show();
                    }
                });
```

```
_progressDialog.setButton(DialogInterface.BUTTON_NEGATIVE, "取消", new
        DialogInterface.OnClickListener() {
        public void onClick(DialogInterface dialog,
            int whichButton)
        {
        Toast.makeText(getBaseContext(),"取消 clicked!", Toast.LENGTH_SHORT).show();
        }
    });
    return _progressDialog;
    }
    return null;
}
```

(2)按 F11 键在 Android 模拟器上对应用程序进行调试。单击按钮即可显示进度条对话框,如图 1-41 所示。注意观察进度条将累计到 100。

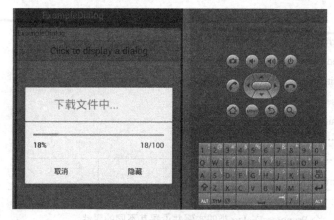

图 1-41 进度条对话框

第 2 章　Android 界面布局

对于网站开发人员来说，网站结构和界面设计是影响浏览用户的第一视觉的关键。而对于 Android 应用开发来说，除了功能强大的应用程序外，屏幕界面效果也是影响程序质量的重要元素。因为消费者永远喜欢界面既美观，功能又强大的软件产品，因此在设计优美界面前，一定要先对屏幕进行布局。

2.1　控件类概述

2.1.1　View 简介

在介绍 Android 的布局管理器之前，有必要让读者了解 Android 平台下的控件类。首先要了解的是 View 类，该类为所有可视化控件的基类，主要提供了控件绘制和事件处理的方法。创建用户界面所使用的控件都继承自 View，如 TextView、Button、CheckBox 等。

关于 View 及其子类的相关属性，既可以在布局 XML 文件中进行设置，也可以通过成员方法在代码中动态设置。View 类常用的属性及其对应方法如表 2-1 所示。

表 2-1　View 类常用属性及其对应方法说明

属性名称	对应方法	描述
android:background	setBackgroundResource(int)	设置背景
android:clickable	setClickable(boolean)	设置界面点击效果
android:visibility	setVisibility(int)	控制界面可见性
android:focusable	setFocusable(boolean)	控制界面可聚焦
android:id	setId(int)	设置界面索引号
android:longClickable	setLongClickable(boolean)	设置界面长单击效果
android:soundEffectsEnabled	setSoundEffectsEnabled(boolean)	设置音效启动效果
android:saveEnabled	setSaveEnabled(boolean)	设置界面初启动性能
android:nextFocusDown	setNextFocusDownId(int)	定义当向下搜索时应该获取焦点的 View，如果该 View 不存在或不可见，则会抛出 RuntimeException 异常
android:nextFocusLeft	setNextFocusLeftId(int)	定义当向左搜索时应该获取焦点的 View
android:nextFocusRight	setNextFocusRightId(int)	定义当向右搜索时应该获取焦点的 View
android:nextFocusUp	setNextFocusUpId(int)	定义当向上搜索时应该获取焦点的 View，如果该 View 不存在或不可见，则会抛出 RuntimeException 异常

说明：任何继承自 View 的子类都将拥有 View 类的以上属性及对应方法。

2.1.2 ViewGroup 简介

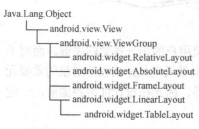

图 2-1 布局管理器的类结构

另外一个需要了解的是 ViewGroup 类，它也是 View 类的子类，但是可以充当其他控件的容器。ViewGroup 的子控件既可以是普通的 View，也可以是 ViewGroup，实际上，这是使用了 Composite 的设计模式。Android 中的一些高级控件如 Galley、GridView 等都继承自 ViewGroup。

与 Java SE 不同，Android 中并没有设计布局管理器，而是为每种不同的布局提供了一个 ViewGroup 的子类，常用的布局及其类结构如图 2-1 所示。

2.2 线性布局

线性布局的形式可以分为两种：第一种是横向线性布局；第二种是纵向线性布局。同时，使用此布局时可以通过设置控件的 Weight 参数控制各个控件在容器中的相对大小。LinearLayout 布局的属性既可以在布局文件（XML）中设置，也可以通过成员方法进行设置。

2.2.1 LinearLayout 简介

表 2-2 给出了 LinearLayout 常用的属性及这些属性的对应设置方法。

表 2-2 LinearLayout 常用属性及对应方法

属性名称	对应方法	描述
android:orientation	setOrientation(int)	设置线性布局的朝向，可取 horizontal
android:gravity	setGravity(int)	设置线性布局的内部元素的布局方式

在线性布局中可使用 gravity 属性来设置控件的对齐方式，gravity 可取的值及说明如表 2-3 所示。

提示：当需要为 gravity 设置多个值时，用"|"分隔即可。

表 2-3 gravity 可取的属性及说明

属性值	描述
top	不改变控件大小，对齐到容器顶部
bottom	不改变控件大小，对齐到容器底部
left	不改变控件大小，对齐到容器左侧
right	不改变控件大小，对齐到容器右侧
center_vertical	不改变控件大小，对齐到容器纵向中央位置

续表

属 性 值	描 述
center_horizontal	不改变控件大小，对齐到容器横向中央位置
center	不改变控件大小，对齐到容器中央位置
fill_vertical	若有可能，纵向拉伸以填满容器
fill_horizontal	若有可能，横向拉伸以填满容器
fill	若有可能，纵向和横向同时拉伸以填满容器

2.2.2 线性布局实例介绍

【实现步骤】

（1）在 Eclipse 中新建一个名为 Example1 的工程，首先打开 res/values 目录下的 strings.xml，在其中输入如下代码。

```xml
<?xml version="1.0" encoding="utf-8"?>
<resources>
    <string name="hello">Example1</string>
    <string name="app_name">Example1</string>
</resources>
```

说明：在 strings.xml 中主要声明了程序中要用到的字符串资源，这样将所有字符串资源统一管理有助于提高程序的可读性及可维护性。

（2）打开 res/layout 目录下的 main.xml，将其中已有的代码替换为如下代码。

```xml
<?xml version="1.0" encoding="utf-8"?>
<LinearLayout xmlns:android="http://schemas.android.com/apk/res/android"
    android:layout_width="fill_parent"
    android:layout_height="fill_parent"
    android:orientation="horizontal">     <!--设置线性布局为水平方向-->
    <Button android:id="@+id/button1"
        android:layout_width="wrap_content"
        android:layout_height="wrap_content"
        android:text="Button1"
        android:layout_weight="1"     <!--设置正比例分配控件范围-->
        />
    <Button android:id="@+id/button2"
        android:layout_width="wrap_content"
        android:layout_height="wrap_content"
        android:text="Button2"
        android:layout_weight="1" />
    <Button android:id="@+id/button3"
        android:layout_width="wrap_content"
        android:layout_height="wrap_content"
```

```
            android:text="Button3"
            android:layout_weight="1" />
        <Button android:id="@+id/button4"
            android:layout_width="wrap_content"
            android:layout_height="wrap_content"
            android:text="Button4"
            android:layout_weight="1" />
        <Button android:id="@+id/button5"
            android:layout_width="wrap_content"
            android:layout_height="wrap_content"
            android:text="Button5"
            android:layout_weight="1" />
</LinearLayout>
```

运行程序，效果如图 2-2 所示。

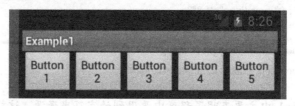

图 2-2 线性布局效果图

在以上代码中，在根 LinearLayout 视图组（ViewGroup）中包含了 5 个 Button，它的子元素是以线性方式（horizontal，水平的）布局的。

2.3 表格布局

本节将要介绍的布局管理器是表格布局，首先将对 TableLayout 类进行简单的介绍，然后通过一个例子来说明表格布局的用法。

2.3.1 TableLayout 简介

TableLayout 类以行和列的形式管理控件，每行为一个 TableRow 对象，也可以为一个 View 对象，当为 View 对象时，该 View 对象将跨越该行的所有列。在 TableRow 中可以添加子控件，每添加一个子控件为一列。

TableLayout 布局中并不会为每一行、每一列或每个单元格绘制边框，每一行可以有 0 或多个单元格，每个单元格为一个 View 对象。TableLayout 中可以有空的单元格，单元格也可以像 HTML 中那样跨越多个列。

在表格布局中，一个列的宽度由该列中最宽的那个单元格指定，而表格的宽度是由父容器指定的。在 TableLayout 中，可以为列设置三种属性。

- Shrinkable,如果一个列被标识为 Shrinkable,则该列的宽度可以进行收缩,以使表格能够适应其父容器的大小。
- Stretchable,如果一个列被标识为 Stretchable,则该列的宽度可以进行拉伸,以使填满表格中空闲的空间。
- Collapsed,如果一个列被标识为 Collapsed,则该列将会被隐藏。

注意:

一个列可以同时具有 Shrinkable 和 Stretchable 属性,在这种情况下,该列的宽度将任意拉伸或收缩以适应父容器。

TableLayout 继承自 LinearLayout 类,除了继承来自父类的属性和方法,TableLayout 类中还包含表格布局所特有的属性和方法。这些属性及对应方法说明如表 2-4 所示。

表 2-4 TableLayout 类常用属性及对应方法说明

属性名称	对应方法	描 述
android:collapseColumns	setColumnCollapsed(int,boolean)	设置指定列号的列为 Collapsed,列号从 0 开始
android:shrinkColumns	setShrinkAllColumns(boolean)	设置指定列号的列为 Shrinkable,列号从 0 开始
android:stretchColumns	setStretchAllColumns(boolean)	设置指定列号的列为 Stretchable,列号从 0 开始

说明: setShrinkAllColumns 和 setStretchAllColumns 实现的功能是将表格中的所有列设置为 Shrinkable 或 Stretchable。

2.3.2 表格布局实例介绍

【实现步骤】

(1) 使用 Eclipse 创建一个名为 Example2 的工程。

(2) 打开 res/values 目录下的 strings.xml,在其中输入如下代码。

```
<?xml version="1.0" encoding="utf-8"?>
<resources>
    <string name="hello">Example2</string>
    <string name="app_name">Example2</string>
</resources>
```

(3) 打开 res/layout 目录下的 main.xml 文件,将其中已有的代码替换为如下代码。

```
<?xml version="1.0" encoding="utf-8"?>
    <TableLayout xmlns:android="http://schemas.android.com/apk/res/android"
        android:layout_width="fill_parent"
        android:layout_height="fill_parent"
        >
        <TableRow
            android:layout_width="wrap_content"
            android:layout_height="fill_parent"
            android:padding="14dip">
```

```xml
<TextView
    android:text="姓名"
    android:gravity="left"
    />
<TextView
    android:text="电话"
    android:gravity="right"/>
</TableRow>

<View
    android:layout_height="2dip"
    android:background="#FFFFFF" />

<TableRow
    android:layout_width="wrap_content"
    android:layout_height="fill_parent"
    android:padding="14dip">
    <TextView
        android:text="AA"
        android:gravity="left"
        />
    <TextView
        android:text="000-555-111"
        android:gravity="right"/>
</TableRow>

<TableRow
    android:layout_width="wrap_content"
    android:layout_height="fill_parent"
    android:padding="14dip">
    <TextView
        android:text="BB"
        android:gravity="left" />
    <TextView
        android:text="222-000-333"
        android:gravity="right"/>
</TableRow>
<TableRow
    android:layout_width="wrap_content"
    android:layout_height="fill_parent"
    android:padding="14dip">
```

```xml
<TextView
    android:text="AB"
    android:gravity="left" />
<TextView
    android:text="222-333-111"
    android:gravity="right"/>
</TableRow>
<TableRow
    android:layout_width="wrap_content"
    android:layout_height="fill_parent"
    android:padding="14dip" >
    <TextView
        android:text="姓名"
        android:gravity="left" />
    <TextView
        android:text="性别"
        android:gravity="right"/>
</TableRow>
<View
    android:layout_height="2dip"
    android:background="#FFFFFF" />
<TableRow
    android:layout_width="wrap_content"
    android:layout_height="fill_parent"
    android:padding="14dip" >
    <TextView
        android:text="AA"
        android:gravity="left" />
    <TextView
        android:text="女"
        android:gravity="right"/>
</TableRow>
<TableRow
    android:layout_width="wrap_content"
    android:layout_height="fill_parent"
    android:padding="14dip">
    <TextView
        android:text="BB"
        android:gravity="left" />
    <TextView
        android:text="男"
```

```
                    android:gravity="right"/>
            </TableRow>
            <TableRow
                android:layout_width="wrap_content"
                android:layout_height="fill_parent"
                android:padding="14dip">
                <TextView
                    android:text="AB"
                    android:gravity="left"    />
            <TextView
                android:text="男"
                android:gravity="right"/>
            </TableRow>
</TableLayout>
```

运行程序，效果如图 2-3 所示。

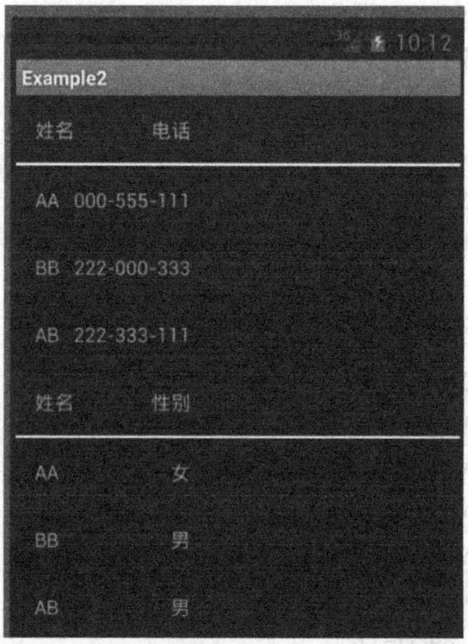

图 2-3 表格布局

2.4 相对布局

本节将要介绍的是相对布局。相对布局比较容易理解，首先介绍 RelativeLayout 类的相关知识，然后通过一个实例来说明相对布局的使用。

2.4.1 RelativeLayout 简介

在相对布局中,子控件的位置是相对兄弟控件或父容器而决定的。出于性能考虑,在设计相对布局时要按照控件之间的依赖关系排列,例如,View A 的位置是由相对于 View B 的位置来决定的,则需要保证在布局文件中 View B 在 View A 的前面。

在进行相对布局时用到的属性很多,首先来看属性值只取 true 或 false 的属性,如表 2-5 所示。

表2-5 相对布局中只取 true 或 false 的属性

属 性 名 称	描 述
android:layout_centerHorizontal	当前控件位于父控件的横向中间位置
android:layout_centerVertical	当前控件位于父控件的纵向中间位置
android:layout_centerInParent	当前控件位于父控件的中央位置
android:layout_alignParentBottom	当前控件底端与父控件底端对齐
android:layout_alignParentLeft	当前控件左侧与父控件左侧对齐
android:layout_alignParentRight	当前控件右侧与父控件右侧对齐
android:layout_alignParentTop	当前控件顶端与父控件顶端对齐
android:layout_alignWithParentIfMissing	参照控件不存在或不可见时参照父控件

接下来再来看属性值为其他控件 id 的属性,如表 2-6 所示。

表2-6 相对布局中取值为其他控件 id 的属性及说明

属 性 名 称	描 述
android:layout_toRightOf	使当前控件用于给出 id
android:layout_toLeftOf	使当前控件用于给出 id
android:layout_above	使当前控件用于给出 id
android:layout_below	使当前控件用于给出 id
android:layout_alignTop	使当前控件的上边界给出 id
android:layout_alignBottom	使当前控件的下边界给出 id
android:layout_alignLeft	使当前控件的左边界给出 id
android:layout_alignRight	使当前控件的右边界给出 id

最后要介绍的是属性值以像素为单位的属性及说明,如表 2-7 所示。

表2-7 相对布局中取值为像素的属性及说明

属 性 名 称	描 述
android:layout_marginLeft	当前控件左侧的留白
android:layout_marginRight	当前控件右侧的留白
android:layout_marginTop	当前控件上方的留白
android:layout_marginBottom	当前控件下方的留白

注意：在进行相对布局时要避免出现循环依赖，例如，设置相对布局在父容器中的排列方式为 WRAP_CONTENT，就不能再将相对布局的子控件设置为 ALIGN_PARENT_BOTTOM。因为这样会造成子控件和父控件相互依赖和参照的错误。

2.4.2 相对布局实例介绍

【实现步骤】

（1）使用 Eclipse 创建一个名为 Example3 的工程。

（2）打开 res/values 目录下的 strings.xml 文件，在其中输入如下代码。

```xml
<resources>
    <string name="app_name">Example3</string>
    <string name="hello_world">Example3</string>
    <string name="menu_settings">Settings</string>
    <string name="title_activity_main">MainActivity</string>
</resources>
```

（3）打开 res/layout 目录下的 main.xml 文件，将其中已有的代码替换为如下代码。

```xml
<?xml version="1.0" encoding="utf-8"?>
<RelativeLayout xmlns:android="http://schemas.android.com/apk/res/android"
    android:layout_width="wrap_content"
    android:layout_height="wrap_content"
    android:gravity="top" >
    <EditText
        android:id="@+id/re_edit_0"
        android:layout_width="wrap_content"
        android:layout_height="wrap_content"
        android:layout_alignParentRight="true"
        android:text="相对布局" />
    <TextView
        android:layout_width="wrap_content"
        android:layout_height="wrap_content"
        android:layout_above="@id/re_edit_0"
        android:background="#FF0000"
        android:text="学习中！！！ "
        android:textColor="#000000"
        android:textSize="18dip" />
    <TextView
        android:id="@+id/textView1"
        android:layout_width="wrap_content"
        android:layout_height="wrap_content"
        android:layout_below="@+id/re_edit_1"
```

```
                    android:layout_marginTop="88dp"
                    android:layout_toLeftOf="@+id/re_edit_1"
                    android:background="#FF0000"
                    android:text="工作中！！！ "
                    android:textColor="#000000"
                    android:textSize="18dip" />
        <EditText
                    android:id="@+id/re_edit_1"
                    android:layout_width="wrap_content"
                    android:layout_height="wrap_content"
                    android:layout_below="@+id/textView1"
                    android:layout_marginTop="93dp"
                    android:layout_toLeftOf="@+id/textView1"
                    android:ems="10"
                    android:text="相对布局" />
        </RelativeLayout>
```

运行程序，效果如图 2-4 所示。

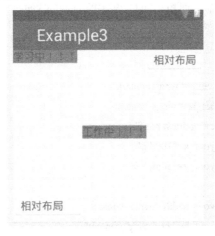

图 2-4 相对布局

2.5 绝对布局

本节要介绍的绝对布局是一种用起来比较费时的布局管理器。首先介绍 AbsoluteLayout 类的相关知识，然后通过一个案例来说明绝对布局的用法。

2.5.1 AbsoluteLayout 简介

所谓绝对布局是指屏幕中所有控件的摆放由开发人员通过设置控件的坐标来指定，控件容器不再负责管理其子控件的位置。由于子控件的位置和布局都通过坐标来指定，因此 AbsoluteLayout 类中并没有开发特有的属性和方法。

2.5.2 绝对布局实例介绍

【实现步骤】

（1）在 Eclipse 中新建一个名为 Example4 的工程。打开 res/values 目录下的 strings.xml 文件，在其中输入如下代码。

```xml
<?xml version="1.0" encoding="utf-8"?>
<resources>
    <string name="hello">Example4</string>
    <string name="app_name">Example4</string>
</resources>
```

（2）把图片 f 及 g（图片 f 及 g 为载入自定义的图片）复制到 res/drawable 目录下。
（3）打开 res/layout 目录下的 main.xml 文件，将其中已有的代码替换为如下代码。

```xml
<?xml version="1.0" encoding="utf-8"?>
<AbsoluteLayout xmlns:android="http://schemas.android.com/apk/res/android"
android:layout_width="fill_parent"
android:layout_height="fill_parent">
    <ImageView
        android:layout_width="wrap_content"
        android:layout_height="wrap_content"
        android:src="@drawable/f"
        android:layout_x="100dip"
        android:layout_y="50dip" />
    <ImageView
        android:layout_width="wrap_content"
        android:layout_height="wrap_content"
        android:src="@drawable/g"
        android:layout_x="50dip"
        android:layout_y="300dip" />
    <TextView
        android:layout_width="wrap_content"
        android:layout_height="wrap_content"
        android:text="当前坐标点 x = 50dip y = 300 dip"
        android:background="#FFFFFF"
        android:textColor="#FF0000"
        android:textSize="18dip"
        android:layout_x="30dip"
        android:layout_y="280dip" />
```

```xml
<TextView
    android:layout_width="wrap_content"
    android:layout_height="wrap_content"
    android:layout_x="35dp"
    android:layout_y="85dp"
    android:background="#FFFFFF"
    android:text="当前坐标点  x = 100dip y = 50dip"
    android:textColor="#FF0000"
    android:textSize="18dip" />
</AbsoluteLayout>
```

运行程序，效果如图 2-5 所示。

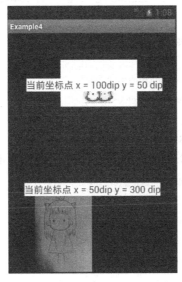

图 2-5 绝对布局

2.6 帧布局

帧布局是最容易理解的一种布局，在本节的内容中，先介绍 FrameLayout 类的相关知识，然后通过一个小实例来说明帧布局的用法。

2.6.1 FrameLayout 简介

FrameLayout 帧布局在屏幕上开辟出了一块区域，在这块区域中可以添加多个子控件，但是所有的子控件都被对齐到屏幕的左上角。帧布局的大小由子控件中尺寸最大的那个子控件来决定。如果子控件一样大，同一时刻只能看到最上面的子控件。

FrameLayout 继承自 ViewGroup，除了继承自父类的属性和方法，FrameLayout 类中包含了自己特有的属性和方法，如表 2-8 所示。

表 2-8　FrameLayout 属性及对应方法

属 性 名 称	对 应 方 法	描　　述
android:foreground	setForeground(Drawable)	设置绘制在所有子控件之上的内容
android:foregroundGravity	setForegroundGravity(int)	设置绘制在所有子控件之上内容的 gravity

提示：在 FrameLayout 中，子控件是通过栈来绘制的，所以后添加的子控件会被绘制在上层。

2.6.2　帧布局实例介绍

【实现步骤】

（1）在 Eclipse 中新建一个名为 Example6 的工程。打开 res/values 目录下的 strings.xml 文件，在其中输入如下代码。

```xml
<?xml version="1.0" encoding="utf-8"?>
<resources>
    <string name="hello">Example6</string>
    <string name="app_name">Example6</string>
</resources>
```

（2）把图片 f 及 g 复制到 res/drawable 目录下。
（3）打开 res/layout 目录下的 main.xml 文件，将其中已有的代码替换为如下代码。

```xml
<?xml version="1.0" encoding="utf-8"?>
<FrameLayout
    xmlns:android="http://schemas.android.com/apk/res/android"
        android:layout_width="fill_parent"
        android:layout_height="fill_parent">
    <ImageView
        android:layout_width="wrap_content"
        android:layout_height="wrap_content"
        android:src="@drawable/f"   />
    <TextView
    android:layout_width="wrap_content"
    android:layout_height="wrap_content"
        android:text="嘻嘻"
        android:background="#FF0000"
        android:textColor="#000000"
        android:textSize="18dip" />
    <ImageView
        android:layout_width="wrap_content"
        android:layout_height="wrap_content"
        android:src="@drawable/g"
        android:layout_gravity="center"   />
```

```
        <EditText
            android:layout_width="wrap_content"
            android:layout_height="wrap_content"
            android:text="幸福快乐"
            android:layout_gravity="center" />
</FrameLayout>
```

运行程序，效果如图 2-6 所示。

图 2-6　帧布局

2.7　列表布局

列表布局是一个 ViewGroup 类，其以列表形式显示它的子视图（view）元素，列表是可滚动的列表。列表元素通过 ListAdapter 自动插入到列表。

2.7.1　ViewGroup 简介

ListAdapter 扩展来自 Adapter，它是 ListView 和数据列表之间的桥梁。ListView 可以显示任何包装在 ListAdapter 中的数据。该类提供了两个公有类型的抽象方法。
- public abstract boolean areAllItemsEnabled ()：表示 ListAdapter 中的所有元素是否是可激活的？如果返回真，即所有的元素是可选择的，即可单击的。
- public abstract boolean isEnabled (int position)：判断指定位置的元素是否是可激活的。

2.7.2　列表布局实例介绍

下面通过一个例子来，创建一个可滚动的列表，并从一个字符串数组读取列表元素。当一个元素被选择时，显示该元素在列表中的位置的消息。

【实现步骤】

(1) 在 Eclipse 中新建一个名为 Example7 的工程。打开 res/values 目录下的 strings.xml 文件,在其中输入如下代码。

```xml
<?xml version="1.0" encoding="utf-8"?>
<resources>
    <string name="hello">Example7</string>
    <string name="app_name">Example7</string>
</resources>
```

(2) 打开 res/layout 目录下的 main.xml 文件,将其中已有的代码替换为如下代码。

```xml
<?xml version="1.0" encoding="utf-8"?>
<TextView xmlns:android="http://schemas.android.com/apk/res/android"
    android:layout_width="fill_parent"
    android:layout_height="fill_parent"
    android:padding="10dp"
    android:textSize="16sp" >
</TextView>
```

(3) 打开项目 src/com.example.example7 目录下的 MainActivity.java 文件,将其中已有的代码替换为如下代码。

```java
package com.example.example7;
import android.app.ListActivity;
import android.os.Bundle;
import android.view.View;
import android.widget.AdapterView;
import android.widget.ArrayAdapter;
import android.widget.ListView;
import android.widget.TextView;
import android.widget.Toast;
import android.widget.AdapterView.OnItemClickListener;
public class MainActivity extends ListActivity
{
    //注意这里 HelloWorld 类不是扩展自 Acitvity,而是扩展自 ListAcitivty
    /** 活动第一次被创建时调用 */
    @Override
    public void onCreate(Bundle savedInstanceState)
    {
        super.onCreate(savedInstanceState);
        setListAdapter(new ArrayAdapter<String>(this, R.layout.main, COUNTRIES));
        ListView lv = getListView();
        lv.setTextFilterEnabled(true);
        lv.setOnItemClickListener(new OnItemClickListener()
        {
```

```
                    public void onItemClick(AdapterView<?> parent, View view, int position, long id)
                    {
                            // 单击时，显示 TextView 的文字
                            Toast.makeText(getApplicationContext(),((TextView) view).getText(),
Toast.LENGTH_SHORT).show();
                    }
                });
        }
        static final String[] COUNTRIES = new String[]
        {
                "1", "2", "3", "4", "5","6", "7", "8", "9", "10","11", "12", "13", "14", "15", "16", "17", "18", "19",
"20","21", "22", "23", "24" };
        }
```

提示：onCreate()函数中并不像往常一样通过 setContentView()为活动（Activity）加载布局文件，替代的是通过 setListAdapter(ListAdapter)自动添加一个 ListView 填充整个屏幕的 ListActivity。在此文件中这个方法以一个 ArrayAdapter 为参数：setListAdapter(new ArrayAdapter<String>(this, R.layout.main, COUNTRIES))，这个 ArrayAdapter 管理填入 ListView 中的列表元素。ArrayAdapter 的构造函数的参数为：this（表示应用程序的上下文 context）、表示 ListViewde 布局文件（这里是 R.layout.main）、插入 ListView 的 List 对象对数组（这里是 COUNTRES）。

setOnItemClickListener（OnItemClickListener）定义了每个元素的单击（on-click）监听器，当 ListView 中的元素被单击时，onItemClick()方法被调用，在这里是一个 Toast 消息——每个元素的位置将显示。

运行程序，单击 6，效果如图 2-7 所示。

图 2-7　视图布局

第 3 章　Widget 组件布局

Widget 是 Android 1.5 版所引进的特性之一。Widget 可让用户在主屏幕界面及时了解程序显示的重要信息。标准的 Android 系统已包含几个 Widget 的示例，如模拟时钟、音乐播放器等。不过这并不能阻止大家开发更加美观、功能更丰富的版本。另外，微博博客、RSS 订阅、股市信息、天气预报这些 Widget 都有流行的可能。

使用 Eclipse 创建一个 Example8 工程作为程序的入口，自动生成的主文件为 MainActivity.java，其代码如下：

```java
package com.example.example8;
import android.os.Bundle;
import android.app.Activity;
import android.view.Menu;
import android.view.MenuItem;
import android.support.v4.app.NavUtils;
public class MainActivity extends Activity {
    @Override
    public void onCreate(Bundle savedInstanceState) {
        super.onCreate(savedInstanceState);
        setContentView(R.layout.main);
    }
    @Override
    public boolean onCreateOptionsMenu(Menu menu)
    {
        getMenuInflater().inflate(R.menu.main, menu);
        return true;
    }
}
```

在上述代码中，关联了一个模板布局文件 main.xml。这样，就可以在里面继续添加需要的控件了，如按钮、列表框、进度条和图片等。

布局文件 main.xml 的代码如下：

```xml
<RelativeLayout xmlns:android="http://schemas.android.com/apk/res/android"
    xmlns:tools="http://schemas.android.com/tools"
    android:layout_width="match_parent"
    android:layout_height="match_parent" >
    <TextView
        android:layout_width="wrap_content"
```

```
            android:layout_height="wrap_content"
            android:layout_centerHorizontal="true"
            android:layout_centerVertical="true"
            android:padding="@dimen/padding_medium"
            android:text="@string/hello_world"
            tools:context=".MainActivity" />
</RelativeLayout>
```

通过以上代码，在手机屏幕中使用了 Widget 组件。执行后不会显示任何信息，这是因为我们还没有在里面添加任何元素。由此可见，Widget 组件只是起了一个"容器"的作用，我们只需要把要显示的屏幕元素添加到这个"容器"中即可。

3.1 Button 控件

Button 控件是一个按钮控件。在项目中应用时，当单击 Button 后会触发一个事件，这个事件会实现用户需要的功能。例如，用户输入一些信息，单击"OK"或"YES"按钮后会实现对应的功能。

3.1.1 setOnClickListener 方法

setOnClickListener 方法用于为按钮绑定一个监听器，用于处理当按钮按下时的动作。在使用时，用户需要重载其内部的 OnClick 方法，并在其中执行相应的动作。该方法使用得最为广泛，主要用于应用程序执行用户动作的场合。其语法格式为：

```
public void setOnClickListener(View.OnClickListener1)
```

通过以下代码来演示单击按钮改变背景颜色。

```
final TextView myTextView=(TextView)findViewById(R.id.myTextView);   //文本框
bt1=(Button)findViewById(R.id.button1);                              //按钮
 bt1.setOnClickListener(listener=new OnClickListener()
{
    //设置监听器
    @Override
    public void onClick(View v)
    {
        // TODO Auto-generated method stub
        Resources res = getResources();                              //资源
        Drawable drawable = res.getDrawable(R.drawable.ic_action_search);
        bg.setBackgroundDrawable(drawable);                          //设置背景图片
        myTextView.setTextColor(Color.RED);                          //设置字体颜色
        myTextView.setText("背景为搜索页面！");                      //设置文本
    }
});
```

在以上代码中，先通过 getWindow 获取窗口对象，然后分别获取文本框和按钮的对象，声明了监听器，接着，通过 setOnClickListener 方法为按钮绑定监听器。在监听器的 OnClick 方法中，重新设置了背景图片、文本内容和字体颜色等。

运行程序，效果如图 3-1 所示。

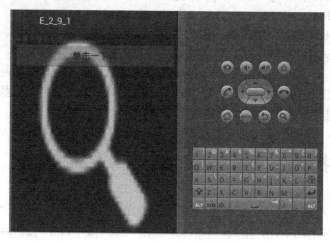

图 3-1　单击来改变背景颜色

3.1.2　setOnLongClickListener 方法

setOnLongClickListener 方法用于为按钮绑定一个监听器，处理当按钮长时间按下时的动作。在使用时，用户需要重载其内部的 OnLongClick 方法，并在其中执行相应的动作。所谓的长按是指单击按钮后长时间不放，主要用于应用程序执行特定用户动作的场合。其语法格式为：

```
public void setOnLongClickListener(View.OnLongClickListener1)
```

通过以下代码来演示怎样实现长按按钮的功能。

```
final Window bg=getWindow();                                    //获取窗口对象
OnLongClickListener longlistener;                               //长按按钮
final Button bt1=(Button)findViewById(R.id.button1);
final TextView myTextView=(TextView)findViewById(R.id.myTextView);
bt1.setOnLongClickListener(longlistener=new OnLongClickListener()
{
    //设置监听器
    @Override
    public boolean onLongClick(View v)
    {
        // TODO:自动生成方法存根
        bt1.setBackgroundColor(Color.RED);                      //设置按钮的背景色
        bt1.setTextColor(Color.BLACK);                          //设置按钮字体颜色
        bt1.setText("长按按钮的操作！");                         //设置按钮文本
        myTextView.setText("按钮的颜色被改变了！");
```

```
                return false;
            }
        });
    }
}
```

在代码中，先是变量声明和对象获取，接着为按钮绑定了长按监听器。在重载内部的 onLongClick 方法时，设置了按钮的背景色、字体颜色和文本内容。

运行程序，效果如图 3-2 所示。

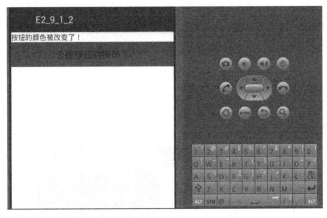

图 3-2　长按按钮的操作效果

3.1.3　setOnFocusChangeListener 方法

setOnFocusChangeListener 方法用于为按钮绑定一个监听器，处理当按钮获得或者失去焦点时的动作。在使用时，用户需要重载其内部的 onFocusChange 方法，并在其中执行相应的动作。所谓的焦点改变，可以是获得焦点或者失去焦点，主要用于应用程序执行待定用户动作的场合。其语法格式为：

```
        public void setOnFocusChangeListener(View.OnLongClickListener1)
```

通过以下代码来演示怎样实现焦点改变的动作。

```
final Window bg=getWindow();                              //获取窗口对象
 OnFocusChangeListener focuslistener;                     //监听器
        final Button bt1=(Button)findViewById(R.id.button1);
        final TextView myTextView=(TextView)findViewById(R.id.myTextView);
        bt1.setOnFocusChangeListener(focuslistener=new OnFocusChangeListener()
        {
            @Override
            public void onFocusChange(View v, boolean hasFocus)
            {
                //设置监听器动作
                // TODO 自动生成方法存根
```

```
                    if(hasFocus)                                    //判断是否获得焦点
                    {
                        Toast.makeText(getApplicationContext(),"按钮获得焦点！",Toast.LENGTH_LONG).show();
                    }
                    else
                    {
                        Toast.makeText(getApplicationContext(),"按钮失去焦点！",Toast.LENGTH_LONG).show();
                    }
                }
            });
        }
    }
```

运行程序，效果如图 3-3 所示。

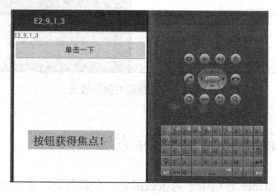

图 3-3　按钮获取焦点

3.1.4　setOnTouchListener 方法

setOnTouchListener 方法用于为按钮绑定一个监听器，处理当按钮被触摸时的动作。在使用时，用户需要重载其内部的 OnTouch 方法，并在其中执行相应的动作。所谓的触摸动作，可以是触摸按下、触摸抬起或触摸移动，主要用于应用程序执行特定用户动作的场合。其语法格式为：

```
        public void setOnTouchListener(View.OnTochListener1)
```

通过以下代码来演示触摸监听器获取触摸点的坐标。

```
        final Button bt1=(Button)findViewById(R.id.button1);              //按钮对象
        final TextView myTextView=(TextView)findViewById(R.id.myTextView); //文本框对象
        bt1.setOnTouchListener(new View.OnTouchListener()
        {
            //设置触摸监听器
```

```
        int px;                                              //触摸点的 x 坐标
        int py;                                              //触摸点的 y 坐标
        @Override
        public boolean onTouch(View v, MotionEvent event)    //重载 onTouch 方法
        {
        // TODO 自动存根法
        switch(event.getAction())
        {
            case MotionEvent.ACTION_DOWN:                    //触摸按下
            px=(int)event.getX();
            py=(int)event.getY();
            myTextView.setText("px="+px+";py="+py);
            break;
            case MotionEvent.ACTION_MOVE:                    //触摸移动
            px=(int)event.getX();
            py=(int)event.getY();
        myTextView.setText("当前触摸点的坐标为: px="+px+", py="+py);
        }
            return false;
         }
     });
   }
 }
```

运行程序，效果如图 3-4 所示。

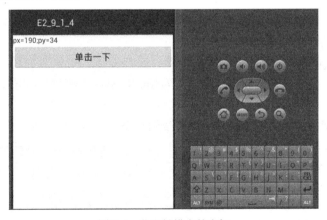

图 3-4　获取触摸点的坐标

3.1.5　setPressed 方法

setPressed 方法用于设置按钮的状态为被按下，主要用于设置按钮初始状态或自动控制按钮动作的场合。其语法格式为：

```
public void setPressd(Boolean pressed)
```
其中，参数 pressed 表示按钮的状态，当其取值为 true 时表示按钮被按下；当取值为 false 时表示按钮未被按下。

通过以下代码来演示如何设置按钮的按下状态。

```java
public class MainActivity extends Activity
{
    /**活动第一次被创建时调用*/
    @Override
    public void onCreate(Bundle savedInstanceState)
    {
        super.onCreate(savedInstanceState);
        setContentView(R.layout.main);
        Button bt1=null;                                        //声明按钮对象
        TextView myTextView=null;                               //声明文本框对象
        bt1=(Button)findViewById(R.id.button1);                 //获取按钮对象
        myTextView=(TextView)findViewById(R.id.myTextView);
    }
}
```

代码中，先声明并获取了按钮及文本框对象，然后通过 setPressed 方法设置按钮被按下的状态。

3.1.6 setClickable 方法

setClickable 方法用于设置按钮是否可以被用户按下，主要用于根据条件有选择地使能按钮功能的场合。

值得注意的是，如果 setClickable 方法之后设置了按钮监听器，那么将重新使按钮处于可按状态。setClickable 方法语法格式为：

```
public void setClickable(Boolean clickable)
```
其中，参数 clickable 代表按钮的状态，当取值为 true 时表示按钮可按；当取值为 false 时，表示按钮不可按。

通过以下代码来演示如何使按钮处于不可按的状态。

```java
final Window bg=getWindow();                                            //获取窗口对象
OnClickListener listener=null;                                          //声明监听器
Button bt1=null;
final TextView myTextView=(TextView)findViewById(R.id.myTextView);      //文本框
bt1=(Button)findViewById(R.id.button1);                                 //按钮
bt1.setOnClickListener(listener=new OnClickListener()
{
    //设置监听器
    @Override
    public void onClick(View v)
```

```java
        {
            // TODO 自动存根法
            Resources res = getResources();                    //资源
            Drawable drawable = res.getDrawable(R.drawable.white);
            bg.setBackgroundDrawable(drawable);                //设置背景图片
            myTextView.setTextColor(Color.RED);                //设置字体颜色
            myTextView.setText("背景已经设置为黄色！");          //设置文本
        }
    });
        bt1.setClickable(false);
    }
}
```

以上代码先通过 getWindow 获取窗口对象，然后分别获取了文本框和按钮的对象，声明了监听器。接着，通过 setOnClickListener 方法为按钮绑定监听器。在监听器的 onClick 方法中，重新设置了背景图片、文本内容和字体颜色等。最后，通过 setClickable 方法设置该按钮不可按。

3.1.7 setLongClickable 方法

setLongClickable 方法用于设置按钮是否可执行长按操作，主要用于根据条件有选择地使能按钮功能的场合。

值得注意的是，如果 setLongClickable 方法之后设置了按钮监听器，那么将重新使按钮处于可长按状态。该方法语法格式为：

```java
        public void setLongClickable(boolean longClickable)
```

其中，参数 longClickable 代表了按钮的状态，当取值为 true 时，表示按钮可长按；当取值为 false 时，表示按钮不可长按。

通过以下代码来演示如何使按钮处于不可长按的状态。

```java
    public void onCreate(Bundle savedInstanceState)
    {
        super.onCreate(savedInstanceState);
        setContentView(R.layout.main);
        final Window bg=getWindow();                           //获取窗口对象
        OnLongClickListener longlistener;                      //长按监听器
        final Button bt1=(Button)findViewById(R.id.button1);
        final TextView myTextView=(TextView)findViewById(R.id.myTextView);
        bt1.setOnLongClickListener(longlistener=new OnLongClickListener()
        {
            //设置监听器
            @Override
            public boolean onLongClick(View v)
            {
```

```
                    //TODO:自动存根法
                       bt1.setBackgroundColor(Color.RED);        //设置按钮的背景色
                       bt1.setTextColor(Color.BLACK);            //设置按钮字体颜色
                       bt1.setText("执行了长按按钮的操作！");      //设置按钮文本
                       myTextView.setText("长按按钮改变了按钮的颜色！");
                       return false;
                   }
               });
                   bt1.setLongClickable(false);
           }
```

在代码中，先是变量声明和对象获取，接着对按钮绑定了长按监听器。在重载内部的 onLongClick 方法时，设置了按钮的背景色、字体颜色和文本内容。最后，通过 setLongClickable 方法设置该按钮不可长按。

3.1.8 Button 控件综合实例

下面通过一个综合例子来实现 Button 用法。下面为工程 Example8 添加代码。

【实现步骤】

（1）修改布局文件 main.xml，在里面添加一个 TextView 和一个 Button。

```
       <LinearLayout xmlns:android="http://schemas.android.com/apk/res/android"
           android:orientation="vertical" android:layout_width="fill_parent"
           android:layout_height="fill_parent">
           <TextView
               android:layout_width="wrap_content"
               android:layout_height="wrap_content"
               android:layout_centerHorizontal="true"
               android:layout_centerVertical="true"
               android:padding="@dimen/padding_medium"
               android:text="@string/Example8"
               />
           <Button android:id="@+id/Click_Button"
               android:layout_width="wrap_content"
               android:layout_height="wrap_content"
               android:text="这是按钮" />
       </LinearLayout>
```

（2）在文件 mainActivity.java 中，先通过 findViewByID()获取 TextView 文本和 Button 按钮的资源，然后为 Button 按钮添加事件监听器 Button.OnClickListence()，最后定义处理事件处理程序。

```
           //获取 TextView 文本和 Button 按钮的资源
               show=(TextView)findViewById(R.id.show_TextView);
```

```
press=(Button)findViewById(R.id.Click_Button);
//为 Button 按钮添加事件监听器 Button.OnClickListener()
press.setOnClickListener(new Button.onClickListener()
{
    @Override
    public void onClick(View v)
    {}
});
//定义处理事件处理程序
press.setOnClickListener(new Button.onClickListener()
{
    @Override
    public void onClick(View v)
    {
        //单击按钮后输入一段文本
        show.setText("单击一下按钮");
    }
});
```

运动后首先显示一个"按钮+文本"样式的界面,当单击界面中的"单击一下按钮"后会执行单击事件。

3.2 TextView 控件

TextView 是一个典型的文本显示控件,该控件只提供了显示的功能,不允许用户对文本进行编辑。Android 系统的 TextView 相当于其他编辑语言中的方框。在程序中,我们主要通过 TextView 来显示文字信息。

3.2.1 append 方法

append 方法用于在 TextView 文本框原有内容的后面再添加显示文本。append 方法通常用于更改文本框显示内容的场合,此文本框内的原有内容不变,仅增加显示字符串。其语法格式为:

    public final void append(CharSequence text)

或

    public void append(CharSequence text,int start,int end)

其中,参数 CharSequence 为需要追加的文本内容;参数 start 为追加文本的起始位置;参数 end 为追加文本的结束位置。

通过以下代码来演示使用两种方式来追加文本框显示的文本。

```java
public class MainActivity extends Activity
{
    /** 活动第一次被创建时调用 */
    @Override
    public void onCreate(Bundle savedInstanceState)
    {
        super.onCreate(savedInstanceState);
        setContentView(R.layout.main);
        TextView myTextView=null;                              //声明变量
        myTextView=(TextView)findViewById(R.id.myTextView);    //通过 ID 获取对象
        myTextView.append("aabbcc");                           //采用第一种方式追加文本
        myTextView.append("AAABBBCCC",2,5);                    //采用第二种方式追加文本
    }
}
```

在代码中，先设置 TextView 变量，然后通过 findViewById 方法获取文本框对象。最后，通过 setText 方法设置需要显示的文本。在例子中可发现使用 setText 方法十分方便，而且支持换行符。

运行程序，效果如图 3-5 所示。

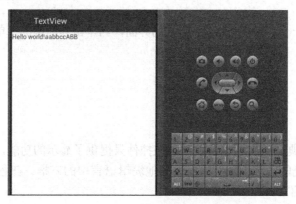

图 3-5　用 append 方法追加文字

3.2.2　addTextChangedListener 方法

addTextChangedListener 方法用于在 TextView 文本框中添加一个文本更改监听器，该方法主要用于监视文本框内容的改变，然后在相应的事件处理函数中执行相应的处理代码。其典型应用与实现文本框和编辑框的文本内容同步。其语法格式为：

```
public void addTextChangedListener(TextWatcher watcher)
```

其中，参数 watcher 为设置的文本内容监听器。

通过以下代码来演示怎样监视文本框内容的改变。

```java
public class MainActivity extends Activity
{
```

```java
/**活动第一次被创建时调用 */
@Override
public void onCreate(Bundle savedInstanceState)
{
    super.onCreate(savedInstanceState);
    setContentView(R.layout.main);
    TextView myTextView=null;                               //声明变量
    TextWatcher watcher=null;
    myTextView=(TextView)findViewById(R.id.myTextView);     //获取对象
    myTextView.setText("11223");                            //设置显示内容
    myTextView.addTextChangedListener(new TextWatcher()
    {
        //添加监听器
        @Override
        public void afterTextChanged(Editable s)
        {
            //文本改变后
            // TODO：自动存根法
        }
        @Override
        public void beforeTextChanged(CharSequence s, int start, int count, int after)
        {
            // TODO 自动生成方法存根
            //文本改变前
        }
        @Override
        public void onTextChanged(CharSequence s, int start, int before, int count)
        {
            // TODO：自动生成方法存根
            //文本改变时
            Toast.makeText(getApplicationContext(), "原字符串共 "+String.valueOf(before)+"个字符。"+"从第"+String.valueOf(start)+"个字符开始更改为字符串："+s+"。共"+String.valueOf(count)+"字符。",
            Toast.LENGTH_LONG).show();                      //显示提示信息
        }
    });
    myTextView.setText("11223445588");                      //重设文本内容
}
```

在以上代码中，先声明了变量，然后通过 findViewById 方法获取文本框对象，并通过 setText 方法来设置显示内容。接着，通过 addTextChargedListener 方法为该文本框添加文本更改监听器 watcher，并重载了该监听器的 3 个事件处理函数。其中，afterTextChanged 函数用于文本改变后，

beforeTextChanged 函数用于文本改变前，onTextChanged 函数用于文本改变时，在此重载的 onTextChanged 函数中显示提示信息。最后，通过 setText 方法来重设内容。此时由于文本框的内容改变，将触发 onTextChanged 函数显示信息。

运行程序，效果如图 3-6 所示。

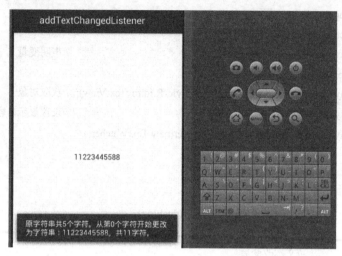

图 3-6　文本更改监听器

3.2.3　setText 方法

setText 方法用于设置一个 TextView 文本框的显示内容，可以直接赋值一个字符串，也可以通过资源 ID 的方法来引用。该方法是设置文本框显示文本最常用的方法。其语法格式为：

```
public final void setText(CharSequence text)
```

其中，参数 CharSequence 为文本字符串的资源 ID；参数 text 为需要设置显示的字符串。

通过以下代码来演示怎样设置文本框字符串。

```
public class MainActivity extends Activity
{
    /* 活动第一次被创建时调用 */
    @Override
    public void onCreate(Bundle savedInstanceState)
    {
        super.onCreate(savedInstanceState);
        setContentView(R.layout.main);
        TextView myTextView=null;                              //设置变量
        myTextView=(TextView)findViewById(R.id.myTextView);    //获取对象
        //设置显示内容
        myTextView.setText("setText 方法：\r\n"+         "+
            "设置的 TextView 文本框用于显示内容，可以直接赋值一个字符串，也可以通过资源 ID 的方式来引用。其是设置文本框显示文本的最常用的方法。");
```

        }
    }

以上代码中，先设置 TextView 变量，然后通过 findViewById 方法获取文本框对象。最后，通过 setText 方法设置需要显示的文本。通过例子可发现使用 setText 方法十分方便，而且支持换行符。

运行程序，效果如图 3-7 所示。

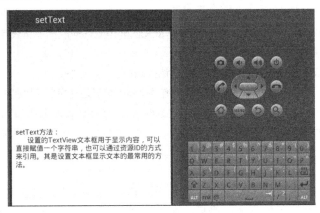

图 3-7　设置文本

3.2.4　setTextSize 方法

setTextSize 方法用于设置一个 TextView 文本框显示内容的字体，可通过指定单位和字体数值大小，也可直接指定 sp(scaled pixel)单位的字体大小。设置文本框显示文本字体大小最常用的是 setTextSize 方法。其语法格式为：

```
public void setTextSize(int unit,float size)
```

或

```
public void setTextSize(float size)
```

其中，参数 size 为字体大小，浮点数；参数 unit 为字体大小的单位，在 Android 中可以取的值包括如下几种。

- COMPLEX_UNIT_DIP：设备独立像素，不同设备有不同的显示效果，与设备硬件有关。一般为了支持 WVGA、HVGA 和 QVGA 等多种分辨率时推荐使用这个，不依赖像素。
- COMPLEX_UNIT_IN：英寸。
- COMPLEX_UNIT_MM：毫米。
- COMPLEX_UNIT_PT：一个标准的长度单位，1pt=1/72 英寸，在印刷业较常见，非常简单易用。
- COMPLEX_UNIT_PX：像素单位，不同设备显示效果相同。
- COMPLEX_UNIT_SP：比较像素单位，主要用于文本字体的显示。

通过以下代码来演示如何为文本框设置字符串的字体大小。

```
public class MainActivity extends Activity
{
```

```
/** 活动第一次被创建时调用 */
@Override
public void onCreate(Bundle savedInstanceState)
{
    super.onCreate(savedInstanceState);
    setContentView(R.layout.main);
    TextView myTextView=null;                                    //声明变量
    myTextView=(TextView)findViewById(R.id.myTextView);          //获得对象
    //设置内容
    myTextView.setText("setTextSize方法：\r\n"+"           "+
        "用 TextView 设置框显示内容的字体,可以通过指定单位和字体数值大小,也可以直接指定 sp（scaled pixel）单位的字体大小。其是设置文本框显示文本字体大小的最常用的方法。");
    myTextView.setTextSize(25);                                  //设置字体
}
```

以上代码中，先设置 TextView 变量，然后通过 findViewById 方法获取文本对象。接着，通过 setText 方法设置需要显示的文本。最后，通过 setTextSize 方法设置字体大小，在此处采用的是 sp 比例像素单位。

运行程序，效果如图 3-8 所示。

图 3-8　设置文本字体大小

3.2.5　setTypeface 方法

setTypeface 方法用于设置画笔的字体样式，可以指定系统自带的字体，也可以使用自定义的字体。该方法是设置画笔显示文本字体最常用的方法。其语法格式为：

public Typeface setTypeface (Typeface typeface)

其中，参数 typeface 为字体样式，具有如下几种取值。

- Typeface.DEFAULT：默认字体。

- Typeface.DEFAULT_BOLD：加粗字体。
- Typeface.MONOSPACE：monospace 字体。
- Typeface.SANS_SERIF：sans 字体。
- Typeface.SERIF：serif 字体。

通过以下代码来演示如何为文本框设置字符串的字体样式。

```
public class MainActivity extends Activity
{
    /* 活动第一次被创建时调用 */
    @Override
    public void onCreate(Bundle savedInstanceState)
    {
        super.onCreate(savedInstanceState);
        setContentView(R.layout.main);
        TextView myTextView=null;                              //声明变量
        myTextView=(TextView)findViewById(R.id.myTextView);    //获取对象
        //设置文本内容
        myTextView.setText("setTypeface 方法： \r\n"+"         "+
            "用于设置 TextView 文本框显示内容的字体样式，可以指定系统自带的字体，也可以使用自定义的字体。该方法是设置文本框显示文本字体的最常用的方法。");
        myTextView.setTextSize(25);                            //设置字体大小
        myTextView.setTextColor(Color.BLUE);                   //设置颜色
        //设置字体样式
        myTextView.setTypeface(Typeface.SANS_SERIF,Typeface.BOLD);
    }
}
```

以上代码中，先设置 TextView 变量，然后通过 findViewById 的方法获取文本框对象。接着，通过 setText 方法设置需要显示的文本，通过 setTextSize 方法设置字体大小，通过 setTextColor 方法来设置字体的颜色。最后通过 setTypeface 方法设置字体样式。

运行程序，效果如图 3-9 所示。

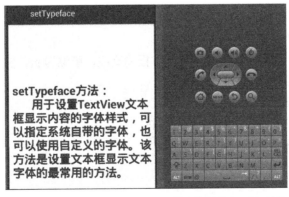

图 3-9　设置字体样式

3.2.6　setTextColor 方法

setTextColor 方法用于设置一个 TextView 文本框显示内容的字体颜色,可以通过 XML 文件的颜色状态列表来设置,或者直接指定颜色。该方法是设置文本框显示文本颜色最常用的方法。其语法格式为:

```
public void setTextColor(ColorStateList colors)
```

或

```
public void setTextColor(int color)
```

其中,参数 colors 为颜色状态列表,这个需要在 XML 文件中定义;参数 color 表示颜色数值。

通过以下代码来演示怎样为文本框设置字符串的字体颜色。

```java
public class MainActivity extends Activity
{
    /* 活动第一次被创建时调用 */
    @Override
    public void onCreate(Bundle savedInstanceState)
    {
        super.onCreate(savedInstanceState);
        setContentView(R.layout.main);
        TextView myTextView=null;                                    //声明变量
        myTextView=(TextView)findViewById(R.id.myTextView);          //获得对象
        //设置文本内容
        myTextView.setText("setTextColor 方法: \r\n"+        "+
                "用于设置一个文本框显示内容的字体颜色,可以通过 XML 文件的颜色状态列表来设置,或者直接指定颜色值。其是设置文本框显示文本颜色的最常用的方法。");
        myTextView.setTextSize(25);                                  //设置字体大小
        myTextView.setTextColor(Color.GRAY);                         //设置字体颜色
    }
}
```

3.2.7　setHeight 方法

setHeight 方法用于设置一个 TextView 文本框的高度,单位为 pt。该方法一般用于设置和更改文本框大小的场合。其语法格式为:

```
public void setHeight(int pixels)
```

其中,参数 pixels 表示文本框的高度,单位为 pt。

通过以下代码来演示怎样设置文本框的高度。

```java
public class MainActivity extends Activity
{
    /* 活动第一次被创建时调用 */
    @Override
```

```
            public void onCreate(Bundle savedInstanceState)
            {
                super.onCreate(savedInstanceState);
                setContentView(R.layout.main);
                TextView myTextView=null;                                    //声明变量
                myTextView=(TextView)findViewById(R.id.myTextView);          //获取对象
                myTextView.setText("setHeight方法：\r\n"+"        "+"用于设置一个文本框的高度,单位是
pt。一般用于设置和更改文本框大小的场合。");
                myTextView.setTextSize(25);                                  //设置字体大小
                myTextView.setTextColor(Color.BLUE);                         //设置字体颜色
                //设置字体样式
                myTextView.setTypeface(Typeface.SANS_SERIF,Typeface.BOLD);
                myTextView. setHeight (450);                                 //设置文本框高度
            }
        }
```

以上代码中，首先设置了 TextView 变量，然后通过 findViewById 方法获取文本框对象。接着，通过 setText 方法设置需要显示的文本，通过 setTextSize 方法设置字体大小，通过 setTextColor 方法来设置字体的颜色，通过 setTypeface 方法设置字体样式，最后，通过 setHeight 方法设置文本框高度为 450。

运行程序，效果如图 3-10 所示。

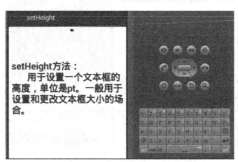

图 3-10　设置文本框的高度

3.2.8　setBackgroundColor 方法

setBackgroundColor 方法用于设置一个文本框的背景颜色，一般通过直接设置颜色（颜色值一定是大写字母）值来表示，是设置文本框背景色最常用的方法。其语法格式为：

```
        public void setBackgroundColor(int color)
```
其中，参数 color 为颜色数值。

通过以下代码来演示怎样为文本框设置背景颜色。

```
        public class MainActivity extends Activity
        {
            /* 活动第一次被创建时调用   */
```

```
            @Override
            public void onCreate(Bundle savedInstanceState)
            {
                super.onCreate(savedInstanceState);
                setContentView(R.layout.main);
                TextView myTextView=null;                                //声明变量
                myTextView=(TextView)findViewById(R.id.myTextView);      //获取对象
                //设置一个文本框
     myTextView.setText("setBackgroundColor 方法：\r\n"+"          "+
             "用于设置一个文本框的背景颜色，一般通过直接设置颜色值来表示。其是设置文本
框背景色的最常用的方法。");
                myTextView.setTextSize(25);                              //设置字体大小
                myTextView.setTextColor(Color.BLACK);                    //设置字体颜色为黑色
                //设置字体样式
                myTextView.setTypeface(Typeface.SANS_SERIF,Typeface.BOLD);
                myTextView.setBackgroundColor(Color.YELLOW);             //设置背景颜色为黄色
            }
        }
```

运行程序，效果如图 3-11 所示。

图 3-11　设置文本框的背景色

3.2.9　getHeight 方法

getHeight 方法用于获取一个文本框的高度，单位为 pt，一般用于设置和更改文本框大小的场合。值得注意的是，该方法在文本框对象绘制之后才能够调用，否则将永远返回 0。其语法格式为：

```
    public void final int getHeight()
```

通过以下代码来演示怎样获取文本框的高度。

```
    public class MainActivity extends Activity
    {
        /* 活动第一次被创建时调用 */
```

```java
@Override
public void onCreate(Bundle savedInstanceState)
{
    super.onCreate(savedInstanceState);
    setContentView(R.layout.main);
    OnClickListener listener=null;                              //声明监听器
    Button bt1=null;                                            //声明按钮变量
        //文本框变量
    final TextView myTextView=(TextView)findViewById(R.id.myTextView);
    bt1=(Button)findViewById(R.id.button1);                     //获取 ID
    myTextView.setText("获取 TextView 的参数信息！");            //设置文本框
    myTextView.setBackgroundColor(Color.YELLOW);                //设置背景颜色为黄色
    bt1.setOnClickListener(listener=new OnClickListener()
    {
        //设置监听器
        @Override
        public void onClick(View v)
        {
            // TODO：自动存根法
            Toast.makeText(getApplicationContext(), "文本框高度为："+myTextView.getHeight(),
                Toast.LENGTH_LONG).show();                      //显示信息
        }
    });
}
```

在以上代码中，先声明了监听器，然后分别声明了按钮变量和文本框变量。接着，通过 setText 方法为 TextView 设置文本。最后，为按钮添加监听器，并在监听器的 onClick 方法中调用 getHeight 方法获取文本框高度，并通过 Toast 视图显示。

运行程序，效果如图 3-12 所示。

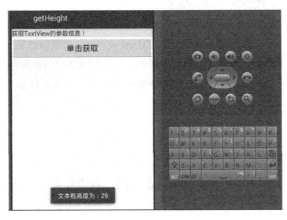

图 3-12　获取文本框高度

3.2.10 getWidth 方法

getWidth 方法用于获取一个 TextView 文本框的宽度，单位为 pt，一般用于设置和更改文本框大小的场合。值得注意的是，该方法应该在文本框对象绘制之后才能够调用，否则将永远返回 0。其语法格式为：

```
public final int getWidth()
```

通过以下代码来演示怎样获取文本框的宽度。

```java
public class MainActivity extends Activity
{
    /* 活动第一次被创建时调用 */
    @Override
    public void onCreate(Bundle savedInstanceState)
    {
        super.onCreate(savedInstanceState);
        setContentView(R.layout.main);
        OnClickListener listener=null;                              //声明监听器
        Button bt1=null;                                            //声明按钮变量
        //文本框变量
        final TextView myTextView=(TextView)findViewById(R.id.myTextView);
        bt1=(Button)findViewById(R.id.button1);                     //获取 ID
        myTextView.setText("获取 TextView 的参数信息！");              //设置文本
        myTextView.setBackgroundColor(Color.YELLOW);
        bt1.setOnClickListener(listener=new OnClickListener()
        {                                                           //设置监听器
            @Override
            public void onClick(View v)
            {
                // TODO：自动存根法
                Toast.makeText(getApplicationContext(),"文本框高度为："+myTextView.getHeight(),
                    Toast.LENGTH_LONG).show();                      //显示信息
                Toast.makeText(getApplicationContext(),"文本框的宽度为："+myTextView.getWidth(),
                    Toast.LENGTH_LONG).show();                      //显示信息
            }
        });
    }
}
```

运行程序，文本框分别获取文本框的高度及宽度，效果如图 3-13 所示。

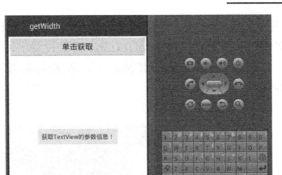

图 3-13　获取文本框的宽度

3.2.11　setPadding 方法

setPadding 方法用于设置一个 TextView 文本框的内边距，单位为 pt。该方法的显示效果相当于文字内容和文本框边缘之间留有空白，一般用于自定义文本框显示样式的场合。其语法格式为：

```
public void setPadding(int left,int top,int right,int bottom)
```

其中，参数 left 用于设置文本框左侧内边距；参数 top 用于设置文本框上部内边距；参数 right 用于设置文本框右侧内边距；参数 bottom 用于设置文本框下部内边距。

通过以下代码来演示怎样设置文本框的内边距。

```
public class MainActivity extends Activity
{
    /* 活动第一次被创建时调用 */
    @Override
    public void onCreate(Bundle savedInstanceState)
    {
        super.onCreate(savedInstanceState);
        setContentView(R.layout.main);
        TextView myTextView=null;                                          //声明变量
        myTextView=(TextView)findViewById(R.id.myTextView);                //获取对象
        myTextView.setText("setPadding 方法：\r\n"+        "+
            "用于设置一个 TextView 文本框的内边距，单位是 pt。一般用于自定义文本框显示样式的场合。");
        myTextView.setTextSize(25);                                        //设置字体大小
        myTextView.setTextColor(Color.BLUE);                               //设置字体颜色
        myTextView.setBackgroundColor(Color.YELLOW);                       //设置背景颜色
        myTextView.setPadding(40, 40, 40, 40);                             //设置文本框内边距
    }
}
```

以上代码中，先设置 TextView 变量，然后通过 findViewById 方法获取文本框对象。接着，通过 setText 方法设置需要显示的文本，通过 setTextSize 方法设置字体大小，通过 setTextColor 方法来设置字体的颜色，通过 setBackgroundColor 方法设置背景颜色为黄色。最后，通过 setPadding 方法设置文本框四边的内边距为 40。

运行程序，效果如图 3-14 所示。

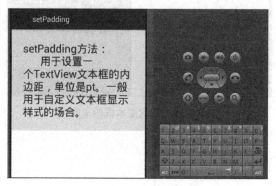

图 3-14　设置文本框内边距

3.2.12　getPaddingLeft 方法

getPaddingLeft 方法用于获取一个 TextView 文本框的左侧内边距，其返回值的单位为 pt，一般用于自定义文本框显示样式的场合。其语法格式为：

```
public int getPaddingLeft()
```

3.2.13　getPaddingTop 方法

getPaddingTop 方法用于获取一个 TextView 文本框的上部内边距，其返回值的单位为 pt，一般用于自定义文本框显示样式的场合。其语法格式为：

```
public int getPaddingTop()
```

3.2.14　getPaddingrRight 方法

getPaddingrRight 方法用于获取一个 TextView 文本框的右侧内边距，其返回值的单位为 pt，一般用于自定义文本框显示样式的场合。其语法格式为：

```
public int getPaddingRight()
```

3.2.15　getPaddingBottonm 方法

getPaddingBottonm 方法用于获取一个 TextView 文本框的下部内边距，其返回值的单位为 pt。其一般用于自定义文本框显示样式的场合。其语法格式为：

```
public int getPaddingBottonm ()
```

通过以下代码来演示怎样获取文本框的内边距。

```
public class MainActivity extends Activity
{
    /** 活动第一次被创建时调用 */
    @Override
    public void onCreate(Bundle savedInstanceState)
    {
        super.onCreate(savedInstanceState);
        setContentView(R.layout.main);
        TextView myTextView=null;                              //声明变量
        myTextView=(TextView)findViewById(R.id.myTextView);    //获取对象
        myTextView.setText("setPadding 方法：\r\n"+"           "+
              "用于设置一个 TextView 文本框的内边距，单位是 pt。一般用于自定义文本框显示样式的场合。");
        myTextView.setTextSize(25);                            //设置字体大小
        myTextView.setTextColor(Color.BLUE);                   //设置字体颜色
        myTextView.setBackgroundColor(Color.YELLOW);           //设置背景颜色
        myTextView.setPadding(35, 50, 35, 50);                 //设置内边距
        //获取左侧内边距
        System.out.println("文本框左侧内边距为："+myTextView.getPaddingLeft());
        //获取上部内边距
        System.out.println("文本框上部内边距为："+myTextView.getPaddingTop());
        //获取右侧内边距
        System.out.println("文本框右侧内边距为："+myTextView.getPaddingRight());
        //获取下部内边距
        System.out.println("文本框下部内边距为："+myTextView.getPaddingBottom());
    }
}
```

运行程序，可在 Eclipse 集成开发环境中输出如图 3-15 所示的效果图。

图 3-15 获取文本框内边距

3.2.16 getCurrentTextColor 方法

getCurrentTextColor 方法用于获取一个文本框当前字体的颜色，其返回值即为颜色的数值代

码，一般用于字体颜色设置和更改的场合。其语法格式为：

```
public final int getCurrentTextColor()
```

其返回值为整型，代表了颜色数值，采用的是 RGB 方式定义。下面介绍一些常用的颜色数值。

- Color.BLACK：对应的颜色值为-16777216(0xff000000)。
- Color.BLUE：对应的颜色值为-16776961(0xff0000ff)。
- Color.CYAN：对应的颜色值为-16711681(0xff00ffff)。
- Color.DKGRAY：对应的颜色值为-12303292(0xff444444)。
- Color.GRAY：对应的颜色值为-7829368(0xff888888)。
- Color.GREEN：对应的颜色值为-16711936(0xff00ff00)。
- Color.LTGRAY：对应的颜色值为-3355444(0xffcccccc)。
- Color.MAGENTA：对应的颜色值为-65281(0xffff00ff)。
- Color.RED：对应的颜色值为-65536(0xffff0000)。
- Color.TRANSPARENT：对应的颜色值为 0(0x00000000)。
- Color.WHITE：对应的颜色值为-1(0xffffffff)。
- Color.YELLOW：对应的颜色值为-256(0xffffff00)。

通过以下代码来演示怎样获取文本框字体的颜色。

```
public class MainActivity extends Activity
{
    /** 活动第一次被创建时调用 */
    @Override
    public void onCreate(Bundle savedInstanceState)
    {
        super.onCreate(savedInstanceState);
        setContentView(R.layout.main);
        TextView myTextView=null;                                    //声明变量
        myTextView=(TextView)findViewById(R.id.myTextView);          //获取控件 ID
        myTextView.setText("获取 TextView 的参数信息！");              //设置文本
        myTextView.setTextColor(Color.CYAN);                         //设置字体颜色
        //获取字体颜色
        System.out.println("字体的颜色是"+myTextView.getCurrentTextColor());
    }
}
```

在以上代码中，先设置 TextView 变量，然后通过 findViewById 方法获取文本框对象。接着，通过 setText 方法设置需要显示的文本，通过 setTextColor 方法来设置字体的颜色为灰色。最后，通过 getCurrentTextColor 方法获取当前字体颜色。

运行程序，输出如图 3-16 所示的信息。

图 3-16　获取字体颜色

3.2.17 getText 方法

getText 方法用于获取一个 TextView 文本框显示的文本内容，一般用于设置和更改文本框显示内容的场合。其语法格式为：

```
public CharSequence getText()
```

通过以下代码来演示怎样获取文本框的内容。

```
public class    MainActivity extends Activity
{
    /**活动第一次被创建时调用 */
    @Override
    public void onCreate(Bundle savedInstanceState)
    {
        super.onCreate(savedInstanceState);
        setContentView(R.layout.main);
        TextView myTextView=null;                                    //声明变量
        myTextView=(TextView)findViewById(R.id.myTextView);          //获取控件 ID
        myTextView.setText("获取 TextView 的参数信息！ ");            //设置文本内容
        //获取文本框的内容
        System.out.println("文本框的内容为："+myTextView.getText());
    }
}
```

以上代码，先声明变量，然后通过 findViewById 方法获取文本框 ID，通过 setText 方法设置显示的内容。最后，调用 getText 方法获取文本框内容并输出。

运行程序，输出如图 3-17 所示的信息。

图 3-17　获取文本框的内容

3.2.18 TextView 控件的综合实例

下面通过一个综合实例来演示文本框的用法。

要在屏幕中显示一段文字，即在手机屏幕中可通过文本框控件 TextView 来显示文本。在使用 TextView 控件时通常需要遵循如下步骤。

（1）导入 TextView 包，实现代码为：

```
Import android.widget.TextView;
```

（2）在文件 mainActivity.java 中声明一个 TextView，实现代码为：

```
Private TextView mTextView01;
```

（3）在文件 main.xml 中定义一个 TextView 对象 TextView01，实现代码为：

```
<TextView android:text="TextView01"
         android:"@+id/TextView01"
         android:layout_width="wrap_content"
         android:layout_height="wrap_content"
         andorid:layout_x="60x"
         android:layout_y="69px">
```

（4）利用 findViewById()方法获取 main.xml 中的 TextView，实现代码为：

```
mTextView01=(TextView) findViewById(R.id.TextView01);
```

（5）设置 TextView 标签内容，实现代码为：

```
String str_2="进入 Android";
mTextView01.setText(str_2);
```

（6）设置文本超级链接，实现代码为：

```
<TextView
         android:"@+id/TextView02"
         android:layout_width="wrap_content"
         android:layout_height="wrap_content"
         andorid:LautoLink="all";
         android:text="Android 开发者:http://developer.android.com/index.html">
</TextView>
```

本示例只须修改 mainActivity.java 文件即可，需在其里面添加 12 个 TextView 对象变量、1 个 LinearLayout 对象变量、1 个 WC 整数变量、1 个 LinearLayout.LayoutParams 变量。实现代码为：

```
//定义使用的对象
private LinearLayout myLayout;
private LinearLayout.LayoutParams layoutP;
private int WC=LinearLayout.LayoutParams.WRAP_CONTENT;
private    TextView    black_TV,blue_TV,dkgray_TV,gray_TV,green_TV,ltgray_TV,magenta_TV,red_TV,transparent_TV, white_TV,yellow_TV;
@Override
public void onCreate(Bundle savedInstanceState)
{
    super.onCreate(savedInstanceState);
    //实例化一个 LinearLayout 布局对象
    myLayout=new LinearLayout(this);
    //设置 LinearLayout 布局背景图片
    myLayout.setBackgroundResource(R.drawable.back);
    //加载主屏布局
    setContentView(myLayout);
    //实例化一个 LinearLayout 布局参数，用来添加 View
    layoutP=new LinearLayout.LayoutParams(WC,WC);
    //构造实例化 TextView 对象
```

```
        constructTextView();
        //把 TextView 添加到 LinearLayout 布局中
        addTextView();
        //设置 TextView 文本颜色
        setTextViewColor();
        //设置 TextView 文本内容
        setTextViewText();
    }
    //设置 TextView 文本内容
    public void setTextViewText()
    {
        black_TV.setText("黑色");
        blue_TV.setText("蓝色");
        gray_TV.setText("灰色");
        green_TV.setText("绿色");
        red_TV.setText("红色");
        transparent_TV.setText("透明");
        white_TV.setText("白色");
        yellow_TV.setText("黄色");
        ltgray_TV.setText("浅灰色");
        dkgray_TV.setText("灰黑色");
        cyan_TV.setText("青绿色");
        magenta_TV.setText("红紫色");
    }
    //设置 TextView 文本颜色
    public void setTextViewColor()
    {
        black_TV.setTextColor(Color.BLACK);
        blue_TV.setTextColor(Color.BLUE);
        gray_TV.setTextColor(Color.GRAY);
        green_TV.setTextColor(Color.GREEN);
        red_TV.setTextColor(Color.RED);
        transparent_TV.setTextColor(Color.TRANSPARENT);
        white_TV.setTextColor(Color.WHITE);
        yellow_TV.setTextColor(Color.YELLOW);
        ltgray_TV.setTextColor(Color.ITGRAY);
        dkgray_TV.setTextColor(Color.DKGRAY);
        cyan_TV.setTextColor(Color.CYAN);
        magenta_TV.setTextColor(Color.MAGENTA);
    }
    //构造实例化 TextView 对象
```

```
public void constructTextView()
{
    black_TV=new TextView(this);
    blue_TV=new TextView(this);
    gray_TV=new TextView(this);
    green_TV=new TextView(this);
    red_TV=new TextView(this);
    transparent_TV=new TextView(this);
    white_TV=new TextView(this);
    yellow_TV=new TextView(this);
    ltgray_TV=new TextView(this);
    dkgray_TV=new TextView(this);
    cyan_TV=new TextView(this);
    magenta_TV=new TextView(this);
}
//把 TextView 添加到 LinearLayout 布局中
public void addTextView()
{
    myLayout.addView(black_TV,layoutP);
    myLayout.addView(blue_TV,layoutP);
    myLayout.addView(gray_TV,layoutP);
    myLayout.addView(green_TV,layoutP);
    myLayout.addView(red_TV,layoutP);
    myLayout.addView(transparent_TV,layoutP);
    myLayout.addView(white_TV,layoutP);
    myLayout.addView(yellow_TV,layoutP);
    myLayout.addView(ltgray_TV,layoutP);
    myLayout.addView(dkgray_TV,layoutP);
    myLayout.addView(cyan_TV,layoutP);
    myLayout.addView(magenta_TV,layoutP);
}
```

如果要设置手机屏幕中的字体,在 Android 手机系统中,可以使用 TextView 来设置屏幕中静态域的字体。在计算机系统中,使用 style 类型来表示字体类型,如表 3-1 所示;使用 Typeface 来表示字体的风格,如表 3-2 所示。

表 3-1　int Style 类型说明

字　　体	说　　明
BOLD	粗体
BOLD_ITALIC	粗斜体
ITALIC	斜体
NORMAL	普通字体

表 3-2　Typeface 类型说明

字体	说明
DEFAULT	默认字体
DEFAULT_BOLD	默认粗体
MONOSPACE	单间隔字体
SANS_SERIF	无衬线字体
SERIF	衬线字体

本实例的实现过程比较简单，通过修改 mainActivity.java 文件即可实现多种字体样式的显示，实现代码如下：

```java
public class TypefaceStudy extends Activity
{
    //Typeface 类指定字体和固有风格
    //该类用于绘制，与可选绘制设置一起使用，如 textSkewX、textScaleX，当绘制（测量）时来
      指定如何显示文本
    //定义实例化一个布局大小，用来添加 TextView
    final int WRAP_CONTENT=ViewGroup.LayoutParams.WRAP_CONTENT;
    //定义 TextView 对象
    private  TextView   bold_TV,bold_italic_TV,default_TV,default_bold_TV,italic_TV,monospace_TV,normal_TV,sans_serif_TV,serif_TV;
    //定义 LinearLayout 布局对象
    private LinearLayout.LayoutParams linearLayoutParams;
    @Override
    public void onCreate(Bundle icicle)
    {
        super.onCreate(icicle);
        //设置 LinearLayout 布局为垂直布局
        linearLayout.setOrientation(LinearLayout.VERTICAL);
        //设置布局管理图
        linearLayout.setBackgroundResource(R.drawable.green);
        //加载 LinearLayout 为主屏布局，显示
        linearLayoutParams=new LinearLayout.LayoutParams(WRAP_CONTENT,WRAP_CONTENT);
        constructTextView();
        setTextSizeOf();
        setTextViewText();
        setStyleOfFont();
        setFontColor();
        toAddTextViewToLayout();
    }
    public void constructTextView()
```

```java
        {
            //实例化 TextView 对象
            bold_TV=new TextView(this);
            bold_italic_TV=new TextView(this);
            default_TV=new TextView(this);
            default_bold_TV=new TextView(this);
            italic_TV=new TextView(this);
            monospace_TV=new TextView(this);
            normal_TV=new TextView(this);
            sans_serif_TV=new TextView(this);
            serif_TV=new TextView(this);
        }
        public void setTextSizeOf()
        {
            //设置绘制的文本大小,该值必须大于 0
            bold_TV.setTextSize(24.0f);
            bold_italic_TV.setTextSize(24.0f);
            default_TV.setTextSize(24.0f);
            default_bold_TV.setTextSize(24.0f);
            italic_TV.setTextSize(24.0f);
            monospace_TV.setTextSize(24.0f);
            normal_TV.setTextSize(24.0f);
            sans_serif_TV.setTextSize(24.0f);
            serif_TV.setTextSize(24.0f);
        }
        public void setTextViewText()
        {
            //设置文本
            bold_TV.setText("BOLD");
            bold_italic_TV.setText("BOLD_ITALIC")
            default_TV.setText("DEFAULT")
            default_bold_TV.setText("DEFAULT_BOLD")
            italic_TV.setText("ITALIC")
            monospace_TV.setText("MONOSPACE")
            normal_TV.setText("NORMAL")
            sans_serif_TV.setText("SNAS")
            serif_TV.setText("SERIF")
        }
        public void setStyleOfFont()
        {
            //设置字体风格
```

```java
        bold_TV.setTypeface(null,Typeface.BOLD);
        bold_italic_TV.setTypeface(null,Typeface.BOLD_ITALIC);
        default_TV.setTypeface(null,Typeface.DEFAULT);
        default_bold_TV.setTypeface(null,Typeface.DEFAULT_BOLD);
        italic_TV.setTypeface(null,Typeface.ITALIC);
        monospace_TV.setTypeface(null,Typeface.MONOSPACE);
        normal_TV.setTypeface(null,Typeface.NORMAL);
        sans_serif_TV.setTypeface(null,Typeface.SNAS);
        serif_TV.setTypeface(null,Typeface.SERIF);
    }
    public void setFontColor()
    {
        //设置文本颜色
        bold_TV.setTextColor(Color.BLACK);
        bold_italic_TV.setTextColor(Color.CYAN);
        default_TV.setTextColor(Color.BULE);
        default_bold_TV.setTextColor(Color.MAGENTA);
        italic_TV.setTextColor(Color.WHITE);
        monospace_TV.setTextColor(Color.RED);
        normal_TV.setTextColor(Color.YELLOW);
        sans_serif_TV.setTextColor(Color.GRAYS);
        serif_TV.setTextColor(Color.LTGRAY);
    }
    public void toAddTextViewToLayout()
    {
        //把 TextView 加入 LinearLayout 布局中
        linearLayout.addView(bold_TV,linearLayouttParams);
        linearLayout.addView(bold_italic_TV,linearLayouttParams);
        linearLayout.addView(default_TV,linearLayouttParams);
        linearLayout.addView(default_bold_TV,linearLayouttParams);
        linearLayout.addView(italic_TV,linearLayouttParams);
        linearLayout.addView(monospace_TV,linearLayouttParams);
        linearLayout.addView(normal_TV,linearLayouttParams);
        linearLayout.addView(sans_TV,linearLayouttParams);
        linearLayout.addView(serif_TV,linearLayouttParams);
    }
}
```

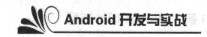

3.3 EditText 控件

编辑框（EditText）是一个典型的文本编辑控件，与其他编程语言中的编辑框类似。EditText 控件允许不仅提供信息显示的功能，还允许用户输入、编辑或者选择文本。在 Android 程序中，EditText 编辑框是使用最为广泛的信息输入接口，也是最重要的人机交互接口，如用户输入框、密码输入框等。

3.3.1 setText 方法

setText 方法用于设置一个 EditText 编辑框的显示内容，可以直接赋值一个字符串，也可以通过资源 ID 的方式来引用。该方法是在程序中设置编辑框显示文本最常用的方法。其语法格式为：

```
public final void setText(int resid)
```
或
```
public final void setText(CharSequence text)
```

其中，参数 resid 为文本字符串的资源 ID；参数 text 为需要设置显示的字符串。

通过以下代码来演示怎样为编辑框设置字符串。

```java
public class MainActivity extends Activity
{
    /** 活动第一次被创建时调用. */
    @Override
    public void onCreate(Bundle savedInstanceState)
    {
        super.onCreate(savedInstanceState);
        setContentView(R.layout.main);
        Button bt1=null;                                          //声明按钮变量
        TextView myTextView=null;                                 //声明文本框变量
        bt1=(Button)findViewById(R.id.button1);                   //获取按钮对象
        myTextView=(TextView)findViewById(R.id.myTextView);       //获取文本框对象
        //编辑框对象
        final EditText myEditText=(EditText)findViewById(R.id.editText1);
        myTextView.setTextColor(Color.RED);                       //设置字体颜色
        myTextView.setBackgroundColor(Color.YELLOW);              //设置背景颜色
        bt1.setOnClickListener(new View.OnClickListener()
        {
            //设置监听器
            @Override
            public void onClick(View v)
```

```
            {
                // TODO: 自动存根法
                myEditText.setText("欢迎来到 EditText!");          //设置文本
            }
        });
    }
}
```

在以上代码中,先声明了获取了文本框、按钮和编辑框对象,然后通过按钮设置监听器,在监听器的 onClick 方法中通过 setText 方法设置 EditText 的文本内容。

运行程序,效果如图 3-18 所示。

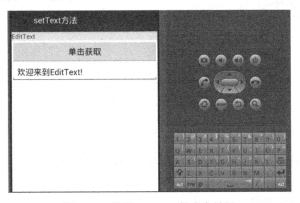

图 3-18 设置 EditText 的内容效果

其中,res/layout 目录下的 main.xml 文件代码为(其他方法此文件设置类似):

```xml
<LinearLayout xmlns:android="http://schemas.android.com/apk/res/android"
    android:orientation="vertical"
    android:layout_width="fill_parent"
    android:layout_height="fill_parent" >
<TextView
    android:layout_width="fill_parent"
    android:layout_height="wrap_content"
    android:text="@string/hello_world"
    android:id="@+id/myTextView"/>
    <Button  android:layout_width="match_parent"  android:layout_height="wrap_content"  android:id="@+id/button1" android:text="@string/btn1"></Button>
    <EditText  android:layout_width="match_parent"  android:text="EditText"  android:layout_height="wrap_content" android:id="@+id/editText1"></EditText>
</LinearLayout>
```

res/layout 目录下的 strings.xml 文件代码为(其他方法此文件设置类似):

```xml
<resources>
    <string name="app_name">E2_9_3_1</string>
    <string name="hello_world">EditText</string>
```

```xml
<string name="menu_settings">Settings</string>
<string name="title_activity_main">setText 方法</string>
<string name="btn1">单击获取</string>
</resources>
```

3.3.2 getText 方法

getText 方法用于获取一个 EditText 编辑框显示的文本内容，一般用于设置和更改编辑框显示内容的场合。其语法格式为：

```
public Editable getText()
```

通过以下代码来演示怎样获取编辑框的内容。

```java
public class MainActivity extends Activity
{
    /** 活动第一次被创建时调用. */
    @Override
    public void onCreate(Bundle savedInstanceState)
    {
        super.onCreate(savedInstanceState);
        setContentView(R.layout.main);
        Button bt1=null;                                              //声明按钮变量
        TextView myTextView=null;                                     //声明文本框变量
        bt1=(Button)findViewById(R.id.button1);                       //获取按钮对象
        myTextView=(TextView)findViewById(R.id.myTextView);           //获取文本框对象
        myTextView.setTextColor(Color.RED);                           //设置字体颜色
        myTextView.setBackgroundColor(Color.YELLOW);                  //设置背景颜色
        //编辑框对象
        final EditText myEditText=(EditText)findViewById(R.id.editText1);
        myEditText.setText("欢迎来到 EditText!");                      //设置显示内容
        bt1.setOnClickListener(new View.OnClickListener()
        {
            //设置监听器
            @Override
            public void onClick(View v)
            {
                // TODO：自动存根法
                Toast.makeText(getApplicationContext(), "编辑框的内容为："+myEditText.getText(), Toast.LENGTH_LONG).show();     //显示信息
            }
        });
    }
}
```

在以上代码中，先通过 findViewById 方法获取文本框 ID，再通过 setText 方法设置显示的内容。最后，在监听器的 onClick 方法中获取编辑框内容并通过 Toast 视图输出。

运行程序，效果如图 3-19 所示。

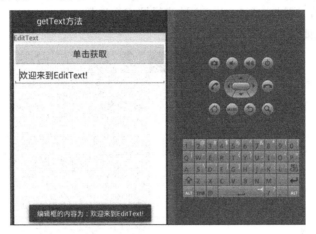

图 3-19　获取编辑框内容

3.3.3　setSelection 方法

setSelection 方法用于设置一个 EditText 编辑框中哪些内容被选择，并将被选择的内容以高亮显示。一般用于设置和更改编辑框显示内容的场合。其语法格式为：

```
public void setSelection(int start,int stop)
```

其中，参数 start 为文本选择的起始字符；参数 stop 为文本选择的终止字符。

通过以下代码来演示怎样设置编辑框的文本内容选择。

```
public class MainActivity extends Activity
{
    /** 活动第一次被创建时调用 */
    @Override
    public void onCreate(Bundle savedInstanceState)
    {
        super.onCreate(savedInstanceState);
        setContentView(R.layout.main);
        Button bt1=null;                                          //声明按钮变量
        TextView myTextView=null;                                 //声明文本框变量
        bt1=(Button)findViewById(R.id.button1);                   //获取按钮对象
        myTextView=(TextView)findViewById(R.id.myTextView);       //获取文本框对象
        //编辑框对象
        final EditText myEditText=(EditText)findViewById(R.id.editText1);
        myEditText.setText("欢迎 EditText!");                      //设置显示内容
        myEditText.setSelection(0,6);                             //设置需要选择的内容
```

            }
        }

以上代码中，先声明变量，然后通过 findViewById 方法获取文本框 ID，通过 setTex 方法设置显示的内容。最后，通过 setSelection 方法来设置文本选择。

运行程序，效果如图 3-20 所示。

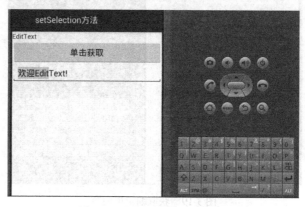

图 3-20　设置编辑框文本选择

3.3.4　setHint 方法

setHint 方法用于设置一个 EditText 的提示信息，当编辑框中的内容为空时将显示该提示信息。该方法一般用于提示用户输入的场合，使程序更具人性化。其语法格式为：

```
public final void setHint(CharSequence hint)
```

其中，参数 hint 为设置的提示信息。

通过以下代码来演示怎样设置编辑框的提示信息。

```java
public class MainActivity extends Activity
{
    /*活动第一次被创建时调用 */
    @Override
    public void onCreate(Bundle savedInstanceState)
    {
        super.onCreate(savedInstanceState);
        setContentView(R.layout.main);
        Button bt1=null;                                                //声明按钮变量
        TextView myTextView=null;                                       //声明文本框变量
        bt1=(Button)findViewById(R.id.button1);                         //获取按钮对象
        myTextView=(TextView)findViewById(R.id.myTextView);             //获取文本框对象
        //编辑框对象
        final EditText myEditText=(EditText)findViewById(R.id.editText1);
        myEditText.setText("欢迎来到 EditText!");                        //设置显示内容
        myEditText.setHint("在此输入所需要字串：");                        //设置提示内容
```

                }
        }

以上代码是通过 setHint 方法来设置提示信息的,当刚开始运行代码并选择编辑框时出现如图 3-21 所示的效果。当单击图 3-20 中的 ⊠ 对文本框中的内容进行删除时,即出现如图 3-22 所示的效果。通过模拟器键盘可输入对应的代码。

图 3-21 设置编辑框提示信息初始效果

图 3-22 编辑框的提示信息

3.2.5 setOnKeyListener 方法

setOnKeyListener 方法用于为 EditText 编辑框设置按键监听器,该监听器用于监听 Android 手机上按键的动作。在程序中需要重载其中的 onKey 方法,值得注意的是,不同的 Android 对于同一个按钮的编码可能会不一样。其语法格式为:

```
public void setOnKeyListener(View.OnKeyListener1)
```

通过以下代码来演示怎样在 EditText 中输入字符的编码。

```java
public class MainActivity extends Activity
{
    /*活动第一次被创建时调用 */
    @Override
    public void onCreate(Bundle savedInstanceState)
    {
        super.onCreate(savedInstanceState);
        setContentView(R.layout.main);
        Button bt1=null;                                           //声明按钮变量
        TextView myTextView=null;                                  //声明文本框变量
        bt1=(Button)findViewById(R.id.button1);                    //获取按钮对象
        myTextView=(TextView)findViewById(R.id.myTextView);        //获取文本框对象
        //编辑框对象
        final EditText myEditText=(EditText)findViewById(R.id.editText1);
        myEditText.setText("进入 EditText!");                      //设置显示内容
        myEditText.setHint("请输入字符串: ");                      //设置提示内容
        myEditText.setOnKeyListener(new View.OnKeyListener()
```

```
                {
                    @Override
                    public boolean onKey(View arg0, int arg1, KeyEvent arg2)
                    {
                        // TODO:自动存根法
                        Toast.makeText(getApplicationContext(), "按键的 keyCode（关键代码）为："+arg1,
                            Toast.LENGTH_LONG).show();          //显示信息
                        return false;
                    }
                });
            }
        }
```

以上代码中通过 setHint 方法来设置提示信息。接着为该编辑框设置 setOnKeyListener 监听器。运行代码，当通过模拟键盘进行按键输入时，可在手机屏幕上输出如图 3-23 所示的信息。

图 3-23　获取 Android 键盘输入

3.3.6　EditText 控件的综合实例

下面使用 EditText 来设置进入 Android 手机的用户名及密码页面。其操作步骤如下：

（1）首先在 Eclipse 环境下，建立一个名为 EditPass 的工程。

（2）打开 res/values 目录下的 strings.xml 文件，代码修改为：

```xml
<resources>
    <string name="app_name">EditPass</string>
    <string name="hello_world">Hello world!</string>
    <string name="menu_settings">Settings</string>
    <string name="title_activity_main">Android_EditText</string>
    <string name="btn1">OK</string>
</resources>
```

（3）打开 res/layout 目录下的 main.xml 文件，代码修改为：

```xml
<?xml version="1.0" encoding="utf-8"?>
<LinearLayout xmlns:android="http://schemas.android.com/apk/res/android"
```

```xml
        android:orientation="vertical"
        android:layout_width="fill_parent"
        android:layout_height="fill_parent">
    <EditText android:id="@+id/et_ShowName"
        android:layout_width="fill_parent"
            android:layout_height="wrap_content"></EditText>
    <Button android:id="@+id/btn_Ok" android:text="OK"
        android:layout_width="fill_parent"
        android:layout_height="wrap_content" />
    <TextView android:id="@+id/tv_Welcome"
        android:layout_width="wrap_content"
            android:layout_height="wrap_content"></TextView>
</LinearLayout>
```

（4）打开 src/com.example.editpass 目录下的 MainActivity.java 文件，代码修改为：

```java
package com.example.editpass;

import android.os.Bundle;
import android.app.Activity;
import android.view.Menu;
import android.view.MenuItem;
import android.support.v4.app.NavUtils;

import android.app.Activity;
import android.os.Bundle;
import android.view.View;
import android.view.View.OnClickListener;
import android.widget.Button;
import android.widget.EditText;
import android.widget.TextView;
public class MainActivity extends Activity
{
    private Button btn_Ok;
    private EditText et_ShowName;
    private TextView tv_Welcome;
    /* 活动第一次被创建时调用*/
    @Override
    public void onCreate(Bundle savedInstanceState)
    {
        super.onCreate(savedInstanceState);
        setContentView(R.layout.main);
        btn_Ok = (Button) findViewById(R.id.btn_Ok);
        et_ShowName = (EditText) findViewById(R.id.et_ShowName);
```

```
            tv_Welcome = (TextView) findViewById(R.id.tv_Welcome);
            btn_Ok.setOnClickListener(new OnClickListener()
            {
                @Override
                public void onClick(View v)
                {
                    //DO：自动存根法
                    tv_Welcome.setText(et_ShowName.getText() + "，欢迎您！");
                }
            });
        }
    }
```

运行程序，再利用 Android 的模拟键盘在编辑框中输入对应的字串并单击"OK"按钮，即可进入如图 3-24 所示的界面。

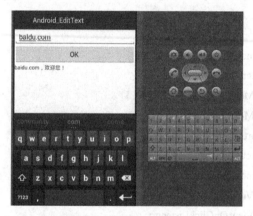

图 3-24　EditText 的应用

3.4　RadioButton 控件

RadioButton 是 Android 系统中的单选按钮，它是一个双状态按钮，可以为选中或不选中。RadioButton 往往需要与 RadioGroup 单选框一组使用。在实际应用中，放在 RadioGroup 之内的所有 RadioButton 每次只能有一个被选中，因此也称为单选按钮组。

3.4.1　setOnCheckedChangedListener 方法

setOnCheckedChangedListener 方法用于设置一个 RadioGroup 单选组按钮的监听器。当其中的单选按钮状态改变时，将触发该监听器。该方法是改变单选按钮组状态的主要方法。其语法格式为：

```
public void setOnCheckedChangedListener(RadioGroup.OnCheckedChangedListener listener)
```

通过以下代码来演示怎样实现自动获取选项的改变。

```java
public class MainActivity extends Activity
{
    /*活动第一次被创建时调用 */
    @Override
    public void onCreate(Bundle savedInstanceState)
    {
        super.onCreate(savedInstanceState);
        setContentView(R.layout.main);
        RadioGroup group=(RadioGroup)findViewById(R.id.radioGroup1);         //获取对象
        final RadioButton CB1=(RadioButton)findViewById(R.id.radio0);        //获取对象
        final RadioButton CB2=(RadioButton)findViewById(R.id.radio1);        //获取对象
        final RadioButton CB3=(RadioButton)findViewById(R.id.radio2);        //获取对象
        final RadioButton CB4=(RadioButton)findViewById(R.id.radio3);        //获取对象
        group.setOnCheckedChangeListener(new RadioGroup.OnCheckedChangeListener()
        {
            //设置监听器
            @Override
            public void onCheckedChanged(RadioGroup group, int checkedId)
            {
                // TODO：自动存根法
                String str="";
                if(checkedId==CB1.getId())                      //检查是哪个单选按钮被选中
                    str=(String) CB1.getText();
                else if(checkedId==CB2.getId())
                    str=(String) CB2.getText();
                else if(checkedId==CB3.getId())
                    str=(String) CB3.getText();
                else if(checkedId==CB4.getId())
                    str=(String) CB4.getText();
                //输出信息
                Toast.makeText(getApplicationContext(),"你最喜欢的运动为："+str,Toast.LENGTH_LONG).show();
            }
        });
    }
}
```

以上代码中，先通过 findViewById 方法获取 RadioButton 和 RadioGroup 对象，然后为 RadioGroup 设置监听器，并重载其中的 onCheckedChanged 方法。此处用于检查是哪个单选按钮被选中，然后输出信息。

运行程序，效果如图 3-25 所示。

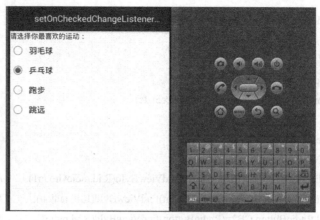

图 3-25 单选按钮组

其中，res/values 目录下的 strings.xml 文件代码修改为：

```xml
<resources>
    <string name="app_name">E_2_9_4_1</string>
    <string name="hello_world">Hello world!</string>
    <string name="menu_settings">Settings</string>
    <string name="title_activity_main">setOnCheckedChangeListener 方法</string>
    <string name="btn1">反向选择</string>
    <string name="tv1">请选择你最喜欢的运动：</string>
    <string name="CB1">羽毛球</string>
    <string name="CB2">乒乓球</string>
    <string name="CB3">跑步</string>
    <string name="CB4">跳远</string>
</resources>
```

res/layout 目录下的 main.xml 文件代码修改为：

```xml
<?xml version="1.0" encoding="utf-8"?>
<LinearLayout xmlns:android="http://schemas.android.com/apk/res/android"
    android:orientation="vertical"
    android:layout_width="fill_parent"
    android:layout_height="fill_parent" android:id="@+id/LL">
    <TextView android:layout_width="match_parent" android:layout_height="wrap_content" android:id="@+id/textView1"
        android:text="@string/tv1"></TextView>
    <RadioGroup android:layout_height="wrap_content"
        android:layout_width="match_parent" android:id="@+id/radioGroup1">
        <RadioButton android:layout_height="wrap_content" android:checked="true" android:id="@+id/radio0" android:text="@string/CB1" ndroid:layout_width="match_parent"></RadioButton>
        <RadioButton android:layout_width="match_parent" android:layout_height="wrap_content" android:text="@string/CB2"
            android:id="@+id/radio1"></RadioButton>
```

```xml
        <RadioButton   android:layout_height="wrap_content"    android:layout_width="match_parent"
android:text="@string/CB3"
          android:id="@+id/radio2"></RadioButton>
            <RadioButton android:layout_height="wrap_content"
android:layout_width="match_parent"         android:text="@string/CB4"
          android:id="@+id/radio3"></RadioButton>
        </RadioGroup>
</LinearLayout>
```

3.4.2 check 方法

check 方法用于设置一个 RadioGroup 单选按钮组中某个单选按钮的选中状态。当指定的 RadioButton 单选按钮被选中后，之前选中的按钮将被取消选中。其语法格式为：

```
public viod check(int id)
```

其中，参数 id 为 RadioButton 单选按钮的 ID。

3.4.3 clearCheck 方法

clearCheck 方法用于清空一个 RadioGroup 单选按钮组的选项状态，此时其中的任何一个 RadioButton 都处于选中状态。该方法往往用于重新让用户选择的场合，常用于"清空"按钮中。其语法格式为：

```
public void clearCheck()
```

通过以下代码来演示怎样实现清空选项。

```java
final RadioGroup group=(RadioGroup)findViewById(R.id.radioGroup1);       //获取对象
final RadioButton CB1=(RadioButton)findViewById(R.id.radio0);            //获取对象
final RadioButton CB2=(RadioButton)findViewById(R.id.radio1);            //获取对象
final RadioButton CB3=(RadioButton)findViewById(R.id.radio2);            //获取对象
final RadioButton CB4=(RadioButton)findViewById(R.id.radio3);            //获取对象
Button bt=(Button)findViewById(R.id.button1);                            //获取对象
group.setOnCheckedChangeListener(new RadioGroup.OnCheckedChangeListener()
{
    //设置监听器
    @Override
    public void onCheckedChanged(RadioGroup group, int checkedId)
    {
        // TODO：自动存根法
        String str="";
        if(checkedId==CB1.getId())                                       //检查是哪个单选按钮被选中
            str=(String) CB1.getText();
        else if(checkedId==CB2.getId())
```

```
            str=(String) CB2.getText();
        else if(checkedId==CB3.getId())
            str=(String) CB3.getText();
        else if(checkedId==CB4.getId())
            str=(String) CB4.getText();
        //输出信息
        Toast.makeText(getApplicationContext(),"你最喜欢的运动为："+str,Toast.LENGTH_LONG).show();
        }
    });

    bt.setOnClickListener(new View.OnClickListener()
    {
        //设置按钮监听器
        @Override
        public void onClick(View v)
        {
            // TODO：自动存根法
            group.clearCheck();                            //清空选项
        }
    });
    }
}
```

在以上代码中，调用 clearCheck 方法来清空选项。运行代码，单击"清空"按钮，效果如图 3-26 所示。

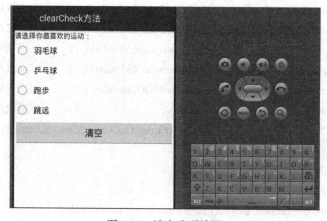

图 3-26　清空选项效果

3.5　CheckBox 控件

CheckBox 是 Android 系统中的复选框控件，这种控件具有双状态，可以选中或不选中。在

Android 系统中，很多需要多项选择的地方均采用了复选框来实现。

CheckBox 类中的很多方法都继承于 TextView 类。

3.5.1 isChecked 方法

isChecked 方法用于检查一个 CheckBox 复选框是否被选中。如果被选中，则返回 true，否则返回 false。该方法常用于获取有多少选中项的时候。其语法格式为：

```
public abstract Boolean isChecked()
```

通过以下代码来演示怎样获取复选框的选中项。

```java
public class MainActivity extends Activity
{
    TextView    textview_button=null;
    private CheckBox    checkbox_Anroid,checkbox_Ios,checkboxWinphone;
    @Override
    protected void onCreate(Bundle savedInstanceState)
    {
        // TODO：自动存根法
        super.onCreate(savedInstanceState);
        setContentView(R.layout.main);
        //通过 id 获取 id 对应的组件
        checkbox_Anroid=(CheckBox)findViewById(R.id.checkbox_android_id);
        checkbox_Ios=(CheckBox)findViewById(R.id.checkbox_ios_id);
        checkboxWinphone =(CheckBox)findViewById(R.id.checkbox_winphone_id);
        //组件单击事件
        checkbox_Anroid.setOnClickListener(checklistener);
        checkbox_Ios.setOnClickListener(checklistener);
        checkboxWinphone.setOnClickListener(checklistener);
    }
    //单击事件
    private OnClickListener    checklistener = new OnClickListener()
    {
        @Override
        public void onClick(View v)
        {
            // TODO Auto-generated method stub
            TextView    txt = (TextView)v;
            switch (txt.getId())
            {
                case R.id.checkbox_android_id:
```

```
                    if(checkbox_Anroid.isChecked())
                    {
                        showToast("你选中的是: "+txt.getText().toString());
                    }else
                      {
                            showToast("你取消的是: "+txt.getText().toString());
                      }
                    break;
                case R.id.checkbox_ios_id:
                    if(checkbox_Ios.isChecked())
                      {
                        showToast("你选中的是: "+txt.getText().toString());
                      }else
                      {
                        showToast("你取消的是: "+txt.getText().toString());
                      }
                    break;
                case R.id.checkbox_winphone_id:
                    if(checkboxWinphone.isChecked())
                      {
                        showToast("你选中的是: "+txt.getText().toString());
                      }else
                      {
                        showToast("你取消的是: "+txt.getText().toString());
                      }
                    break;
                default:
                    break;
            }
        }
    };
    private void showToast(String str)
    {
        // TODO: 自动存根法
        Toast.makeText(getApplicationContext(),str, Toast.LENGTH_LONG).show();
    }
}
```

运行程序，效果如图 3-27 所示。

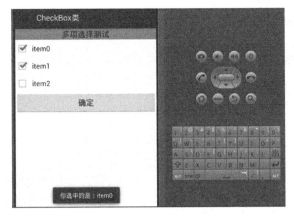

图 3-27 复选框的选中项

3.5.2 setChecked 方法

setChecked 方法用于设置一个 CheckBox 复选框是否被选中。该方法是程序代码中设置复选框状态的最主要的方法,如某些初始化选择项的场合。其语法格式为:

public abstract void setChecked(Boolean checked)

其中,参数 checked 为复选框的状态,true 表示选中,false 表示不选中。

3.5.3 toggle 方法

toggle 方法用于更改一个 CheckBox 复选框的状态。如果该复选框为选中,则将更改为未选中;如果该复选框未选中,则将更改为选中。该方法通常用于反向选择多个项目的场合。其语法格式为:

public abstract void toggle()

通过以下代码来演示怎样实现反向选择的功能。

```
public class MainActivity extends Activity
{
    /* 第一次调用活动 */
    @Override
    public void onCreate(Bundle savedInstanceState)
    {
        super.onCreate(savedInstanceState);
        setContentView(R.layout.main);
        final CheckBox CB1=(CheckBox)findViewById(R.id.checkBox1);   //获取对象
        final CheckBox CB2=(CheckBox)findViewById(R.id.checkBox2);   //获取对象
        final CheckBox CB3=(CheckBox)findViewById(R.id.checkBox3);   //获取对象
        final CheckBox CB4=(CheckBox)findViewById(R.id.checkBox4);   //获取对象
        Button bt=(Button)findViewById(R.id.button1);                //获取对象
        CB2.setChecked(true);                                        //设为选中
```

```
bt.setOnClickListener(new View.OnClickListener()
{
    //设置监听器
    @Override
    public void onClick(View v)
    {
        //：自动存根法
        String str="你选择的项为：";
        CB1.toggle();                              //反向选择
        CB2.toggle();                              //反向选择
        CB3.toggle();                              //反向选择
        CB4.toggle();                              //反向选择
        if(CB1.isChecked())                        //判断复选框 1 是否被选中
            str=str+CB1.getText();
        if(CB2.isChecked())                        //判断复选框 2 是否被选中
            str=str+CB2.getText();
        if(CB3.isChecked())                        //判断复选框 3 是否被选中
            str=str+CB3.getText();
        if(CB4.isChecked())                        //判断复选框 4 是否被选中
            str=str+CB4.getText();
        //显示结果
        Toast.makeText(getApplicationContext(),str,Toast.LENGTH_LONG).show();
    }
});
}
```

在以上代码中，先通过 findViewById 方法获取 4 个复选框和 1 个按钮对象，然后设置第二个复选框为选中状态。接着，为按钮绑定监听器。在该监听器内部，通过 toggle 方法实现反向选择，并输出选择的结果。

运行代码，选择 item1 及 item2 项，并单击"反向选择"按钮后，得到如图 3-28 所示的效果。

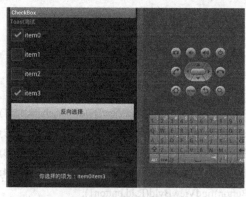

图 3-28　反向选择效果

3.5.4 setOnCheckedChangeListener 方法

setOnCheckedChangeListener 方法用于设置一个 CheckBox 复选框的监听器。当按钮的状态改变时，将触发该监听器。该方法是获知 CheckBox 复选框状态改变的主要方法。其语法格式为：

public void setOnCheckedChangeListener(CompoundButton.OnCheckedChangeListener listener)

通过如下代码来演示怎样实现自动获取选项的改变。

```java
public class MainActivity extends Activity
{
    /* 活动第一次被调用 */
    @Override
    public void onCreate(Bundle savedInstanceState)
    {
        super.onCreate(savedInstanceState);
        setContentView(R.layout.main);
        final CheckBox CB1=(CheckBox)findViewById(R.id.checkBox1);        //获取对象
        final CheckBox CB2=(CheckBox)findViewById(R.id.checkBox2);        //获取对象
        final CheckBox CB3=(CheckBox)findViewById(R.id.checkBox3);        //获取对象
        final CheckBox CB4=(CheckBox)findViewById(R.id.checkBox4);        //获取对象
        CB1.setOnCheckedChangeListener(new CompoundButton.OnCheckedChangeListener()
        {
            @Override
            public void onCheckedChanged(CompoundButton buttonView, boolean isChecked)
            {
                // TODO：自动存根法
                if(isChecked)                                              //判断状态
                    Toast.makeText(getApplicationContext(),CB1.getText()+"被选中",
                Toast.LENGTH_LONG).show();
                else
                    Toast.makeText(getApplicationContext(),CB1.getText()+"被取消",
                Toast.LENGTH_LONG).show();
            }
        });
        CB2.setOnCheckedChangeListener(new CompoundButton.OnCheckedChangeListener()
        {
            @Override
            public void onCheckedChanged(CompoundButton buttonView, boolean isChecked)
            {
                // TODO：自动存根法
                if(isChecked)                                              //判断状态
                    Toast.makeText(getApplicationContext(),CB2.getText()+"被选中",
```

```java
                    Toast.LENGTH_LONG).show();
                else
                    Toast.makeText(getApplicationContext(),CB2.getText()+"被取消",
                Toast.LENGTH_LONG).show();
            }
        });
        CB3.setOnCheckedChangeListener(new CompoundButton.OnCheckedChangeListener()
        {
            @Override
            public void onCheckedChanged(CompoundButton buttonView, boolean isChecked)
            {
                // TODO Auto-generated method stub
                if(isChecked)                                      //判断状态
                    Toast.makeText(getApplicationContext(),CB3.getText()+"被选中",
                Toast.LENGTH_LONG).show();
                else
                    Toast.makeText(getApplicationContext(),CB3.getText()+"被取消",
                Toast.LENGTH_LONG).show();
            }
        });
        CB4.setOnCheckedChangeListener(new CompoundButton.OnCheckedChangeListener()
        {
            @Override
            public void onCheckedChanged(CompoundButton buttonView, boolean isChecked)
            {
                // TODO：自动存根法
                if(isChecked)                                      //判断状态
                    Toast.makeText(getApplicationContext(),CB4.getText()+"被选中",
                Toast.LENGTH_LONG).show();
                else
                    Toast.makeText(getApplicationContext(),CB4.getText()+"被取消",
                Toast.LENGTH_LONG).show();
            }
        });
    }
}
```

在以上代码中，为 4 个复选框都设置了状态更改监听器。在监听器内部，判断选中状态，并通过 Toast 视图显示结果。

运行程序，选中"item2"项，出现如图 3-29 所示的效果。

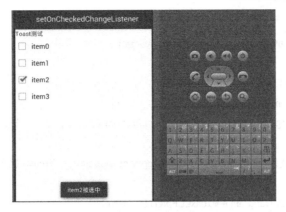

图 3-29　设置监听器

3.6　Toast 控件

Toast 是一种提供给用户简洁信息的视图。Toast 类帮助你创建和显示该信息。该视图以浮于应用程序之上的形式呈现给用户。因为它并不获得焦点，即使用户正在输入也不会受到影响。它的目标是尽可能以不显眼的方式使用户看到你提供的信息。有两个例子就是音量控制和设置信息保存成功。使用该类最简单的方法就是调用一个静态方法，让它来构造你需要的一切并返回一个新的 Toast 对象。

3.6.1　cancel 方法

cancel 方法用于关闭 Toast 视图。如果视图已经显示则将其关闭，还没有显示则不再显示。一般不需要调用该方法。正常情况下，视图会在超过存续期间后消失。其语法格式为：

```
public void cancel()
```

3.6.2　getDuration 方法

getDuration 方法用于获取 Toast 视图显示的持续时间。getDuration 方法通常用于获取 Toast 视图持续时间，然后根据应用程序的需要进行修改的场合。该方法的返回值为 int 类型，可以取以下两种值。

- LENGTH_LONG：较长的视图持续时间。
- LENGTH_SHORT：较短的视图持续时间，其为系统默认值。

getDuration 的语法格式为：

```
public int getDuration()
```

通过以下代码来演示怎样使用 getDuration 方法获取 Toast 视图的持续时间。

```
public class MainActivity extends Activity
{
    /* 第一次调用活动 */
```

```
@Override
public void onCreate(Bundle savedInstanceState)
{
    super.onCreate(savedInstanceState);
    setContentView(R.layout.main);
    Toast mg1 = Toast.makeText(getApplicationContext(), "Toast 提示信息 1", Toast.LENGTH_LONG);
    Toast mg2 = Toast.makeText(getApplicationContext(), "Toast 提示信息 2", Toast.LENGTH_SHORT);
    if(mg1.getDuration()==Toast.LENGTH_LONG)
    {
        System.out.println("Toast 提示信息 1 的持续时间为 LENGTH_LONG!");
    }
    else
    {
        System.out.println("Toast 提示信息 1 的持续时间为 LENGTH_SHORT!");
    }
    if(mg2.getDuration()==Toast.LENGTH_LONG)
    {
        System.out.println("Toast 提示信息 2 的持续时间为 LENGTH_LONG!");
    }
    else
    {
        System.out.println("Toast 提示信息 2 的持续时间为 LENGTH_SHORT!");
    }
}
```

代码中，声明了两个 Toast 对象 mg1 和 mg2，分别设置持续时间为 Toast.LENGTH_LONG 和 Toast.LENGTH_SHORT。接着调用 getDuration 方法来获取 Toast 对象的持续时间，并对其进行判断，分别输出显示。

运行程序在 Eclipse 环境中输出如图 3-30 所示的信息。

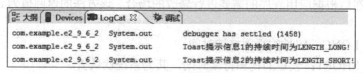

图 3-30　获取视图的持续时间

3.6.3　getGravity 方法

getGravity 方法用于获取 Toast 视图的提示信息在屏幕上显示的位置，通常用于显示和设置

Toast 视图在手机屏幕上显示位置的场合。其返回值类型为 int，为 Android 内置的 Gravity 标准常量。常量取值如下：
- BOTTOM：底部，常量值为 80(0x00000050)。
- CENTER：水平方向和垂直方向的中部，常量值为 17(0x00000011)。
- CENTER_HORIZONTAL：水平方向的中部，常量值为 1(0x00000001)。
- CENTER_VERTICAT：垂直方向的中部，常量值为 16(0x00000010)。
- LEFT：左部，常量值为 3(0x00000003)。
- RIGHT：左部，常量值为 5(0x00000005)。
- TOP：顶部，常量值为 48(0x000000030)。

值得注意的是，返回值可以为多个 Gravity 常量的组合。例如，81 代表了 Toast 视图位于屏幕的底部，且是水平方向上的中部。其语法格式为：

```
public int getGravity()
```

通过以下代码来演示怎样使用 getGravity 方法来获取一个默认 Toast 视图的显示位置。

```
public class MainActivity extends Activity
{
    /** 第一次调用活动 */
    @Override
    public void onCreate(Bundle savedInstanceState)
    {
        super.onCreate(savedInstanceState);
        setContentView(R.layout.main);
        Toast msg1 = Toast.makeText(getApplicationContext(), "Toast 提示信息 2", Toast.LENGTH_SHORT);
        System.out.println("Toast 提示信息 2 的显示位置是"+msg1.getGravity());
    }
}
```

运行程序，在 Eclipse 环境下输出如图 3-31 的信息。

图 3-31 getGravity 获取视图显示位置

从以上结果可看到一个 Toast 对象的默认显示位置为 81，即 Gravity.BOTTOM 和 Gravity.GENTER_HORIZONTAL 的组合，位于屏幕的底部和水平方向上的中部。调用 getGravity 方法获得的是显示位置的组合形式。

3.6.4 getHorizontalMargin 方法与 getVerticalMargin 方法

getHorizontalMargin 方法与 getVerticalMargin 方法分别用于获取 Toast 视图的提示信息在屏幕的水平方向及垂直方向的页边空白。它们的返回值为 float 类型，是容器的边缘与提示信息的横（纵）向空白（与容器宽度的比）。它们的调用格式为：

```
public float getHorizontalMargin ()
public float getVerticalMargin ()
```

通过以下代码来演示怎样获取 Toast 视图的页边空白值。

```
public class MainActivity extends Activity
{
    /* 第一次调用活动 */
    @Override
    public void onCreate(Bundle savedInstanceState)
    {
        super.onCreate(savedInstanceState);
        setContentView(R.layout.main);
        Toast mg = Toast.makeText(getApplicationContext(), "Toast 提示信息 2", Toast.LENGTH_SHORT);
        mg.setMargin(0.25f, 0.25f);
        System.out.println("Toast 提示信息 2 的水平方向的页边空白为"+mg.getHorizontalMargin());
        System.out.println("Toast 提示信息 2 的垂直方向的页边空白为"+mg.getVerticalMargin());
        mg.show();
    }
}
```

运行程序，在 Eclipse 环境下输出如图 3-32 的信息。

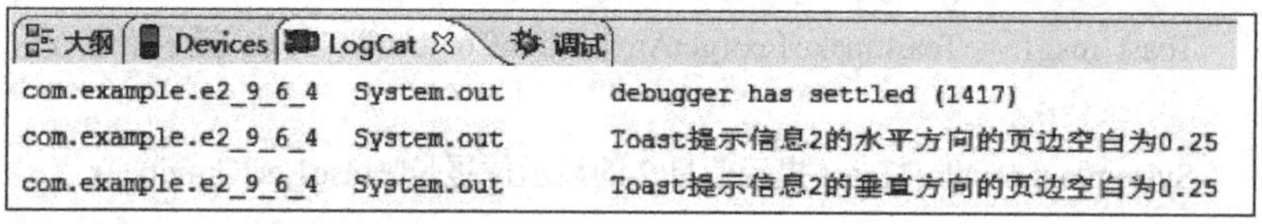

图 3-32　获取视图的页边空白

3.6.5　makeText 方法

makeText 方法用于生成一个包含文本视图的标准 Toast 对象。通过该方法显示 Toast 视图的文本提示内容和持续时间。应用程序往往通过 Toast 对象来显示一个简短的提示信息，如音量调节或文件保存成功的提示信息。其语法格式为：

```
public static Toast makeText(Context context,CharSequence text,int duration)
```

其中，参数 context 表示使用的上下文，通常是应用程序或者 Activity 对象；参数 text 表示要显示的文本，可以是已格式化的文本，也可通过字符串资源 ID 来引用；参数 duration 表示持续的时间，可以取值为 Toast.LENGTH_SHORT 或 Toast.LENGTH_LONG。

通过以下代码来演示怎样生成标准的 Toast 视图对象。

```
public class MainActivity extends Activity
{
    /*第一次调用活动*/
```

```
    @Override
    public void onCreate(Bundle savedInstanceState)
    {
        super.onCreate(savedInstanceState);
        setContentView(R.layout.main);
        Toast mg;
        mg = Toast.makeText(getApplicationContext(), "Toast 提示信息 2", Toast.LENGTH_LONG);
        mg.show();
    }
}
```

运行程序，效果如图 3-33 所示。

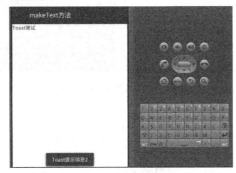

图 3-33　生成标准 Toast 对象

3.6.6　setView 方法

setView 方法用于设置 Toast 视图中要显示的 View 对象，该方法可以灵活地显示自定义的 Toast 视图，其中可以包含图片、文字等。在自定义 Toast 风格时，setView 方法是比较常用的一种手段。其语法格式为：

```
public void setView(View view)
```

其中，参数 view 为需要显示的 View 对象。

通过以下代码来演示怎样设置图片形式的 Toast 弹出信息。

```
public class MainActivity extends Activity
{
    /*第一次调用活动*/
    @Override
    public void onCreate(Bundle savedInstanceState)
    {
        super.onCreate(savedInstanceState);
        setContentView(R.layout.main);
        Toast toast = new Toast(getApplicationContext());
        ImageView view = new ImageView(getApplicationContext());
        view.setImageResource(R.drawable.fg);
```

```
            toast.setView(view);
            toast.show();
        }
    }
```
运行程序，效果如图 3-34 所示。

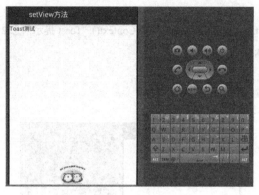

图 3-34　设置 View 对象

3.6.7　getView 方法

getView 方法用于获取一个 Toast 视图的 View 对象，通常用于一些用户自定义风格的 Toast 视图显示中，其返回类型为 View 类型的对象。其语法格式为：

```
public View getView()
```
通过以下代码来演示怎样在 Toast 视图中同时显示图片和文字。
```
    public class MainActivity extends Activity
    {
        /* 第一次调用活动*/
        @Override
        public void onCreate(Bundle savedInstanceState)
        {
            super.onCreate(savedInstanceState);
            setContentView(R.layout.main);
            Toast toast = Toast.makeText(getApplicationContext(), "图文共同显示", Toast.LENGTH_LONG);                                                    //创建 Toast
            //创建布局，水平布局
            LinearLayout mLayout = new LinearLayout(getApplicationContext());
            mLayout.setOrientation(LinearLayout.HORIZONTAL);
            //用于显示图像的 ImageView
            ImageView mImage = new ImageView(getApplicationContext());
            mImage.setImageResource(R.drawable.fg);
            View toastView = toast.getView();             //获取显示文字的 Toast View
            mLayout.addView(mImage);                      //添加到布局
```

```
            mLayout.addView(toastView);                    //添加到布局
            toast.setView(mLayout);                        //设置 View 对象
            toast.show();
        }
    }
```

以上代码中，先创建了一个 Toast 对象，显示的文字为"图文共同显示"。然后创建一个布局，将用户显示图像的 ImageView 和 Toast 文字信息都添加到布局中，其中 Toast 文字信息是通过 getView 方法来实现的。最后，使用 setView 方法将其设置给 toast，并显示该提示信息。

运行程序，效果如图 3-35 所示。

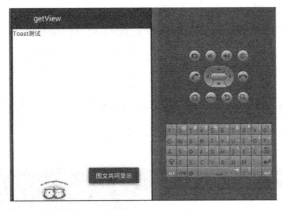

图 3-35　图片与文字共同显示效果

3.6.8　setGravity 方法

setGravity 方法用于设置提示信息在手机屏幕上的显示位置，经常用于自定义 Toast 视图显示位置的场合。其语法格式为：

```
        public void setGravity(int gravity,int xOffset,int yOffset)
```

其中，参数 gravity 为 Android 内置的 Gravity 标准常量；xOffset 参数为 Toast 视图在 x 方向上的偏移，大于 0 向右移，小于 0 向左移；yOffset 参数为 Toast 视图在 y 方向上的偏移，大于 0 向下移，小于 0 向上移。

通过以下代码来演示怎样将 Toast 视图显示在指定的位置上。

```
        public class MainActivity extends Activity
        {
            /* 第一次调用活动*/
            @Override
            public void onCreate(Bundle savedInstanceState)
            {
                super.onCreate(savedInstanceState);
                setContentView(R.layout.main);
                Toast toast = Toast.makeText(getApplicationContext()," setGravity 方法设置显示位置",
Toast.LENGTH_SHORT);
```

```
            //创建布局，水平布局
            LinearLayout mLayout = new LinearLayout(getApplicationContext());
            mLayout.setOrientation(LinearLayout.HORIZONTAL);
            //用于显示图像的 ImageView
            ImageView mImage = new ImageView(getApplicationContext());
            mImage.setImageResource(R.drawable.fg);
            View toastView = toast.getView();                //获取显示文字的 Toast View
            mLayout.addView(mImage);                         //添加到布局
            mLayout.addView(toastView);                      //添加到布局
            toast.setView(mLayout);                          //设置 View 对象
            toast.setGravity(Gravity.CENTER_VERTICAL, -1, 1);//设置显示位置
            toast.show();
        }
    }
```

在以上代码中，使用 setGrvaity 方法来设置显示位置位于手机屏幕垂直方向的中部、左下部，x 设为-1，即向左方向偏移；y 设为 1，即向下偏移。

运行程序，效果如图 3-36 所示。

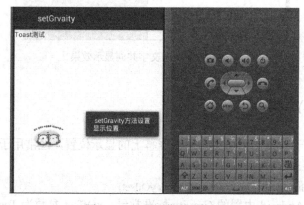

图 3-36 设置图像显示位置

3.6.9 getXOffset 方法与 getYOffset 方法

getXOffset 方法与 getYOffset 方法均用于获取 Toast 视图相对于参照位置的水平方向（垂直方向）偏移像素量，其返回值均为 int 类型。这两种方法常与 setGavity 方法一起使用，用来自定义 Toast 视图的显示位置。它们的调用格式为：

```
        public int getXOffset()
        public int getYOffset()
```

通过以下代码来演示怎样获取 Toast 视图显示位置。

```
        public class MainActivity extends Activity
        {
            /*第一次调用活动 */
```

```
        @Override
        public void onCreate(Bundle savedInstanceState)
        {
            super.onCreate(savedInstanceState);
            setContentView(R.layout.main);
            Toast msg = Toast.makeText(getApplicationContext(), "设置偏移量显示位置", Toast.LENGTH_LONG);
            msg.setGravity(Gravity.CENTER, 285, 150);                    //设置显示位置
            //获取偏移
            System.out.println("X Offset="+msg.getXOffset()+" Y Offset="+msg.getYOffset());
            msg.show();
        }
```

以上代码中，先创建一个 Toast 对象，接着调用 setGravity 方法来设置显示位置为水平和垂直方向的中部，并设置水平方向的偏移量为 285，垂直方向的偏移量为 150。然后通过 getXOffset 方法和 getYOffset 方法获取偏移量。

运行程序，Android 模拟器效果如图 3-37 所示，并在 Eclipse 环境下输出如图 3-38 的信息。

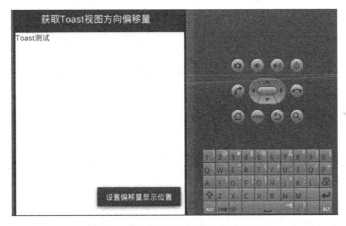

图 3-37　自定义偏移量的显示位置

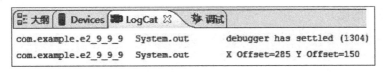

图 3-38　获取偏移量信息

3.6.10　setDuration 方法

setDuration 方法用于设置 Toast 视图显示的持续时间，通常用于获取 Toast 视图的持续时间，接着根据应用程序的需要进行修改的场合。其语法格式为：

```
        public void setDuration(int duration)
```

其中，参数 duration 为 Toast 视图的持续时间，具有如下两种取值。
- LENGTH_LONG：较长的视图持续时间。
- LENGTH_SHORT：较短的视图持续时间，为系统默认值。

通过以下代码演示怎样使用 setDuration 方法来设置较长时间的 Toast 视图显示。

```java
public class MainActivity extends Activity
{
    /** 第一次调用活动 */
    @Override
    public void onCreate(Bundle savedInstanceState)
    {
        super.onCreate(savedInstanceState);
        setContentView(R.layout.main);
        Toast toast = new Toast(getApplicationContext());
        ImageView view = new ImageView(getApplicationContext());
        view.setImageResource(R.drawable.icon);
        toast.setView(view);
        toast.setDuration(Toast.LENGTH_LONG);
        toast.show();
    }
}
```

以上代码中与 setView 方法得到的图像一样，只不过显示的时间不同。

3.6.11 setMargin 方法

setMargin 方法用于设置 Toast 视图提示信息的页边空白，常用于设置自定义 Toast 视图显示位置的场合。其语法格式为：

```
public void setMargin(float horizontalMargin,float verticalMargin)
```

其中，参数 horizontalMargin 为水平方向的页边空白，是容器的边缘与提示信息的横向空白（与容器宽度的比）；参数 verticalMargin 为垂直方向的页边空白，是容器的边缘与提示信息的纵向空白（与容器宽度的比）。

通过以下代码来演示设置 Toast 视图的页边空白值。

```java
public class MainActivity extends Activity
{
    /* 第一次调用活动*/
    @Override
    public void onCreate(Bundle savedInstanceState)
    {
        super.onCreate(savedInstanceState);
        setContentView(R.layout.main);
        Toast msg1 = Toast.makeText(getApplicationContext(), "设置 Toast 视图的页边空白",
```

```
Toast.LENGTH_LONG);
            msg1.setMargin(-0.15f, -0.25f);
            msg1.show();
        }
    }
```

运行程序，效果如图 3-39 所示。

图 3-39　页边空白设置

3.6.12　setText 方法

setText 方法用于设置 Toast 视图提示信息显示的文本内容，常用于设置自定义 Toast 视图显示内容的场合。其语法格式为：

    public void setText(CharSequence s)

其中，参数 s 为 Toast 视图提示信息显示的文本内容，也可通过引用字符串资源 ID 来表示。

通过以下代码来演示怎样设置 Toast 视图的显示内容。

```
    public class MainActivity extends Activity
    {
        /* 第一次调用活动 */
        @Override
        public void onCreate(Bundle savedInstanceState)
        {
            super.onCreate(savedInstanceState);
            setContentView(R.layout.main);
            Toast msg = Toast.makeText(getApplicationContext(), "setText 方法", Toast.LENGTH_SHORT);
            msg.setText("自定义文本！ ");
            msg.show();
        }
    }
```

运行程序,效果如图 3-40 所示。

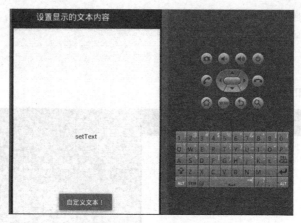

图 3-40　显示文本内容效果

3.6.13　show 方法

show 方法用于按照指定的设置来显示 Toast 视图提示信息。在最终显示 Toast 视图提示信息时,必须调用该方法。其语法格式为:

```
public void show
```

通过以下代码来演示怎样显示 Toast 视图提示信息。

```
public class MainActivity extends Activity
{
    /* 第一次调用活动 */
    @Override
    public void onCreate(Bundle savedInstanceState)
    {
        super.onCreate(savedInstanceState);
        setContentView(R.layout.main);
        Toast msg;
        msg = Toast.makeText(getApplicationContext(),"带图片的 Toast", Toast.LENGTH_LONG);
        msg.setGravity(Gravity.CENTER, 0, 0);            //设置显示位置
        LinearLayout toastView = (LinearLayout) msg.getView();
        //图片
        ImageView imageCodeProject = new ImageView(getApplicationContext());
        imageCodeProject.setImageResource(R.drawable.fg);
        toastView.addView(imageCodeProject, 0);          //添加 View
        msg.show();                                      //显示提示信息
    }
}
```

运行程序,效果如图 3-41 所示。

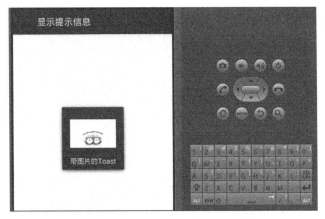

图 3-41　显示 Toast 提示信息

3.6.14　Toast 控件的综合实例

下面通过一个综合例子来演示 Toast 类的用法，该例子包含 5 种效果。其具体操作步骤如下。
（1）在 Eclipse 环境下，新建一个名为 ExampleToast 的工程。
（2）打开 res/layout 目录下的 main.xml 文件，代码修改为：

```
<?xml version="1.0" encoding="utf-8"?>
<LinearLayout xmlns:android="http://schemas.android.com/apk/res/android"
    android:orientation="vertical" android:layout_width="fill_parent"
    android:layout_height="fill_parent" android:padding="5dip" android:gravity="center">
    <Button android:layout_height="wrap_content"
      android:layout_width="fill_parent" android:id="@+id/btnSimpleToast"
      android:text="默认"></Button>
    <Button android:layout_height="wrap_content"
      android:layout_width="fill_parent" android:text="自定义显示位置"
      android:id="@+id/btnSimpleToastWithCustomPosition"></Button>
    <Button android:layout_height="wrap_content"
      android:layout_width="fill_parent" android:id="@+id/btnSimpleToastWithImage"
      android:text="带图片"></Button>
    <Button android:layout_height="wrap_content"
      android:layout_width="fill_parent" android:text="完全自定义"
      android:id="@+id/btnCustomToast"></Button>
    <Button android:layout_height="wrap_content"
      android:layout_width="fill_parent" android:text="其他线程"
      android:id="@+id/btnRunToastFromOtherThread"></Button>
</LinearLayout>
```

（3）选中 res 目录并右击，在弹出的快捷菜单中选择"新建"菜单下的"文件"选项，弹出"新建文件"对话框，如图 3-42 所示。在下侧的"文件名"中输入"custom.xml"，并单击"完成"按钮即可完成自定义"custom.xml"文件。

图 3-42 自定义"custom.xml"文件

custom.xml 文件的代码修改为:

```xml
<?xml version="1.0" encoding="utf-8"?>
<LinearLayout
    xmlns:android="http://schemas.android.com/apk/res/android"
    android:layout_height="wrap_content" android:layout_width="wrap_content"
    android:background="#ffffffff" android:orientation="vertical"
    android:id="@+id/llToast" >
    <TextView
      android:layout_height="wrap_content"
      android:layout_margin="1dip"
      android:textColor="#ffffffff"
      android:layout_width="fill_parent"
      android:gravity="center"
      android:background="#bb000000"
      android:id="@+id/tvTitleToast" />
    <LinearLayout
      android:layout_height="wrap_content"
      android:orientation="vertical"
      android:id="@+id/llToastContent"
      android:layout_marginLeft="1dip"
      android:layout_marginRight="1dip"
      android:layout_marginBottom="1dip"
      android:layout_width="wrap_content"
```

```xml
        android:padding="15dip"
        android:background="#44000000" >
        <ImageView
          android:layout_height="wrap_content"
          android:layout_gravity="center"
          android:layout_width="wrap_content"
          android:id="@+id/tvImageToast" />
        <TextView
          android:layout_height="wrap_content"
          android:paddingRight="10dip"
          android:paddingLeft="10dip"
          android:layout_width="wrap_content"
          android:gravity="center"
          android:textColor="#ff000000"
          android:id="@+id/tvTextToast" />
    </LinearLayout>
</LinearLayout>
```

（4）打开 res/values 目录下的 strings.xml 文件，代码修改为：

```xml
<resources>
    <string name="app_name">EaxmpleToast</string>
    <string name="hello_world">Hello world!</string>
    <string name="menu_settings">Settings</string>
    <string name="title_activity_main">Toast 的综合例子</string>
</resources>
```

（5）打开 src/com.example.eaxmpletoast 目录下的 MainActivity.java 文件，代码修改为：

```java
package com.example.eaxmpletoast;

import android.app.Activity;
import android.os.Bundle;
import android.os.Handler;
import android.view.Gravity;
import android.view.LayoutInflater;
import android.view.View;
import android.view.ViewGroup;
import android.view.View.OnClickListener;
import android.widget.ImageView;
import android.widget.LinearLayout;
import android.widget.TextView;
import android.widget.Toast;

public class MainActivity extends Activity implements OnClickListener
```

```java
{
    Handler handler = new Handler();
    @Override
    public void onCreate(Bundle savedInstanceState)
    {
        super.onCreate(savedInstanceState);
        setContentView(R.layout.main);
        findViewById(R.id.btnSimpleToast).setOnClickListener(this);
        findViewById(R.id.btnSimpleToastWithCustomPosition).setOnClickListener(this);
        findViewById(R.id.btnSimpleToastWithImage).setOnClickListener(this);
        findViewById(R.id.btnCustomToast).setOnClickListener(this);
        findViewById(R.id.btnRunToastFromOtherThread).setOnClickListener(this);
    }
    public void showToast()
    {
        handler.post(new Runnable()
        {
            @Override
            public void run()
            {
                Toast.makeText(getApplicationContext(), "来自其他线程！",
                    Toast.LENGTH_SHORT).show();
            }
        });
    }
    @Override
    public void onClick(View v)
    {
        Toast toast = null;
        switch (v.getId())
        {
        case R.id.btnSimpleToast:
            Toast.makeText(getApplicationContext(), "默认 Toast 样式",
            Toast.LENGTH_SHORT).show();
            break;
        case R.id.btnSimpleToastWithCustomPosition:
            toast = Toast.makeText(getApplicationContext(),
            "自定义位置 Toast", Toast.LENGTH_LONG);
            toast.setGravity(Gravity.CENTER, 0, 0);
            toast.show();
            break;
```

```java
case R.id.btnSimpleToastWithImage:
    toast = Toast.makeText(getApplicationContext(),
    "带图片的 Toast", Toast.LENGTH_LONG);
    toast.setGravity(Gravity.CENTER, 0, 0);
    LinearLayout toastView = (LinearLayout) toast.getView();
    ImageView imageCodeProject = new ImageView(getApplicationContext());
    imageCodeProject.setImageResource(R.drawable.fg);
    toastView.addView(imageCodeProject, 0);
    toast.show();
    break;
case R.id.btnCustomToast:
    LayoutInflater inflater = getLayoutInflater();
    View layout = inflater.inflate(R.layout.custom,
        (ViewGroup) findViewById(R.id.llToast));
    ImageView image = (ImageView) layout
        .findViewById(R.id.tvImageToast);
    image.setImageResource(R.drawable.fg);
    TextView title = (TextView) layout.findViewById(R.id.tvTitleToast);
    title.setText("Attention");
    TextView text = (TextView) layout.findViewById(R.id.tvTextToast);
    text.setText("完全自定义 Toast");
    toast = new Toast(getApplicationContext());
    toast.setGravity(Gravity.RIGHT | Gravity.TOP, 12, 40);
    toast.setDuration(Toast.LENGTH_LONG);
    toast.setView(layout);
    toast.show();
    break;
case R.id.btnRunToastFromOtherThread:
    new Thread(new Runnable()
    {
        public void run()
        {
            showToast();
        }
    }).start();
    break;
    }
    }
}
```

运行程序，得到初始界面如图 3-43 所示。

单击界面中的"完全自定义"按钮，得到如图 3-44 所示的效果。

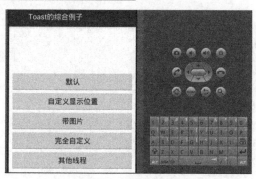

图 3-43 初始界面

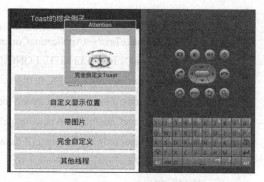

图 3-44 "完全自定义"界面

单击不同按钮，会得到不同效果的界面。

3.7 ImageView 控件

ImageView 类可以加载各种来源的图片（如资源或图片库），需要计算图像的尺寸，以便它可以在其他布局中使用，并提供缩放和着色（渲染）等各种显示选项。

3.7.1 setAdjustViewBounds 方法

设置 setAdjustViewBounds 属性为真可以在 ImageView 调整边界时保持图片的纵横比例。（需要与 maxWidth、maxHeight 一起使用，否则单独使用没有效果。）

1．maxWidth

maxWidth 用于设置该视图支持的最大高度的可选参数，只有 setAdjustViewBounds(boolean) 为真时有效。要设置图像最大尺寸为 100×100，并保持原始比率，操作步骤如下：

（1）设置 adjustViewBounds 为真；

（2）设置 maxWidth 和 maxHeight 为 100；

（2）设置宽、高的布局参数为 WRAP_CONTENT。

注意，如果原始图像较小，即使设置了该参数，图像仍然要比 100×100 小。如果要设置图片为固定大小，需要在布局参数中指定大小，并使用 setScaleType(ImageView.ScaleType)函数来检测如何将图片调整到适当的大小。

其语法格式为：

```
public void setMaxHeight (int maxHeight)
```

2．maxHeight

maxHeight 用于设置该视图支持的最大宽度的可选参数，只有 setAdjustViewBounds(boolean) 为真时有效。要设置图像最大尺寸为 100×100，并保持原始比率，操作步骤如下：

（1）设置 adjustViewBounds 为真；

（2）设置 maxWidth 和 maxHeight 为 100；

（3）设置宽、高的布局参数为 WRAP_CONTENT。

注意：如果原始图像较小，即使设置了该参数，图像仍然要比 100×100 小。如果要设置图片为固定大小，需要在布局参数中指定大小，并使用 setScaleType(ImageView.ScaleType)函数来检测如何将图片调整到适当的大小。

其语法格式为：

> public void setMaxWidth (int maxWidth)

3.7.2 setScaleType 方法

setScaleType 方法用于控制图像应该如何缩放和移动，以使图像与 ImageView 一致。其语法格式为：

> public void setScaleType (ImageView.ScaleType scaleType)

其中，参数 scaleType 为填充方法，在 Android 系统中提供了如下几种填充方式。

（1）ScaleType.CENTER：按原图大小显示图片，但图片宽、高大于控件的宽、高时，截图图片中间部分显示。

（2）ScaleType.CENTER_CROP：按比例放大原图，直至某边等于控件的宽或高。

（3）ScaleType.CENTER_INSDE：当原图宽、高等于控件的宽、高时，按原图大小居中显示。反之将原图缩放到控件的宽、高居中显示。

（4）ScaleType.FIT_CENTER：按比例拉伸图片，拉伸后图片的高度为控件的高度，且显示在该控件的左边。

（5）ScaleType.FIT_XY：拉伸图片（不按比例）以填充控件的宽高。

（6）ScaleType.MATRIX：按照指定的参数拉伸图片以填充控件。

3.7.3 setSelected 方法

setSelected 方法用于改变视图的选中状态，视图有选中和未选中两个状态。注意，选择状态不同于焦点。典型的选中的视图是像 ListView 和 GridView 这样的 AdapterView 中显示的内容，选中的内容会显示为高亮。其语法格式为：

> public void setSelected (boolean selected)

其中，参数 selected 设置为真，将视图设为选中状态；如果设置为假时，视图设为未选中状态。

3.7.4 setImageURI 方法

setImageURI 方法用于设置指定的 URI 为该 ImageView 显示的内容。其语法格式为：

> public void setImageURI (Uri uri)

其中，参数 uri 为图像的 URI。

该操作读取位图，并在 UI 线程中解码，因此可能导致反应迟缓。如果反应迟缓，可以考虑用 setImageDrawable(Drawable)、setImageBitmap(Bitmap)或 SetImageDrawable(Bm)代替。

其中，参数 Bm 为设置的位图；参数 Drawable 为设置的可绘制对象；参数 Bitmap 为设置的二值图像。

3.7.5 setAdjustViewBounds 方法

setAdjustViewBounds 方法用于设置当需要在 ImageView 调整边框时保持可绘制对象的比例时，将该值设为真。其语法格式为：

    public void setAdjustViewBounds (boolean adjustViewBounds)

其中，参数 adjustViewBounds 是否调整边框，以保持可绘制对象的原始比例。

3.7.6 setAlpha 方法

setAlpha 方法用于为 ImageView 图片控件设置图片透明度。值得注意的是，该方法是针对图片资源本身的。如果两个 ImageView 对象显示的是同一个图片资源，则设置其中一个 ImageView 对象的透明度，另外一个会有相同效果。其语法格式为：

    public void setAlpha (int alpha)

其中，参数 alpha 为透明度，其取值范围为 0～255，255 为不透明。

3.7.7 setImageResource 方法

setImageResource 方法用于为 ImageView 图片控件设置图片资源，该资源以指定 ID 的形式来表示。这是设置图片显示的一个主要方法。在使用时，先将图片资源保存在 res 文件夹下，并赋予 ID 即可在程序中使用。其语法格式为：

    public void setImageResource(int resId)

其中，参数 resId 为图片资源 ID。

3.7.8 ImageView 控件综合实例

下面通过实例来演示实现 ImageView 几种不同形式的图片显示效果。

【实现步骤】

（1）在 Eclipse 环境下，建立名为 E3_7 的工程。

（2）打开 res/layout 目录下的 main.xml 文件，代码修改为：

```
<?xml version="1.0" encoding="utf-8"?>
<LinearLayout xmlns:android="http://schemas.android.com/apk/res/android"
    android:layout_width="fill_parent"
    android:layout_height="fill_parent"
    android:orientation="vertical" >
    <TextView
        android:layout_width="fill_parent"
        android:layout_height="wrap_content"
        android:text="scaleType:fitXY"/>
    <ImageView
```

```
            android:id="@+id/image1"
            android:adjustViewBounds="true"
            android:layout_width= "300dp"
            android:layout_height="100dp"
            android:src= "@drawable/ss"
            android:scaleType="fitXY"
            android:background="#F00"/>
        <TextView
            android:layout_width="fill_parent"
            android:layout_height="wrap_content"
            android:text="scaleType:center"/>
         <ImageView
            android:id="@+id/image2"
            android:adjustViewBounds="true"
            android:layout_width= "300dp"
            android:layout_height="100dp"
            android:src= "@drawable/gg"
            android:scaleType="center"
            android:background="#F00"/>
        <TextView
            android:layout_width="fill_parent"
            android:layout_height="wrap_content"
            android:text="scaleType:fitEnd"/>
         <ImageView
            android:id="@+id/image3"
            android:adjustViewBounds="true"
            android:layout_width= "300dp"
            android:layout_height="100dp"
            android:src= "@drawable/ss"
            android:scaleType="fitEnd"
            android:background="#F00"/>
        <TextView
            android:layout_width="fill_parent"
            android:layout_height="wrap_content"
            android:text="scaleType:fitCenter"/>
         <ImageView
            android:id="@+id/image4"
            android:adjustViewBounds="true"
            android:layout_width= "300dp"
            android:layout_height="80dp"
            android:src= "@drawable/gg"
            android:scaleType="fitCenter"
```

```
            android:background="#F00"/>
    </LinearLayout>
```

（3）打开 res/values 目录下的 strings.xml 文件，代码修改为：

```xml
<resources>
    <string name="app_name">E3_7</string>
    <string name="hello_world">Hello world!</string>
    <string name="menu_settings">Settings</string>
    <string name="title_activity_main">ImageView 用法</string>
</resources>
```

（4）打开 src/com.example.e3_7 目录下的 MainActivity.java 文件，代码修改为：

```java
public class MainActivity extends Activity
{
    /**第一次调用活动 */
    @Override
    public void onCreate(Bundle savedInstanceState)
    {
        super.onCreate(savedInstanceState);
        setContentView(R.layout.main);
        ImageView image11=(ImageView)findViewById(R.id.image1);          //获取对象
        image11.setImageResource(R.drawable.ss);                          //设置图片
        ImageView image22=(ImageView)findViewById(R.id.image2);          //获取对象
        image22.setImageResource(R.drawable.gg);                          //设置图片
        ImageView image33=(ImageView)findViewById(R.id.image3);          //获取对象
        image33.setImageDrawable(getResources().getDrawable(R.drawable.ss)); //设置图片
        ImageView image44=(ImageView)findViewById(R.id.image4);          //获取对象
        image44.setImageResource(R.drawable.gg);                          //设置图片
        image44.setAlpha(95);                                             //设置图片透明度
    }
}
```

运行程序，效果如图 3-45 所示。

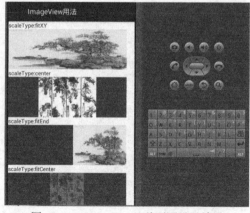

图 3-45 ImageView 几种不同显示效果

3.8 ProgressBar 控件

ProgressBar 是 Android 系统中的进度条，主要用于显示一些操作的进度。特别在执行一些比较耗时间的操作时，以动态的效果告诉用户目前的进展，使之有更好的用户体验。在 Android 手机中，应用程序开启时加载某些资源、从网络下载文件和进行大量数据计算的场合都使用 ProgressBar。

3.8.1 ProgressBar 相关方法

ProgressBar 类中常用的方法主要有以下几种。

1. setMax 方法

setMax 方法用于设置一个 ProgressBar 进度条的最大值，在默认情况下 ProgressBar 进度条的最大值为 100。该方法主要用于一些需要自定义进度条的场合。其语法格式为：

```
public synchronized void setMax(int max)
```

其中，参数 max 为设置的进度条最大值。此时，该进度条的范围为 0~max。

2. setProgress 方法

setProgress 方法用于设置一个 ProgressBar 进度条的当前主要进度值，该值应该小于 setMax 方法设置的最大值。该方法主要用于进度条显示的场合，一般放置在一个线程中，跟随所进行的操作来更新显示进度。其语法格式为：

```
public synchronized void setProgress(int progress)
```

其中，参数 progress 为设置的当前进度值。

3. setSecondaryProgress 方法

setSecondaryProgress 方法用于设置一个 ProgressBar 进度条的当前次要进行值，该值应该小于 setMax 方法设置的最大值。该方法主要用于多重进度显示的场合，例如，在多个文件下载的场合，主要进度显示所有文件的下载进度，次要进度显示当前文件的下载进度。该方法一般放置在一个线程中，跟随所进行的操作来更新显示进度。其语法格式为：

```
public synchronized void setSecondaryProgress(int secondaryProgress)
```

其中，参数 secondaryProgress 为设置的当前次要进度值。

3.8.2 ProgressBar 相关类型

ProgressBar 中分为两种类型：一种是横向的，另一种是旋转型的（默认）。

旋转型的 ProgressBar 需要使用 style 属性为其设置需要的大小，分为大、中、小三种。如果不设置将默认显示中等大小。

其中，

● Widget.ProgressBar.Small：用于显示很小的旋转进度。

- Widget.ProgressBar.Large：用于显示较大的旋转进度。
- Widget.ProgressBar.Inverse：将反色显示中等大小的旋转进度。
- Widget.ProgressBar.Small.Inverse：将反色显示较小的旋转进度。
- Widget.ProgressBar.Large.Inverse：将反色显示较大的旋转进度。

下面通过示例来综合演示两种不同类型的进度条。

1. 同时显示横向及模型进度条

【实现步骤】

（1）在 Eclipse 环境下，建立名为 E3_8_1 的工程。

（2）打开 res/layout 目录下的 main.xml 文件，代码修改为：

```xml
<?xml version="1.0" encoding="utf-8"?>
<LinearLayout xmlns:android="http://schemas.android.com/apk/res/android"
    android:layout_width="fill_parent"
    android:layout_height="fill_parent"
    android:orientation="vertical" >
 <Button
    android:layout_width="match_parent"
    android:layout_height="wrap_content"
    android:id="@+id/button1"
    android:text="@string/str_button1"/>
<ProgressBar
    android:id="@+id/firstBar"
    style="?android:attr/progressBarStyleHorizontal"
    android:layout_width="fill_parent"
    android:layout_height="wrap_content"
    android:max="200"
    android:visibility="gone"/>
<ProgressBar
    android:id="@+id/second"
    style="?android:attr/progressBarStyle"
    android:layout_width="wrap_content"
    android:layout_height="wrap_content"
    android:visibility="gone" />
</LinearLayout>
```

其中，style="?android:attr/progressBarStyleHorizontal"用来定义条件进度条；style="?android:attr/progressBarStyle"用来定义圆形进度条。

（3）打开 res/values 目录下的 strings.xml 文件，代码修改为：

```xml
<resources>
    <string name="app_name">E3_8_1</string>
    <string name="hello_world">Hello world!</string>
    <string name="menu_settings">Settings</string>
```

```xml
<string name="title_activity_main">两种不同类型进度条</string>
<string name="str_button1">请单击</string>
</resources>
```

（4）打开 src/com.example.e3_8_1 目录下的 MainActivity.java 文件，代码修改为：

```java
public class MainActivity extends Activity
{
    //声明变量
    ProgressBar firstProgressBar;
    ProgressBar secondProgressBar;
    Button myButton;
    int i = 0;
    /*第一次调用活动 */
    @Override
    public void onCreate(Bundle savedInstanceState)
    {
        super.onCreate(savedInstanceState);
        setContentView(R.layout.main);
        //根据控件的 ID 来取得代表控件的对象
        firstProgressBar = (ProgressBar)findViewById(R.id.firstBar);
        secondProgressBar = (ProgressBar)findViewById(R.id.second);
        myButton = (Button)findViewById(R.id.button1);
        myButton.setOnClickListener(new MyButton());
    }
    class MyButton implements OnClickListener
    {
        @Override
        public void onClick(View v)
        {
            if(i == 0)
            {
                //通过 setVisibility(View.VISIBLE)来显示进度条
                firstProgressBar.setVisibility(View.VISIBLE);
                secondProgressBar.setVisibility(View.VISIBLE);
            }
            else if(i < 200)
            {
                //通过 setProgress()来进度条的进程
                firstProgressBar.setProgress(i);
                //通过 setSecondaryProgress()来取得进度信息
                firstProgressBar.setSecondaryProgress(i + 10);
                secondProgressBar.setProgress(i);
```

```
                }
                else
                {
                    firstProgressBar.setVisibility(View.GONE);
                    secondProgressBar.setVisibility(View.GONE);
                }
                i = i + 10;
            }
        }
    }
```

运行程序，并单击图中的按钮，效果如图 3-46 所示。

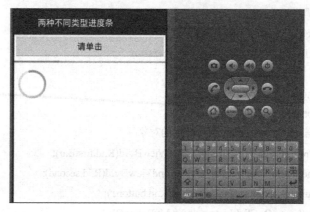

图 3-46　进度条效果

2．进一步实现进度条效果

本实例先使用控件 ProgressBar 实现了进度条，然后使用 Handler 访问了新进程 Activity 中的 Widget，并将运行状态在屏幕中显示出来。通过 Handler 对象和 Message 对象，将进程里的状态往外传递，最后由 Activity 的 Handle 事件来取得运行状态。

【实现步骤】

（1）在 Eclipse 环境下，建立名为 E3_8_2 的工程。

（2）打开 res/layout 目录下的 main.xml 文件，代码修改为：

```xml
<?xml version="1.0" encoding="utf-8"?>
<LinearLayout
    xmlns:android="http://schemas.android.com/apk/res/android"
    android:background="@drawable/white"
    android:orientation="vertical"
    android:layout_width="fill_parent"
    android:layout_height="fill_parent">
    <TextView
        android:id="@+id/TextView1"
        android:layout_width="fill_parent"
```

```xml
        android:layout_height="wrap_content"
        android:textColor="@drawable/blue"
        android:text="@string/hello_world" />
    <ProgressBar
        android:id="@+id/ProgressBar1"
        android:layout_width="wrap_content"
        android:layout_height="wrap_content"
        android:visibility="gone" />
    <Button
        android:id="@+id/Button1"
        android:layout_width="wrap_content"
        android:layout_height="wrap_content"
        android:text="@string/str_button1" />
</LinearLayout>
```

（3）打开 res/values 目录下的 strings.xml 文件，代码修改为：

```xml
<resources>
    <string name="app_name">E3_8_2</string>
    <string name="hello_world">Hello world!</string>
    <string name="menu_settings">Settings</string>
    <string name="title_activity_main">进度条实例</string>
    <string name="str_button1">请单击</string>
    <string name="progress_start">正在加载中....</string>
    <string name="progress_done">运行完毕！</string>
</resources>
```

（4）选择 res 目录下的 value 并右击，在弹出的快捷菜单中选择"新建"菜单下的"文件"选项，在"新建文件"界面左下侧的"文件名"中输入"color.xml"，即可完成 color.xml 文件的新建，其代码修改为：

```xml
<?xml version="1.0" encoding="utf-8"?>
<resources>
    <drawable name="darkgray">#801080</drawable>
    <drawable name="white">#F01FFF</drawable>
    <drawable name="blue">#0011FF</drawable>
</resources>
```

（5）打开 src/com.example.e3_8_2 目录下的 MainActivity.java 文件，代码修改为：

```java
public class MainActivity extends Activity
{
    private TextView mTextView01;
    private Button mButton01;
    private ProgressBar mProgressBar01;
    public int intCounter=0;
    /* 自定义 Handler 信息代码，作为识别事件处理 */
```

```java
protected static final int GUI_STOP_NOTIFIER = 0x108;
protected static final int GUI_THREADING_NOTIFIER = 0x109;
/** 第一次调用活动 */
@Override
public void onCreate(Bundle savedInstanceState)
{
    super.onCreate(savedInstanceState);
    setContentView(R.layout.main);
    mButton01 = (Button)findViewById(R.id.Button1);
    mTextView01 = (TextView)findViewById(R.id.TextView1);
    /* 设置 ProgressBar widget 对象 */
    mProgressBar01 = (ProgressBar)findViewById(R.id.ProgressBar1);
    /* 调用 setIndeterminate 方法赋值 indeterminate 模式为 false */
    mProgressBar01.setIndeterminate(false);
    /* 当单击按钮后，开始运行线程工作 */
    mButton01.setOnClickListener(new Button.OnClickListener()
    {
        @Override
        public void onClick(View v)
        {
            // TODO：自动存根法
            /* 单击按钮让 ProgressBar 显示 */
            mTextView01.setText(R.string.progress_start);
            /* 将隐藏的 ProgressBar 显示出来 */
            mProgressBar01.setVisibility(View.VISIBLE);
            /* 指定 Progress 为最多 100 */
            mProgressBar01.setMax(100);
            /* 初始 Progress 为 0 */
            mProgressBar01.setProgress(0);
            /* 起始一个运行线程 */
            new Thread(new Runnable()
            {
                public void run()
                {
                    /* 默认 0 至 9，共运行 10 次的循环叙述 */
                    for (int i=0;i<10;i++)
                    {
                        try
                        {
                            /* 成员变量，用以识别加载进度 */
                            intCounter = (i+1)*20;
```

```java
                    /* 每运行一次循环，即暂停 1 秒 */
                    Thread.sleep(1000);
                    /* 当 Thread 运行 5 秒后显示运行结束 */
                    if(i==4)
                    {
                        /* 以 Message 对象，传递参数给 Handler */
                        Message m = new Message();
                        /* 以 what 属性指定 User 自定义 */
                        m.what = MainActivity.GUI_STOP_NOTIFIER;
                        MainActivity.this.myMessageHandler.sendMessage(m);
                        break;
                    }
                    else
                    {
                        Message m = new Message();
                        m.what = MainActivity.GUI_THREADING_NOTIFIER;
                        MainActivity.this.myMessageHandler.sendMessage(m);
                    }
                }
                catch(Exception e)
                {
                    e.printStackTrace();
                }
            }
        }
    }).start();
    }
});
}
/* Handler 建构之后，会聆听传来的信息代码 */
Handler myMessageHandler = new Handler()
{
    // @Override
    public void handleMessage(Message msg)
    {
        switch (msg.what)
        {
            /* 当取得识别为离开运行线程时所取得的信息 */
            case MainActivity.GUI_STOP_NOTIFIER:

                /* 显示运行终了 */
```

```java
            mTextView01.setText(R.string.progress_done);

            /* 设置 ProgressBar Widget 为隐藏 */
            mProgressBar01.setVisibility(View.GONE);
            Thread.currentThread().interrupt();
            break;
          /* 当取得识别为持续在运行线程当中时所取得的信息 */
          case MainActivity.GUI_THREADING_NOTIFIER:
            if(!Thread.currentThread().isInterrupted())
            {
              mProgressBar01.setProgress(intCounter);
              /* 将显示进度显示于 TextView 当中 */
              mTextView01.setText
              (
                getResources().getText(R.string.progress_start)+
                "("+Integer.toString(intCounter)+"%)\n"+
                "Progress:"+
                Integer.toString(mProgressBar01.getProgress())+
                "\n"+"Indeterminate:"+
                Boolean.toString(mProgressBar01.isIndeterminate())
              );
            }
            break;
          }
          super.handleMessage(msg);
        }
      };
    }
```

运行程序,并单击按钮,效果如图 3-47 所示。

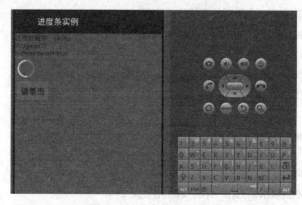

图 3-47 进度条实例效果

3.9 Spinner 控件

Spinner 是 Android 系统提供的下拉列表控件,该控件每次只能选择所有项中的一项。在使用 Spinner 下拉列表前,需要为该控件设置容纳可选数据的适配器。在 Android 手机中,该控件经常用于给用户提供多个选项,可选择其中一个的下拉列表。

3.9.1 setAdapter 方法

setAdapter 方法用于为 Spinner 下拉列表设置适配器,也为 Spinner 提供一个数据源。在使用 Spinner 下拉列表时,必须先为其提供可选数据源,然后才可供用户进行选择。该方法是使用 Spinner 下拉列表的最主要方法。其语法格式为:

```
public void setAdapter(SpinnerAdapter adapter)
```

其中,参数 adapter 为数据适配器,使用时必须先初始化适配器。

3.9.2 setPrompt 方法

setPrompt 方法用于为 Spinner 下拉列表设置标题。通过标题可以使读者更清楚该控件的用途。该方法不是使用 Spinner 下拉列表必需的步骤,但值得为用户提供更好的体验。其语法格式为:

```
public void setPrompt(CharSequence prompt)
```

其中,参数 prompt 为字符串,即在 Spinner 下拉列表中显示的标题内容。

通过以下代码来演示怎样实现适配器及设置其标题。

```java
public class MainActivity extends Activity
{
    /* 第一次调用活动*/
    private List< String> list = new ArrayList< String>();
    private TextView myTextView;
    private Spinner mySpinner;
    private ArrayAdapter< String> adapter;
    private Animation myAnimation;
    @Override
    public void onCreate(Bundle savedInstanceState)
    {
        super.onCreate(savedInstanceState);
        setContentView(R.layout.main);
        //第一步:添加一个下拉列表项的 list,这里添加的项就是下拉列表的菜单项
        list.add("北京");
        list.add("上海");
```

```java
        list.add("深圳");
        list.add("南京");
        list.add("重庆");
        myTextView = (TextView)findViewById(R.id.TextView_Show);
        mySpinner = (Spinner)findViewById(R.id.spinner_City);
        //第二步：为下拉列表定义一个适配器，这里就用到前面定义的 list
        adapter = new ArrayAdapter< String>(this,android.R.layout.simple_spinner_item, list);
        //第三步：为适配器设置下拉列表下拉时的菜单样式
        adapter.setDropDownViewResource(android.R.layout.simple_spinner_dropdown_item);
        //第四步：将适配器添加到下拉列表上
        mySpinner.setAdapter(adapter);
        mySpinner.setPrompt("请选择你所在城市");
        //第五步：为下拉列表设置各种事件的响应，这个事响应菜单被选中
        mySpinner.setOnItemSelectedListener(new Spinner.OnItemSelectedListener()
        {
            public void onItemSelected(AdapterView< ?> arg0, View arg1, int arg2, long arg3)
            {
                // TODO ：自动存根法
                /* 将所选 mySpinner 的值带入 myTextView 中*/
                myTextView.setText("你所在地区是： "+ adapter.getItem(arg2));
                /* 将 mySpinner 显示*/
                arg0.setVisibility(View.VISIBLE);
            }
            public void onNothingSelected(AdapterView< ?> arg0)
            {
                // TODO ：自动存根法
                myTextView.setText("NONE");
                arg0.setVisibility(View.VISIBLE);
            }
        });
        /*下拉菜单弹出的内容选项触屏事件处理*/
        mySpinner.setOnTouchListener(new Spinner.OnTouchListener()
        {
            public boolean onTouch(View v, MotionEvent event)
            {
                /* 将 mySpinner 隐藏，不隐藏也可以，看自己爱好*/
                v.setVisibility(View.INVISIBLE);
                return false;
            }
        });
        /*下拉菜单弹出的内容选项焦点改变事件处理*/
```

```
            mySpinner.setOnFocusChangeListener(new Spinner.OnFocusChangeListener()
            {
                public void onFocusChange(View v, boolean hasFocus)
                {
                    v.setVisibility(View.VISIBLE);
                }
            });
    }
}
```

运行程序，效果如图 3-48 所示。

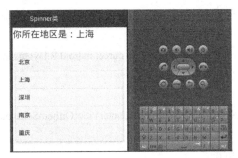

图 3-48　下拉列表

3.9.3　setPromptId 方法

setPromptId 方法用于为 Spinner 下拉列表设置标题。通过标题可以使读者更清楚该控件的用途。该方法通过指定字符串 ID 的形式设置 Spinner 下拉列表标题，这是设置标题的第二种方法。一般推荐采用这种方法，这在多种语言支持的场合非常有用。其语法格式为：

```
        public void setPromptId(int promptId)
```
其中，参数 promptId 为字符串 ID，指向 Spinner 下拉列表中显示的标题内容。

3.9.4　setOnItemSelectedListener 方法

setOnItemSelectedListener 方法用于设置 Spinner 下拉列表的监听器。当用户选择一个项目之后，便会触发该监听器。该方法是感知用户选择的主要方法，使用非常广泛。其语法格式为：

```
        setOnItemSelectedListener(AdapterView.OnItemSelectedListener listener)
```
通过以下代码来演示怎样设置 Spinner 监听器。

```
    public class MainActivity extends Activity
    {
        /* 第一次调用活动 */
        @Override
        public void onCreate(Bundle savedInstanceState)
```

```java
        {
            super.onCreate(savedInstanceState);
            setContentView(R.layout.main);
            Button button=(Button)findViewById(R.id.button1);                    //获取对象
            final Spinner s1=(Spinner)findViewById(R.id.spinner1);               //获取对象
            String[] mCountries={"天津","北京","广东","香港","湖南","山西"};     //字符串数组
            ArrayList<String> allcountries=new ArrayList<String>();              //list 对象
            for(int i=0;i<mCountries.length;i++)                                 //添加数据
            {
                allcountries.add(mCountries[i]);
            }
            ArrayAdapter<String> aspnCountries = new ArrayAdapter<String> (this, android.R.layout.simple_spinner_item, allcountries);     //初始化适配器
            aspnCountries.setDropDownViewResource(android.R.layout.simple_spinner_dropdown_item);
            s1.setAdapter(aspnCountries);                                        //设置适配器
            s1.setPromptId(R.string.prompt);                                     //指定标题 ID
            s1.setOnItemSelectedListener(new AdapterView.OnItemSelectedListener()
            {
                //设置监听器
                @Override
                public void onItemSelected(AdapterView<?> arg0, View arg1, int arg2, long arg3)
                {
                    // TODO Auto-generated method stub
                    Toast.makeText(MainActivity.this, "您选择的是："+arg0.getItemAtPosition(arg2).toString(),
                            Toast.LENGTH_LONG).show();                            //显示选择项
                }
                @Override
                public void onNothingSelected(AdapterView<?> arg0)
                {
                    // TODO：自动存根法
                    Toast.makeText(MainActivity.this, "您没有选择任何选项！", Toast.LENGTH_LONG).show();
                }
            });
            button.setOnClickListener(new View.OnClickListener()
            {
                //设置监听器
                @Override
                public void onClick(View v)
                {
```

```
                    Toast.makeText(getApplicationContext(),"Spinner 标题为："+s1.getPrompt(),
                    Toast.LENGTH_LONG).show();                              //获取标题
                }
            });
        }
    }
```
运行程序，效果如图 3-49 所示。

图 3-49 设置监听器

3.10 AutoCompleteTextView 控件

　　AutoCompleteTextView 是一个可编辑的文本，可自动完成用户的输入。建议列表显示在一个下拉菜单，用户可以从中选择一项，以完成输入。建议列表是从一个数据适配器获取的数据。它有三个重要的方法：clearListSelection()——清除选中的列表项；dismissDropDown()——如果存在关闭下拉菜单；getAdapter()——获取适配器。

3.10.1 setAdapter 方法

　　setAdapter 方法用于为 AutoCompleteTextView 自动完成文本框设置适配器，也为 AutoCompleteTextView 提供一个数据源。在使用 AutoCompleteTextView 自动完成文本框时，必须先为其提供可选数据源，然后才可供用户进行选择。其语法格式为：
```
        public void setAdapter(T adapter)
```
其中，参数 adapter 为数据适配器，使用时必须先初始化适配器。

　　通过以下代码来演示怎样设置自动文本的适配器。
```
        public class MainActivity extends Activity
        {
            final String[] COUNTRIES=new String[] {"Afghanistan", "Australia","America","Canada", "Chile", "China"};
            @Override
            public void onCreate(Bundle savedInstanceState)
```

```
        {
            super.onCreate(savedInstanceState);
            setContentView(R.layout.main);
            AutoCompleteTextView textView = (AutoCompleteTextView) findViewById(R.id.edit);
            ArrayAdapter<String> adapter = new ArrayAdapter<String>(this,
                    android.R.layout.simple_dropdown_item_1line, COUNTRIES);
            textView.setAdapter(adapter);
        }
    }
```

运行程序,效果如图 3-50 所示。

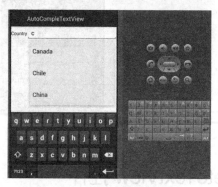

图 3-50 适配器的设置

3.10.2 setThreshold 方法

setThreshold 方法用于为 AutoCompleteTextView 自动完成文本框设置匹配字符数。当用户输入到指定的字符数时,才开始在适配器中寻找匹配的项目。该方法可简化匹配的项目,防止出现过多无关的项目。其调用格式为:

```
public void setThreshold(int threshold)
```

其中,参数 threshold 为匹配字符数。

3.10.3 setCompletionHint 方法

setCompletionHint 方法用于为 AutoCompleteTextView 自动完成文本框设置提示信息。当与用户输入的字符具有匹配项时,显示匹配项,并显示该提示信息,以提示用户。其语法格式为:

```
public void setCompletionHint(CharSequence hint)
```

其中,参数 hint 为字符串提示信息。

通过以下代码来演示怎样设置 AutoCompleteTextView 的提示信息。

```
public class MainActivity extends Activity
    {
        final String[] COUNTRIES = new String[] {"Afghanistan", "Australia","America","Canada", "Chile", "China"};
```

```java
@Override
public void onCreate(Bundle savedInstanceState)
{
    super.onCreate(savedInstanceState);
    setContentView(R.layout.main);
    AutoCompleteTextView textView = (AutoCompleteTextView) findViewById(R.id.edit);
    ArrayAdapter<String> adapter = new ArrayAdapter<String>(this,
            android.R.layout.simple_dropdown_item_1line, COUNTRIES);
    textView.setAdapter(adapter);
    textView.setCompletionHint("单击选择符合的一项");          //设置提示信息
}
```

运行程序，效果如图 3-51 所示。

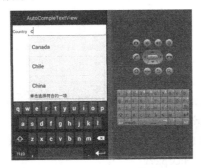

图 3-51 设置提示信息

设置标准除了可以在 MainActivity.java 代码中设置外，也可在 main.xml 代码中设置，其设置代码如下：

```xml
<?xml version="1.0" encoding="utf-8"?>
<LinearLayout xmlns:android="http://schemas.android.com/apk/res/android"
    android:orientation="horizontal" android:layout_width="fill_parent"
    android:layout_height="wrap_content">
    <TextView
        android:layout_width="wrap_content"
        android:layout_height="wrap_content"
        android:text="Country" />
    <AutoCompleteTextView
        android:id="@+id/edit"
        android:layout_width="fill_parent"
        android:layout_height="wrap_content"
        android:completionHint="单击选择符合的一项"
        android:completionThreshold="1"/>
</LinearLayout>
```

3.10.4 setDropDownBackgroundResource 方法

setDropDownBackgroundResource 方法用于为 AutoCompleteTextView 自动完成文本框设置背景资源，该资源以指定 ID 的形式来表示。在使用时，先将图片资源保存在 res 文件夹下，并赋予 ID。然后，即可在程序中使用。其语法格式为：

```
public void setDropDownBackgroundResource(int id)
```

其中，参数 id 为背景图片的 ID。

通过以下代码来演示怎样设置 AutoCompleteTextView 的背景颜色。

```java
public class MainActivity extends Activity
{
    final String[] COUNTRIES = new String[] {"Afghanistan", "Australia","America","Canada", "Chile", "China"};
    @Override
    public void onCreate(Bundle savedInstanceState)
    {
        super.onCreate(savedInstanceState);
        setContentView(R.layout.main);
        AutoCompleteTextView textView = (AutoCompleteTextView) findViewById(R.id.edit);
        ArrayAdapter<String> adapter = new ArrayAdapter<String>(this,
                android.R.layout.simple_dropdown_item_1line, COUNTRIES);
        textView.setAdapter(adapter);
        textView.setCompletionHint("单击选择符合的一项");          //设置提示信息
        textView.setDropDownBackgroundResource(R.drawable.fr);    //设置背景图片
    }
}
```

运行程序，效果如图 3-52 所示。

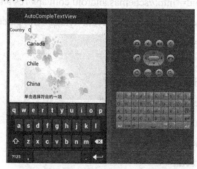

图 3-52 设置背景颜色图

3.10.5 setDropDownBackgroundDrawable 方法

setDropDownBackgroundDrawable 方法用于为 AutoCompleteTextView 自动完成文本框设置背景资源，该资源是指定的 Drawable。这是另外一种设置背景资源的方法。其语法格式为：

```
        public void setDropDownBackgroundDrawable(Drawable d)
```
其中，参数 d 为 Drawable 背景资源。

通过以下代码来演示怎样设置 AutoCompleteTextView 的背景颜色。

```
public class MainActivity extends Activity
{
    final String[] COUNTRIES = new String[] {"Afghanistan", "Australia","America","Canada", "Chile", "China"};
    @Override
    public void onCreate(Bundle savedInstanceState)
    {
        super.onCreate(savedInstanceState);
        setContentView(R.layout.main);
        AutoCompleteTextView textView = (AutoCompleteTextView) findViewById(R.id.edit);
        ArrayAdapter<String> adapter = new ArrayAdapter<String>(this, android.R.layout. simple_dropdown_item_1line, COUNTRIES);
        textView.setAdapter(adapter);
        textView.setCompletionHint("单击选择符合的一项");          //设置提示信息
        //设置背景图片
        textView.setDropDownBackgroundDrawable(getResources().getDrawable(R.drawable.fr));
    }
}
```

3.10.6 MultiAutoCompleteTextView 类

MultiAutoCompleteTextView 是一个继承自 AutoCompleteTextView 的可编辑的文本视图，能够对用户输入的文本进行有效地扩充提示，而不需要用户输入整个内容。（用户输入一部分内容，剩下的部分系统就会给予提示）。

用户必须提供一个 MultiAutoCompleteTextView.Tokenizer 用来区分不同的子串。其重要方法主要有以下几个。

- enoughToFilter()：当文本长度超过阈值时过滤。
- performValidation()：代替验证整个文本，这个子类方法验证每个单独的文字标记。
- setTokenizer(MultiAutoCompleteTextView.Tokenizer t)：用户正在输入时，Tokenizer 设置将用于确定文本相关范围内。

通过以下一个例子来说明 AutoCompleteTextView 类与 MultiAutoCompleteTextView 类的综合使用。

【实现步骤】

（1）在 Eclipse 环境下建立名为 E3_9_6 的工程。

（2）打开 res/layout 目录下的 main.xml 文件，其代码修改为：

```
<?xml version="1.0" encoding="utf-8"?>
<LinearLayout
```

```
xmlns:android="http://schemas.android.com/apk/res/android"
    android:orientation="vertical"
    android:layout_width="fill_parent"
    android:layout_height="fill_parent">
    <AutoCompleteTextView
    android:id="@+id/autoCompleteTextView"
        android:layout_width="fill_parent"
        android:layout_height="wrap_content"/>
    <MultiAutoCompleteTextView
    android:id="@+id/multiAutoCompleteTextView"
        android:layout_width="fill_parent"
        android:layout_height="wrap_content"/>
</LinearLayout>
```

（3）打开 src/com.example.e3_9_6 目录下的 MainActivity.java 文件，代码修改为：

```
package com.example.e3_9_6;
import android.app.Activity;
import android.os.Bundle;
import android.widget.ArrayAdapter;
import android.widget.AutoCompleteTextView;
import android.widget.MultiAutoCompleteTextView;
public class MainActivity extends Activity
{
    //几个城市
    private static final String[] cities=new String[]
        {"shenzhen","guangzhou","foshan","shanghai","beijing","shanxi","wuhang","xiangang","gansu"};
    private AutoCompleteTextView autoCompleteTextView=null;
    private MultiAutoCompleteTextView multiAutoCompleteTextView=null;
    @Override
    public void onCreate(Bundle savedInstanceState)
    {
        super.onCreate(savedInstanceState);
        setContentView(R.layout.main);    autoCompleteTextView=(AutoCompleteTextView)findViewById(R.id.autoCompleteTextView);
        multiAutoCompleteTextView=(MultiAutoCompleteTextView)findViewById(R.id.multiAutoCompleteTextView);
        //创建适配器
        ArrayAdapter<String> adapter=new ArrayAdapter<String>(this,android.R.layout.simple_dropdown_item_1line,cities);
        AutoCompleteTextView.setAdapter(adapter);
        //设置输入多少字符后提示，默认值为2
        AutoCompleteTextView.setThreshold(2);
```

```
        multiAutoCompleteTextView.setAdapter(adapter);
        multiAutoCompleteTextView.setThreshold(2);
        //用户必须提供一个 MultiAutoCompleteTextView.Tokenizer 用来区分不同的子串
        multiAutoCompleteTextView.setTokenizer(new MultiAutoCompleteTextView.CommaTokenizer());
    }
}
```

运行程序，效果如图 3-53 所示。

图 3-53 AutoCompleteTextView 与 MultiAutoCompleteTextView 综合使用

3.11 AnalogClock 控件

在 Android 系统中有一个专门的时钟对象 AnalogClock，通过此对象可以在屏幕中实现一个时钟的效果。在本实例屏幕的上方显示一个时钟效果界面，并在其下方放置一个 TextView 以显示一个时钟效果。

在具体实现上，使用到如下三个对象。

（1）android.os.Handler：通过产生的 Thread 对象在进程内同步调用方法 System.currentTimeMillis()，这样可以取得系统时间。

（2）java.lang.Thread：联系 Activity 与 Thread 的桥梁。

（3）android.os.Message：使用 Message 对象通知 Handler 对象，在收到 Message 对象后将时间变量的值显示在 TextView 中，这样即实现了数字时钟功能。

【实现步骤】

（1）在 Eclipse 环境下，建立名为 E3_10 的工程。

（2）打开 res/layout 目录下的 main.xml 文件，其代码修改为：

```
<?xml version="1.0" encoding="utf-8"?>
<LinearLayout
    android:id="@+id/widget27"
    android:layout_width="fill_parent"
    android:layout_height="fill_parent"
    xmlns:android="http://schemas.android.com/apk/res/android"
    android:orientation="vertical"
    android:background="@drawable/fr">
```

```xml
<AnalogClock
    android:id="@+id/myAnalogClock"
    android:layout_width="wrap_content"
    android:layout_height="wrap_content"
    android:layout_gravity="center_horizontal">
</AnalogClock>
<TextView
    android:id="@+id/myTextView"
    android:layout_width="wrap_content"
    android:layout_height="wrap_content"
    android:text="TextView"
    android:textSize="20sp"
    android:textColor="@drawable/white"
    android:layout_gravity="center_horizontal">
</TextView>
</LinearLayout>
```

（3）选择 res 目录下的 values 并右击，在弹出的快捷菜单中选择"新建"菜单下的"文件"选项，在"新建文件"界面左下侧的"文件名"中输入"color.xml"，即可完成 color.xml 文件的新建，其代码修改为：

```xml
<?xml version="1.0" encoding="utf-8"?>
<resources>
    <drawable name="white">#FF1F0F</drawable>
</resources>
```

（4）打开 src/com.example.e3_10 目录下的 MainActivity.java 文件，代码修改为：

```java
package com.example.e3_10;

import android.app.Activity;
import android.os.Bundle;
/*这里我们需要使用 Handler 类与 Message 类来处理运行线程*/
import android.os.Handler;
import android.os.Message;
import android.widget.AnalogClock;
import android.widget.TextView;
/*需要使用 Java 的 Calendar 与 Thread 类来取得系统时间*/
import com.example.e3_10.R;
import java.util.Calendar;
import java.lang.Thread;

public class MainActivity extends Activity
{
    /*声明一常数作为判别信息使用*/
```

```java
protected static final int GUINOTIFIER = 0x1234;
/*声明两个 widget 对象变量*/
private TextView mTextView;
public AnalogClock mAnalogClock;
/*声明与时间相关的变量*/
public Calendar mCalendar;
public int mMinutes;
public int mHour;
/*声明关键 Handler 与 Thread 变量*/
public Handler mHandler;
private Thread mClockThread;
/** 第一次调用活动 */
public void onCreate(Bundle savedInstanceState)
{
  super.onCreate(savedInstanceState);
  setContentView(R.layout.main);
  /*通过 findViewById 取得两个 widget 对象*/
  mTextView=(TextView)findViewById(R.id.myTextView);
  mAnalogClock=(AnalogClock)findViewById(R.id.myAnalogClock);
  /*通过 Handler 来接收运行线程所传递的信息并更新 TextView*/
  mHandler = new Handler()
  {
    public void handleMessage(Message msg)
    {
      /*这里是处理信息的方法*/
      switch (msg.what)
      {
        case MainActivity.GUINOTIFIER:
        /* 在这要处理 TextView 对象 Show 时间的事件 */
          mTextView.setText(mHour+" : "+mMinutes);
          break;
      }
      super.handleMessage(msg);
    }
  };
  /*通过运行线程来持续取得系统时间*/
  mClockThread=new LooperThread();
  mClockThread.start();
}
/*改写一个 Thread Class 用来持续取得系统时间*/
class LooperThread extends Thread
```

```
            {
                public void run()
                {
                    super.run();
                    try
                    {
                        do
                        {
                            /*取得系统时间*/
                            long time = System.currentTimeMillis();
                            /*通过 Calendar 对象来取得小时与分钟*/
                            final Calendar mCalendar = Calendar.getInstance();
                            mCalendar.setTimeInMillis(time);
                            mHour = mCalendar.get(Calendar.HOUR);
                            mMinutes = mCalendar.get(Calendar.MINUTE);
                            /*让运行线程休息一秒*/
                            Thread.sleep(1000);
                            /*重要关键程序:取得时间后发出信息给 Handler*/
                            Message m = new Message();
                            m.what = MainActivity.GUINOTIFIER;
                            MainActivity.this.mHandler.sendMessage(m);
                        }while(MainActivity.LooperThread.interrupted()==false);
                        /*当系统发出中断信息时停止本循环*/
                    }
                    catch(Exception e)
                    {
                        e.printStackTrace();
                    }
                }
            }
        }
```

运行程序，效果如图 3-54 所示。

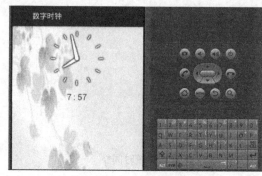

图 3-54　数字时钟效果图

3.12 DatePicker 与 TimePicker 控件

DatePicker 继承自 FrameLayout 类，日期选择控件的主要功能是向用户提供包含年、月、日的日期数据并允许用户对其修改。如果要捕获用户修改日期选择控件中的数据事件，需要为 DatePicker 添加 OnDateChangedListener 监听器。

TimePicker 也继承自 FrameLayout 类。时间选择控件向用户显示一天中的时间（可以为 24 小时，也可以为 AM/PM 制），并允许用户进行选择。如果要捕获用户修改时间数据的事件，便需要为 TimePicker 添加 OnTimeChangedListener 监听器。

3.12.1 DatePicker 控件

DatePicker 类中提供了许多方法用于用户更改或者选择日期。

1．init 方法

init 方法用于初始化一个 DatePicker 日期选择控件，通过该方法可以设置初始显示的年份、月份和日数，还可以设置相应的日期更改监听器。在使用一个 DatePicker 日期选择控件时，该方法必须是首次调用的。其语法格式为：

    public void init(int year,int monthOfYear,int dayOfMonth,DatePicker.OnDateChangedListener onDateChangedListener)

其中，参数 year 为初始年份；参数 monthOfYear 为初始月份，值得注意的是，这里实际显示的月份是此处设置的数值加 1；参数 dayOfMonth 为初始日数；参数 onDateChangedListener 为日期更改监听器。

2．onDateChangedListener 方法

onDateChangedListener 方法用于为 DatePicker 日期选择控件设置日期更改监听器。当用户在界面之上更改了日期后，将触发监听器，此时需要重载其中的 onDateChanged 方法。其语法格式为：

    public void OnDateChangedListener(OnDateChangedListener l)

3．updateDate 方法

updateDate 方法用于更新一个 DatePicker 日期选择控件，通过该方法可以设置显示的年份、月份和日数。值得注意的是，调用该方法将触发日期更改监听器。其语法格式为：

    public void updateDate(int year,int month,int dayOfMonth)

其中，参数 year 为年份；参数 month 为月份，值得注意的是，这里实际显示的月份是此处设置的数值加 1；参数 dayOfMonth 为日数。

3.12.2 TimerPicker 控件

TimerPicker 类为用户提供时间显示功能，还允许用户更改或者选择时间。

1. setCurrentHour 方法

setCurrentHour 方法用于设置一个 TimePicker 时间选择控件，显示小时。其语法格式为：

```
public void setCurrentHour(Integer currentHour)
```

其中，参数 currentHour 为显示的小时数。

2. setCurrentMinute 方法

setCurrentMinute 方法用于设置一个 TimePicker 时间选择控件，显示分钟。其语法格式为：

```
public void setCurrentMinute(Integer currentMinute)
```

3. setIs24HourView 方法

setIs24HourView 方法用于设置一个 TimePicker 时间选择控件是否按照 24 小时进制显示，在默认情况下该控件采用上午/下午模式显示。其语法格式为：

```
public void setIs24HourView(Boolean is24HourView)
```

其中，参数 is24HourView 为布尔型数据，当取值为 true 时表示采用 24 小时制，当取值为 false 时不采用 24 小时制。

4. setOnTimeChangedListener 方法

setOnTimeChangedListener 方法用于为 TimePicker 时间选择控件设置时间更改监听器。当用户在界面之上更改了时间后，将触发该监听器，此时需要重载其中的 onTimeChanged 方法。其语法格式为：

```
public void setOnTimeChangedListener(OnDateChangedListener 1)
```

3.12.3 DatePicker 与 TimePicker 控件综合实例

下面例子介绍模拟日期与时间选择控件的用法。

【实现步骤】

（1）在 Eclipse 环境下，建立名为 E3_11 的工程。

（2）打开 res/layout 目录下的 main.xml 文件，其代码修改为：

```xml
<?xml version="1.0" encoding="utf-8"?>
<LinearLayout
    xmlns:android="http://schemas.android.com/apk/res/android"
    android:orientation="vertical"
    android:layout_width="fill_parent"
    android:layout_height="fill_parent"
    >
<DatePicker
    android:id="@+id/datePicker"
    android:layout_width="wrap_content"
    android:layout_height="wrap_content"
    android:layout_gravity="center_horizontal"/>
```

```xml
<EditText
    android:id="@+id/dateEt"
    android:layout_width="fill_parent"
    android:layout_height="wrap_content"
    android:cursorVisible="false"
    android:editable="false"/>
<TimePicker
    android:id="@+id/timePicker"
    android:layout_width="wrap_content"
    android:layout_height="wrap_content"
    android:layout_gravity="center_horizontal"/>
    <EditText
    android:id="@+id/timeEt"
    android:layout_width="fill_parent"
    android:layout_height="wrap_content"
    android:cursorVisible="false"
    android:editable="false"/>
</LinearLayout>
```

(3) 打开 src/com.example.e3_11 目录下的 MainActivity.java 文件，代码修改为：

```java
package com.example.e3_11;

import java.util.Calendar;
import android.app.Activity;
import android.os.Bundle;
import android.widget.DatePicker;
import android.widget.EditText;
import android.widget.TimePicker;
import android.widget.DatePicker.OnDateChangedListener;
import android.widget.TimePicker.OnTimeChangedListener;

public class MainActivity extends Activity
{
    private EditText dateEt=null;
    private EditText timeEt=null;
    @Override
    public void onCreate(Bundle savedInstanceState)
    {
        super.onCreate(savedInstanceState);
        setContentView(R.layout.main);
        dateEt=(EditText)findViewById(R.id.dateEt);
        timeEt=(EditText)findViewById(R.id.timeEt);
```

```
DatePicker datePicker=(DatePicker)findViewById(R.id.datePicker);
TimePicker timePicker=(TimePicker)findViewById(R.id.timePicker);
Calendar calendar=Calendar.getInstance();
int year=calendar.get(Calendar.YEAR);
int monthOfYear=calendar.get(Calendar.MONTH);
int dayOfMonth=calendar.get(Calendar.DAY_OF_MONTH);
datePicker.init(year,monthOfYear, dayOfMonth, new OnDateChangedListener()
{
    public void onDateChanged(DatePicker view, int year,int monthOfYear, int dayOfMonth)
    {
        dateEt.setText("您选择的日期是: "+year+"年"+(monthOfYear+1)+"月"+dayOfMonth+"日。");
    }
});
    timePicker.setOnTimeChangedListener(new OnTimeChangedListener()
    {
        public void onTimeChanged(TimePicker view, int hourOfDay, int minute)
        {
            timeEt.setText("您选择的时间是: "+hourOfDay+"时"+minute+"分。");
        }
    });
    }
}
```

运行程序,效果如图 3-55 所示。

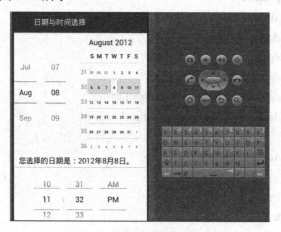

图 3-55　日期与时间选择效果

3.13　SeekBar 控件

SeekBar 是 Android 系统中的滑块控件,主要用于接收用户输入。特别是在一些简化用户输入的场合，可直观地改变用户输入值，使之有更好的用户体验。在 Android 手机中，音量调节、屏幕亮度调节等都使用了 SeekBar 滑块。

setOnSeekBarChangeListener 方法为 SeekBar 的常用方法。该方法用于设置一个 SeekBar 类滑块监听器，在该监听器内部可监听滑块的移动，同时提供用户在 SeekBar 上开始和停止触摸手势时的通知。其语法格式为：

public void setOnSeekBarChangeListener(SeekBar.OnSeekBarChangeListener 1)

通过以下代码来演示怎样使用 SeekBar 滑块。

【实现步骤】

（1）在 Eclipse 环境下，建立名为 E3_12 的工程。

（2）打开 res/layout 目录下的 main.xml 文件，代码修改为：

```xml
<menu xmlns:android="http://schemas.android.com/apk/res/android">
    <item android:id="@+id/menu_settings"
        android:title="@string/menu_settings"
        android:orderInCategory="100"
        android:showAsAction="never"
        android:background="@drawble/white" />
</menu>
```

（3）打开 src/com.example.e3_12 目录下的 MainActivity.java 文件，代码修改为：

```java
package com.example.e3_12;

import android.app.Activity;
import android.os.Bundle;
import android.widget.SeekBar;
import android.widget.TextView;

public class MainActivity extends Activity
{
    /*第一次调用活动 */
    @Override
    public void onCreate(Bundle savedInstanceState)
    {
        super.onCreate(savedInstanceState);
        setContentView(R.layout.main);
        this.setTitle("SeekBar");
        SeekBar seekBar=(SeekBar)this.findViewById(R.id.seekBar1);
        seekBar.setMax(1000);
        seekBar.setOnSeekBarChangeListener(seekBarListener);
    }

    private SeekBar.OnSeekBarChangeListener seekBarListener=new SeekBar.OnSeekBarChangeListener()
```

```java
        {
            //结束滚动时调用的方法
            public void onStopTrackingTouch(SeekBar seekBar)
            {
                // TODO：自动存根法
                System.out.println("Stop:"+seekBar.getProgress());
                TextView txt=(TextView)MainActivity .this.findViewById(R.id.textView1);
                txt.setText("Stop:"+seekBar.getProgress());
            }
            //开始滚动时调用的方法
            public void onStartTrackingTouch(SeekBar seekBar)
            {
                TextView txt=(TextView)MainActivity .this.findViewById(R.id.textView1);
                txt.setText("Start:"+seekBar.getProgress());
            }
            //进度条发生变化时调用的方法
            public void onProgressChanged(SeekBar seekBar, int progress, boolean fromUser)
            {
                TextView txt=(TextView)MainActivity .this.findViewById(R.id.textView1);
                txt.setText("Changed:"+seekBar.getProgress());
            }
        };
    }
```

运行程序，效果如图 3-56 所示。

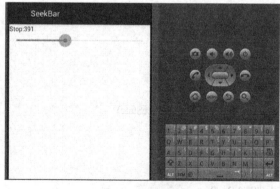

图 3-56 滑块创建

3.14 RatingBar 控件

评分组件 RatingBar 的功能是为用户提供一个评分操作的模式。在日常应用中，经常见到评分系统，用户可以对某个产品或某个观点进行评分处理。

3.14.1 RatingBar 类方法

下面介绍一些常用 RatingBar 类方法。

1. setMax 方法

setMax 方法用于设置一个 RatingBar 星级滑块的最大值,评分等级的范围为 0~max。其语法格式为:

> public synchronized void setMax(int max)

其中,参数 max 为设置的星级滑块最大值。

2. setNumStars 方法

setNumStars 方法用于设置一个 RatingBar 星级滑块的星形数量,该方法主要用于一些需要自定义星级滑块的场合。为了能够正常显示它们,建议将当前控件的布局宽度设置为 wrap content。其语法格式为:

> public void setNumStars(int numStars)

其中,参数 numStars 为设置的星期显示数量。

3. setRating 方法

setRating 方法用于设置一个 RatingBar 星级滑块的显示分数,也即是获取的星形数量。其语法格式为:

> public void setRating(float rating)

其中,参数 rating 为设置的分数,浮点型数据。

4. setStepSize 方法

setStepSize 方法用于设置一个 RatingBar 星级滑块的步长,也即是每次更改的最小长度。例如,假如每次可更改半个星星,则需要将该值设为 0.5。其语法格式为:

> public void setStepSize(float stepSize)

其中,参数 setSize 为设置的步长,浮点型数据。

5. setOnRatingBarChangeListener 方法

setOnRatingBarChangeListener 方法用于设置一个 RatingBar 星级滑块的监听器,当用户更改了星级滑块后,将触发该监听器。其语法格式为:

> public void setOnRatingBarChangedListener(RaingBar.OnRatingBarChangeListener listener)

3.14.2 RatingBar 控件综合实例

通过以下代码来演示怎样实现星级滑块。
【实现步骤】
(1) 在 Eclipse 环境下,建立名为 E3_13 的工程。
(2) 打开 res/layout 目录下的 main.xml 文件,代码修改为:

```xml
<?xml version="1.0" encoding="utf-8"?>
<LinearLayout
    xmlns:android="http://schemas.android.com/apk/res/android"
    android:orientation="vertical"
    android:paddingLeft="10dip"
    android:layout_width="wrap_content"
    android:layout_height="wrap_content">
    <RatingBar
        android:id="@+id/rbOne"
        android:layout_width="wrap_content"
        android:layout_height="wrap_content"
        android:numStars="3"
        android:rating="2.5"/>
    <RatingBar
        android:id="@+id/rbTwo"
        android:layout_width="wrap_content"
        android:layout_height="wrap_content"
        android:numStars="5"
        android:rating="2.25"/>
    <LinearLayout
        android:layout_width="match_parent"
        android:layout_height="wrap_content"
        android:layout_marginTop="10dip">
        <TextView android:id="@+id/textView"
            android:layout_width="wrap_content"
            android:layout_height="wrap_content" />
        <RatingBar android:id="@+id/rbThree"
            style="?android:attr/ratingBarStyleSmall"
            android:layout_marginLeft="5dip"
            android:layout_width="wrap_content"
            android:layout_height="wrap_content"
            android:layout_gravity="center_vertical" />
    </LinearLayout>
    <RatingBar
        android:id="@+id/rbFour"
        style="?android:attr/ratingBarStyleIndicator"
        android:layout_marginLeft="5dip"
        android:layout_width="wrap_content"
        android:layout_height="wrap_content"
        android:layout_gravity="center_vertical" />
</LinearLayout>
```

(3)打开 src/com.example.e3_13 目录下的 MainActivity.java 文件,代码修改为:

```java
import android.app.Activity;
import android.os.Bundle;
import android.util.Log;
import android.widget.RatingBar;
import android.widget.RatingBar.OnRatingBarChangeListener;
import android.widget.TextView;

public class RatingBarActivity extends Activity implements OnRatingBarChangeListener
{
    private RatingBar smallRatingBar = null;
    private RatingBar indicatorRatingBar = null;
    private TextView ratingText = null;
    /*   第一次调用活动 */
    @Override    public void onCreate(Bundle savedInstanceState)
    {
        super.onCreate(savedInstanceState);
        setContentView(R.layout.main);
        ratingText = (TextView) findViewById(R.id.textView);
        indicatorRatingBar = (RatingBar) findViewById(R.id.rbFour);
        smallRatingBar = (RatingBar) findViewById(R.id.rbThree);
        // 在布局中定义不同星级等级
        ((RatingBar)findViewById(R.id.rbOne)).setOnRatingBarChangeListener(this);
        ((RatingBar)findViewById(R.id.rbTwo)).setOnRatingBarChangeListener(this);
    }
    @Override
    public void onRatingChanged(RatingBar ratingBar, float rating, boolean fromUser)
    {
        // 返回所示星级的数量
        final int numStars = ratingBar.getNumStars();
        ratingText.setText(" 欢迎程度 " + rating + "/" + numStars);
        if (indicatorRatingBar.getNumStars() != numStars)
        {
            indicatorRatingBar.setNumStars(numStars);
            smallRatingBar.setNumStars(numStars);
        }
        //获取当前等级评级
            if (indicatorRatingBar.getRating() != rating)
            {
                Log.d("sxp","rating " + rating);
                indicatorRatingBar.setRating(rating);
```

```
                    smallRatingBar.setRating(rating);
                }
                // 获取此评级栏的步长
                final float ratingBarStepSize = ratingBar.getStepSize();
                if (indicatorRatingBar.getStepSize() != ratingBarStepSize)
                {
                    Log.d("sxp","ratingBarStepSize " + ratingBarStepSize);
                    indicatorRatingBar.setStepSize(ratingBarStepSize);
                    smallRatingBar.setStepSize(ratingBarStepSize);
                }
            }
        }
```

运行程序，效果如图 3-57 所示。

图 3-57　星级滑块实现

3.15　Tab 控件

Tab（标签）控件的功能是在屏幕内实现多个标签栏样式的效果，当单击某个标签栏时，会打开对应的界面。

下面通过一个示例来演示怎样实现标签。

【实现步骤】

（1）在 Eclipse 环境下，建立一个名为 E3_14 的工程。

（2）打开 res/layout 目录下的 main.xml 文件，代码修改为：

```
<?xml version="1.0" encoding="utf-8"?>
<FrameLayout xmlns:android="http://schemas.android.com/apk/res/android"
    android:layout_width="fill_parent"
    android:layout_height="fill_parent">
    <!--tab1 的布局 -->
    <LinearLayout android:id="@+id/tab1"
        android:layout_width="fill_parent"
        android:layout_height="fill_parent"
```

```xml
        androidrientation="vertical" >
    <EditText android:id="@+id/widget34"
        android:layout_width="fill_parent"
        android:layout_height="wrap_content"
         android:text="EditText"
        android:textSize="18sp">
    </EditText>
    <Button android:id="@+id/widget30"
        android:layout_width="wrap_content"
        android:layout_height="wrap_content"
        android:text="Button">
    </Button>
</LinearLayout>
<!--tab2 的布局 -->
<LinearLayout android:id="@+id/tab2"
    android:layout_width="fill_parent"
    android:layout_height="fill_parent"
    androidrientation="vertical"  >
    <AnalogClock android:id="@+id/widget36"
        android:layout_width="wrap_content"
        android:layout_height="wrap_content">
    </AnalogClock>
</LinearLayout>
<!--tab3 的布局 -->
<LinearLayout android:id="@+id/tab3"
    android:layout_width="fill_parent"
    android:layout_height="fill_parent"
    androidrientation="vertical">
    <RadioGroup android:id="@+id/widget43"
        android:layout_width="166px" android:layout_height="98px"
        androidrientation="vertical">
        <RadioButton android:id="@+id/widget44"
            android:layout_width="wrap_content"
            android:layout_height="wrap_content"
            android:text="RadioButton">
        </RadioButton>
        <RadioButton android:id="@+id/widget45"
            android:layout_width="wrap_content"
            android:layout_height="wrap_content"
            android:text="RadioButton">
        </RadioButton>
```

```
            </RadioGroup>
        </LinearLayout>
    </FrameLayout>
```

（3）打开 src/com.example.e3_14 目录下的 MainActivity.java 文件，代码修改为：

```java
package com.example.e3_14;

import android.app.TabActivity;
import android.os.Bundle;
import android.view.LayoutInflater;
import android.widget.TabHost;
public class MainActivity extends TabActivity
{
    public void onCreate(Bundle savedInstanceState)
    {
        super.onCreate(savedInstanceState);
        setTitle("Tab 测试");
        TabHost tabHost = getTabHost();
        LayoutInflater.from(this).inflate(R.layout.main,tabHost.getTabContentView(), true);
        tabHost.addTab(tabHost.newTabSpec("tab1").setIndicator("tab1").setContent(R.id.tab1));
        tabHost.addTab(tabHost.newTabSpec("tab3").setIndicator("tab2").setContent(R.id.tab2));
        tabHost.addTab(tabHost.newTabSpec("tab3").setIndicator("tab3").setContent(R.id.tab3));
    }
}
```

运行程序，效果如图 3-58 所示。

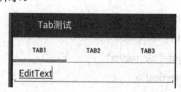

图 3-58 标签 TAB1 测试效果

图 3-58 中的三个标签分别链接不同效果，若单击 TAB2，得到如图 3-59 所示的效果，单击 TAB3 得到如图 3-60 所示的效果。

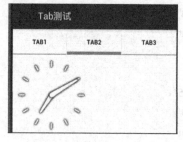

图 3-59 标签 TAB2 测试效果

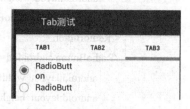

图 3-60 标签 TAB3 测试效果

3.16 Gallery 控件

Gallery 是 Android 系统中的相簿控件,非常华丽,支持滑动操作。其相当于一个水平滚动的图片列表,主要用于相册、图片浏览和选择等功能的应用程序中。

3.16.1 Gallery 类方法

本节介绍 Gallery 类中的主要方法。

1. setAdapter 方法

setAdapter 方法用于为 Gallery 相簿设置适配器,也为 Gallery 提供一个数据源。在使用 Gallery 相簿时,必须先为其提供可选数据源,然后才可供用户进行操作。其语法格式为:

```
public void setAdapter(SpinnerAdapter adapter)
```

其中,参数 adapter 为数据适配器,使用时必须先初始化适配器。

2. setSpacing 方法

setSpacing 方法用于设置 Gallery 相簿中各个图片间的间距。通过适当设置间距,可防止图片的堆叠,直到美化界面的效果。其语法格式为:

```
public vod setSpacing(int spacing)
```

其中,参数 spacing 为图片之间的间距,以像素为单位。

通过以下代码来演示怎样实现 Gallery 相簿适配器及设置图片间的间距。

```
public class MainActivity extends Activity
{
    /* 第一次调用活动*/
    @Override
    public void onCreate(Bundle savedInstanceState)
    {
        super.onCreate(savedInstanceState);
        setContentView(R.layout.main);
        final int[] images={R.drawable.f1,R.drawable.f2,
            R.drawable.f3,R.drawable.f4,R.drawable.f5};            //图片资源数组
        final Gallery g1=(Gallery)findViewById(R.id.gallery1);     //获取对象
        BaseAdapter adapter=new BaseAdapter()
        {
            //初始化适配器
            @Override
            public int getCount()
            {
                // TODO: 自动存根法
                return images.length;                               //返回图片个数
```

```
            }
            @Override
            public Object getItem(int position)
            {
                return null;
            }
            @Override
            public long getItemId(int position)
            {
                return 0;
            }
            @Override
            public View getView(int position, View convertView, ViewGroup parent)
            {
                //获取 View
                ImageView iv=new ImageView(MainActivity.this);
                iv.setImageResource(images[position]);          //设置图片资源
                //设置填充方式
                iv.setScaleType(ImageView.ScaleType.FIT_XY);
                iv.setLayoutParams(new Gallery.LayoutParams(230, 300));    //布局
                return iv;
            }
        };
        g1.setSpacing(50);                                       //设置图片间距
        g1.setAdapter(adapter);                                  //设置适配器
    }
}
```

运行程序,效果如图 3-61 所示。

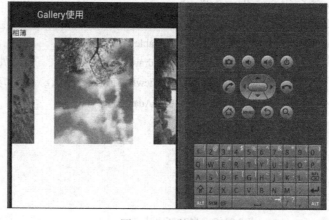

图 3-61 相簿效果

3. setOnItemClickListener 方法

setOnItemClickListener 方法用于设置 Gallery 相簿的监听器。当用户单击某张图片之后，即会触发该监听器。其语法格式为：

```
public void setOnItemClickListener(AdapterView.OnItemClickListener listener)
```

通过以下代码来演示怎样设置相簿监听器。

```java
public class MainActivity extends Activity
{
    /*第一次调用活动 */
    @Override
    public void onCreate(Bundle savedInstanceState)
    {
        super.onCreate(savedInstanceState);
        setContentView(R.layout.main);
        final int[] images={R.drawable.f1,R.drawable.f2,
            R.drawable.f3,R.drawable.f4,R.drawable.f5};          //图片资源数组
        final Gallery g1=(Gallery)findViewById(R.id.gallery1);   //获取对象
        BaseAdapter adapter=new BaseAdapter()
        {
            //初始化适配器
            @Override
            public int getCount()
            {
                // TODO：自动存根法
                return images.length;                             //返回图片个数
            }
            @Override
            public Object getItem(int position)
            {
                return null;
            }
            @Override
            public long getItemId(int position)
            {
                return 0;
            }
            @Override
            public View getView(int position, View convertView, ViewGroup parent)
            {
                // 获取 View
                ImageView iv=new ImageView(MainActivity.this);
```

```
                    iv.setImageResource(images[position]);        //设置图片资源
                    //设置填充方式
                    iv.setScaleType(ImageView.ScaleType.FIT_XY);
                    //布局
                    iv.setLayoutParams(new Gallery.LayoutParams(230, 300));
                    return iv;
                }
            };
            g1.setSpacing(60);                                    //设置图片间距
            g1.setAdapter(adapter);                               //设置适配器
            g1.setOnItemClickListener(new AdapterView.OnItemClickListener()
            {
                //设置监听器
                @Override
            public void onItemClick(AdapterView<?> arg0, View arg1, int arg2, long arg3)
                {
                    Toast.makeText(getApplicationContext(), "您选择的为第"+arg2+"张图片",
                            Toast.LENGTH_LONG).show();            //显示选择项
                }
            });
        }
    }
```

运行程序，效果如图 3-62 所示。

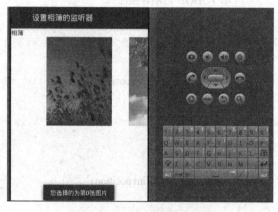

图 3-62　设置相簿监听器

在 Android 的相簿中，第一张图片的 Id 为 0，第二张图片的 Id 为 1，以此类推，第 N 张图片的 Id 为 N−1。

4．setGravity 方法

setGravity 方法用于设置 Gallery 相簿图片的对齐方式，可通过 Gravity 中的常量来指定。其语法格式为：

```
public void setGravity(int gravity)
```

其中,参数 gravity 为图片对齐方式,在程序中可以通过引用 Gravity 常量来表示。

5. setAnimationDuration 方法

setAnimationDuration 方法用于设置动画切换时间,仅限于动画开始时生效。通过调整图片的切换时间可以适应不同用户的操作需求,使程序具有更好的用户体验。其语法格式为:

```
public void setAnimationDuration(int animationDurationMillis)
```

其中,参数 animationDurationMillis 为图片的切换时间,以毫秒为单位。

通过以下代码来演示怎样设置相簿的动画切换时间。

```java
public class MainActivity extends Activity
{
    /**第一次调用活动 */
    @Override
    public void onCreate(Bundle savedInstanceState)
    {
        super.onCreate(savedInstanceState);
        setContentView(R.layout.main);
        final int[] images={R.drawable.f1,R.drawable.f2,
            R.drawable.f3,R.drawable.f4,R.drawable.f5};        //图片资源数组
        final Gallery g1=(Gallery)findViewById(R.id.gallery1); //获取对象
        BaseAdapter adapter=new BaseAdapter()
        {
            //初始化适配器
            @Override
            public int getCount()
            {
                // TODO：自动存根法
                return images.length;                          //返回图片个数
            }
            @Override
            public Object getItem(int position)
            {
                // TODO：自动存根法
                return null;
            }
            @Override
            public long getItemId(int position)
            {
                // TODO：自动存根法
                return 0;
            }
```

```
                @Override
                public View getView(int position, View convertView, ViewGroup parent)
                {
                    // TODO：自动存根法                                              //获取 View
                    ImageView iv=new ImageView(MainActivity.this);
                    iv.setImageResource(images[position]);                          //设置图片资源
                    iv.setScaleType(ImageView.ScaleType.FIT_XY);                    //设置填充方式
                    iv.setLayoutParams(new Gallery.LayoutParams(250, 280));         //布局
                    return iv;
                }
            };
            g1.setGravity(Gravity.CENTER);                                          //设置图片居中对齐
            g1.setAnimationDuration(100);                                           //设置动画切换时间
            g1.setAdapter(adapter);                                                 //设置适配器
        }
    }
```

运行程序，效果如图 3-63 所示。

图 3-63　相簿对齐及动画切换效果

在以上代码中，先初始化了资源数组，然后初始化 BaseAdapter 适配器，在其中设置图片资源的填充方式及布局等信息。接着，通过 setGravity 设置了相簿的对齐方式，通过 setAnimationDuration 方法设置动画切换时间，通过 setAdapter 设置相簿适配器。

3.16.2　Gallery 控件综合实例

本实例用于在屏幕中实现一个相簿功能。

【实现步骤】

（1）在 Eclipse 环境下，建立一个名为 E3_15_4 的工程。

（2）先将 5 张图片导入到 res/drawable 目录下的 drawable-hdpi 文件中。

（3）打开 res/drawable 目录下的 main.xml 文件，其代码修改为：

```xml
<?xml version="1.0" encoding="utf-8"?>
<Gallery xmlns:android="http://schemas.android.com/apk/res/android"
    android:id="@+id/mygallery"
    android:layout_width="fill_parent"
    android:layout_height="wrap_content"/>
```

（4）在 res/values 目录下新建一个名为 color.xml 的文件，其用于设置定义 layout 外部资源的样式，并且设置为随着滑动而改变 layout 背景图的效果，其代码为：

```xml
<?xml version="1.0" encoding="utf-8"?>
<resources>
    <declare-styleable name="Gallery">
        <attr name="android:galleryItemBackground" />
    </declare-styleable>
</resources>
```

（5）打开 res/values 目录下的 strings.xml 文件，其代码修改为：

```xml
<resources>
    <string name="app_name">E3_15_4</string>
    <string name="hello_world">Hello world!</string>
    <string name="menu_settings">Settings</string>
    <string name="my_gallery_text_pre">照片</string>
    <string name="my_gallery_text_post">编号</string>
    <string name="title_activity_main">Gallery 综合实例</string>
</resources>
```

（6）打开 src/com.example.e3_15_4 目录下的 MainActivity.java 文件，代码修改为：

```java
package com.example.e3_15_4;
import com.example.e3_15_4.R;
import android.app.Activity;
import android.os.Bundle;
/* 本范例需使用到的 class */
import android.content.Context;
import android.content.res.TypedArray;
import android.view.View;
import android.view.ViewGroup;
import android.widget.AdapterView;
import android.widget.BaseAdapter;
import android.widget.Gallery;
import android.widget.ImageView;
import android.widget.Toast;
import android.widget.AdapterView.OnItemClickListener;
public class MainActivity extends Activity
{
    /* Called when the activity is first created */
```

```java
@Override
public void onCreate(Bundle savedInstanceState)
{
    super.onCreate(savedInstanceState);
    setContentView(R.layout.main);
    /*通过 findViewById 取得*/
    Gallery g = (Gallery) findViewById(R.id.mygallery);
    /* 添加一个 ImageAdapter 并设置给 Gallery 对象 */
    g.setAdapter(new ImageAdapter(this));

    /* 设置一个 ItemClickListener 并 Toast 被单击图片的位置 */
    g.setOnItemClickListener(new OnItemClickListener()
    {
        public void onItemClick
        (AdapterView<?> parent, View v, int position, long id)
        {
            Toast.makeText
            (MainActivity.this, getString(R.string.my_gallery_text_pre)
            + position+ getString(R.string.my_gallery_text_post),
            Toast.LENGTH_SHORT).show();
        }
    });
}
/* 改写 BaseAdapter，自定义一个 ImageAdapter class */
public class ImageAdapter extends BaseAdapter
{
    /*声明变量*/
    int mGalleryItemBackground;
    private Context mContext;

    /*ImageAdapter 的构造器*/
    public ImageAdapter(Context c)
    {
        mContext = c;
        /* 使用在 res/values/attrs.xml 中的<declare-styleable>定义
        * 的 Gallery 属性.*/
        TypedArray a = obtainStyledAttributes(R.styleable.Gallery);

        /*取得 Gallery 属性的 Index id*/
        mGalleryItemBackground = a.getResourceId
        (R.styleable.Gallery_android_galleryItemBackground, 0);
```

```java
    /*让对象的 styleable 属性能够反复使用*/
    a.recycle();
}
/* 覆盖的方法 getCount，返回图片数目 */
public int getCount()
{
    return myImageIds.length;
}
/* 覆盖的方法 getItemId，返回图像的数组 id */
public Object getItem(int position)
{
    return position;
}
public long getItemId(int position)
{
    return position;
}
/* 覆盖的方法 getView，返回一 View 对象 */
public View getView
(int position, View convertView, ViewGroup parent)
{
    /*产生 ImageView 对象*/
    ImageView i = new ImageView(mContext);
    /*设置图片给 ImageView 对象*/
    i.setImageResource(myImageIds[position]);
    /*重新设置图片的宽高*/
    i.setScaleType(ImageView.ScaleType.FIT_XY);
    /*重新设置 Layout 的宽高*/
    i.setLayoutParams(new Gallery.LayoutParams(150, 120));
    /*设置 Gallery 背景图*/
    i.setBackgroundResource(mGalleryItemBackground);
    /*返回 ImageView 对象*/
    return i;
}
/*建构一 Integer array 并取得预加载 Drawable 的图片 id*/
private Integer[] myImageIds =
{
    R.drawable.f1,
    R.drawable.f2,
    R.drawable.f3,
```

```
                R.drawable.f4,
                R.drawable.f5,
            };
        }
    }
```

运行程序，效果如图 3-64 所示。

图 3-64　实现相簿效果

在 MainActivity.java 文件中，ImageAdapter 继承于 BaseAdapter，对未实现的方法进行重写构造，通过 Gallery 中的 OnItemClick()方法来响应图片滑动及 Layout 宽和高的设置。

3.17　ToggleButton 控件

可以认为 ToggleButton 控件是一个开关，每单击一次会在"开"和"关"这两种状态之间切换，同时切换的还有相应状态的文字和图片。

3.17.1　ToggleButton 类方法

ToggleButton 类继承于 Button 类，其中有很多方法都类似，在此只介绍 ToggleButton 类中特有的方法。

1. setTextOff 方法

setTextOff 方法用于设置一个 ToggleButton 状态开关按钮未选中时显示的文本。在首次使用状态开关按钮时，需要设置该内容。其语法格式为：

```
public void setTextOff(CharSequence textOff)
```

其中，参数 textOff 为字符串，也可以引用字符串资源。

2. setTextOn 方法

setTextOn 方法用于设置一个 ToggleButton 状态开关按钮选中时显示的文本。在首次用状态开关按钮时，需要设置该内容。其语法格式为：

```
public void setTextOn(CharSequence textOff)
```

其中，参数 textOff 为字符串，也可以引用字符串资源。

3. setChecked 方法

setChecked 方法用于设置一个 ToggleButton 状态开关按钮的状态，是选中还是不选中。该方法是更改 ToggleButton 状态形状按钮的主要方法。其语法格式为：

  public void setChecked(Boolean checked)

其中，参数 checked 为按钮状态，当其取值为 true 时表示选中，其取值为 false 时表示未选中。

4．setBackgroundDrawable 方法

setBackgroundDrawable 方法用于设置一个 ToggleButton 状态开关按钮的背景，可以指定相应的显示图片资源。其语法格式为：

  public void setBackgroundDrawable(Drawable d)

其中，参数 d 为图片资源。

3.17.2　ToggleButton 类实现

本实例通过 ToggleButton 类来控制开灯与关灯。

【实现步骤】

（1）在 Eclipse 环境下，建立一个名为 E3_16 的工程。

（2）打开 res/layout 目录下的 main.xml 文件，代码修改为：

```xml
<?xml version="1.0" encoding="utf-8"?>
<LinearLayout
xmlns:android="http://schemas.android.com/apk/res/android"
 android:layout_width="fill_parent"
 android:layout_height="fill_parent"
 android:orientation="vertical">
    <ImageView android:layout_width="wrap_content"
        android:layout_height="wrap_content"
        android:src="@drawable/bulb_off"
        android:id="@+id/imageView"
        android:layout_gravity="center_horizontal" >
    </ImageView>
    <ToggleButton android:layout_width="140dip"
        android:layout_height="wrap_content"
        android:textOn="开灯"
       android:textOff="关灯"
        android:id="@+id/toggleButton"
        android:layout_gravity="center_horizontal"></ToggleButton>
</LinearLayout>
```

（3）打开 src/com.example.3_16 目录下的 MainActivity.java 文件，代码修改为：

```java
public class MainActivity extends Activity
{
    private ImageView imageView=null;
    private ToggleButton toggleButton=null;
```

```
@Override
public void onCreate(Bundle savedInstanceState)
{
    super.onCreate(savedInstanceState);
    setContentView(R.layout.main);
    imageView=(ImageView) findViewById(R.id.imageView);
    toggleButton=(ToggleButton)findViewById(R.id.toggleButton);
    toggleButton.setOnCheckedChangeListener(new OnCheckedChangeListener()
    {
        public void onCheckedChanged(CompoundButton buttonView,boolean isChecked)
        {
            toggleButton.setChecked(isChecked);
            imageView.setImageResource(isChecked?R.drawable.bulb_on:R.drawable.bulb_off);
        }
    });
}
```

运行程序，效果如图 3-65 所示。

图 3-65 初始界面

当单击界面中的"开灯"按钮时，效果如图 3-66 所示。

图 3-66 ToggleButton 类效果

第 4 章 Android 菜单

Menu 控件显示应用程序的自定义菜单。命令、子菜单和分隔符条都可包括在菜单之中。每一个创建的菜单至多有四级子菜单。

4.1 Menu 菜单

大部分的应用程序都包括两种人机互动方式：一种是直接通过 GUI 的 View，其可以满足大部分的交互操作；另外一种是应用 Menu，当按下"Menu"按钮后，即弹出与当前活动状态下的应用程序相匹配的菜单。这两种方式相比较都有各自的优势，而且可以很好地相辅相成。即便用户可以由主界面完成大部分操作，但是适当地扩展 Menu 功能可以更好地完善应用程序，至少用户可以通过排列整齐的按钮清晰地了解当前模式下可以使用的功能。

Android 提供了三种菜单类型，分别为 MenuItem、SubMenu、ContextMenu。

MenuItem 是通过"home"键来显示的；ContextMenu 需要在 View 上按 2s 后显示。这两种 Menu 都可以加入子菜单，子菜单不能嵌套子菜单。MenuItem 最多只能在屏幕最下面显示 6 个菜单选项，称为 Icon Menu，Icon Menu 不能有 Checkable 选项。多于 6 的菜单会以 More Icon Menu 来调出，称为 Expanded Menu。MenuItem 通过 Activity 的 onCreateOptionsMenu 生成，这个函数只会在 Menu 第一次生成时调用。任何想改变 MenuItem 的想法只能通过 onPrepareOptionsMenu 来实现，这个函数会在 Menu 显示前调用。onOptionsItemSelected 用来处理选中的菜单项。

ContextMenu 是跟某个具体的 View 绑定在一起的，在 Activity 中用 Register ForContextMenu 来为某个 View 注册 ContextMenu。ContextMenu 在显示前都会调用 onCreateContextMenu 来生成 Menu。onContextItemSelected 用来处理选中的菜单项。

Android 还提供了对菜单项进行分组的功能，可以把相似功能的菜单项分在同一个组，这样就可以通过调用 setGroupCheckable，setGroupEnabled，setGroupVisible 来设置菜单属性，而无须单独设置。

4.1.1 Menu 菜单方法

Menu 类中提供了几种常用方法用于实现其相关功能。

1. onCreateOptionsMenu 方法

onCreateOptionsMenu 方法用于初始化选项菜单，在应用程序第一次显示选项菜单时调用。onCreateOptionsMenu 方法主要设置菜单项。其语法格式为：

```
public Boolean onCreateOptions(Menu menu)
```

其中，参数 menu 表示菜单，通过该变量可以为应用程序添加各个菜单。

2. add 方法

add 方法用于向菜单中添加菜单项，其返回值为 MenuItem 类型。add 方法是设置选项菜单最主要的方法，通常在 onCreateOptionsMenu 方法中调用。其语法格式为：

```
public abstract MenuItem add(int groupId,int itemId,int order,CharSequence title)
public abstract MenuItem add(int groupId,int itemId,int order,int titleRes)
public abstract MenuItem add (CharSequence title)
public abstract MenuItem add (int titleRes)
```

其中，参数 groupId 为菜单项所在的组 ID。通过对菜单项分组，可以进行批量处理，也可以赋值 NONE 来表示不使用组 ID；参数 itemId 为菜单项的唯一标识 ID；参数 order 为菜单项的顺序，也可赋值为 NONE；参数 title 为菜单项所显示的菜单名称；参数 titleRes 为菜单项所显示的菜单名称，通过引用字符串 ID 来表示。

3. addSubMenu 方法

addSubMenu 方法用于向菜单中添加菜单项，其返回值为 SubMenu 对象。这是设置选项子菜单最主要的方法，通常在 onCreateOptionsMenu 方法中调用。其语法格式为：

```
public abstract SubMenu addSubMenu(int groupId,int itemId,int order,CharSequence title)
public abstract SubMenu addSubMenu(int groupId,int itemId,int order,int titleRes)
public abstract SubMenu addSubMenu(CharSequence title)
public abstract SubMenu addSubMenu(int titleRes)
```

其中，参数 groupId 为子菜单项所在的组 ID，通过对子菜单项分组，可以进行批量处理，也可以赋值 NONE 来表示不使用组 ID；参数 itemId 为子菜单项唯一标识 ID；参数 order 为子菜单项的顺序，也可以赋值为 NONE；参数 titleRes 为子菜单项所显示的名称，通过引用字符串 ID 来表示。

通过以下代码来演示怎样添加菜单和子菜单。

```java
public class MainActivity extends Activity
{
    // 定义字体大小菜单项的标识
    final int FONT_9 = 0x111;
    final int FONT_11 = 0x112;
    final int FONT_13 = 0x113;
    final int FONT_15 = 0x114;
    final int FONT_17 = 0x115;
    // 定义普通菜单项的标识
    final int PLAIN_ITEM = 0x11b;
    // 定义字体颜色菜单项的标识
    final int FONT_BLUE = 0x116;
    final int FONT_YELLOW = 0x117;
    final int FONT_RED = 0x118;
    private EditText edit;
```

```java
@Override
public void onCreate(Bundle savedInstanceState)
{
    super.onCreate(savedInstanceState);
    setContentView(R.layout.main);
    edit = (EditText) findViewById(R.id.txt);
}
@Override
public boolean onCreateOptionsMenu(Menu menu)
{
    // 向 menu 中添加字体大小的子菜单
    SubMenu fontMenu = menu.addSubMenu("字体大小");
    // 设置菜单的图标
    fontMenu.setIcon(R.drawable.pn);
    // 设置菜单头的图标
    fontMenu.setHeaderIcon(R.drawable.pn);
    // 设置菜单头的标题
    fontMenu.setHeaderTitle("选择字体大小");
    fontMenu.add(0, FONT_9, 0, "9 号字体");
    fontMenu.add(0, FONT_11, 0, "11 号字体");
    fontMenu.add(0, FONT_13, 0, "13 号字体");
    fontMenu.add(0, FONT_15, 0, "15 号字体");
    fontMenu.add(0, FONT_17, 0, "17 号字体");
    // 向 menu 中添加普通菜单项
    menu.add(0, PLAIN_ITEM, 0, "普通菜单项");
    // 向 menu 中添加文字颜色的子菜单
    SubMenu colorMenu = menu.addSubMenu("字体颜色");
    colorMenu.setIcon(R.drawable.cn);
    // 设置菜单头的图标
    colorMenu.setHeaderIcon(R.drawable.cn);
    // 设置菜单头的标题
    colorMenu.setHeaderTitle("文字颜色");
    colorMenu.add(0, FONT_BLUE, 0, "蓝色");
    colorMenu.add(0, FONT_YELLOW, 0, "粉色");
    colorMenu.add(0, FONT_RED, 0, "红色");
    return super.onCreateOptionsMenu(menu);
}
@Override
// 菜单项被单击后的回调方法
public boolean onOptionsItemSelected(MenuItem mi)
{
```

```java
            //判断单击的是哪个菜单项,并有针对性地做出响应
            switch (mi.getItemId())
            {
                case FONT_9:
                    edit.setTextSize(9 * 2);
                    break;
                case FONT_11:
                    edit.setTextSize(11 * 2);
                    break;
                case FONT_13:
                    edit.setTextSize(13 * 2);
                    break;
                case FONT_15:
                    edit.setTextSize(15 * 2);
                    break;
                case FONT_17:
                    edit.setTextSize(17 * 2);
                    break;
                case FONT_BLUE:
                    edit.setTextColor(Color.BLUE);
                    break;
                case FONT_YELLOW:
                    edit.setTextColor(Color.YELLOW);
                    break;
                case FONT_RED:
                    edit.setTextColor(Color.RED);
                    break;
                case PLAIN_ITEM:
                    Toast toast = Toast.makeText(MainActivity.this, "您单击了菜单项", Toast.LENGTH_SHORT);
                    toast.show();
                    break;
            }
            return true;
        }
    }
```

以上代码中添加了三个菜单,三个菜单中有两个为子菜单,而且程序还为子菜单设置了图标、标题等。运行程序,单击模拟器中的"Menu"按钮,得到如图 4-1 所示的效果。

图 4-1 所示的菜单中,"字体大小"包含子菜单,"普通菜单项"只是一个菜单项,"字体颜色"也包含子菜单。如果单击"字体大小"菜单,将看到屏幕上显示如图 4-2 所示的子菜单。

由于程序重写了 onOptionItemSelected(MenuItem mi)方法,因此当用户单击指定菜单项时,

程序可以为菜单项的单击事件提供响应。

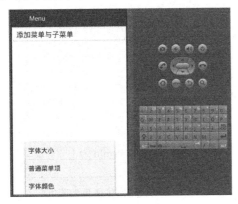

图 4-1 初始页面

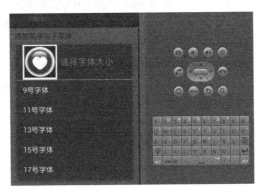

图 4-2 子菜单

4．onOptionsItemSelected 方法

onOptionsItemSelected 方法用于设置当选项菜单中某个选项被选中时执行的操作。该方法传入的是 MenuItem 对象，通过该对象可以获知哪个菜单项被选中，这是执行菜单处理的主要方法。其语法格式为：

```
public Boolean onOptionsItemSelected(MenuItem item)
```

其中，参数 item 为传入的菜单项，程序中通过该参数可以获知哪个菜单项被选中。

通过以下代码来演示怎样设置选项菜单的处理函数。

```java
package iflab.test;

public class MainActivity extends Activity
{
    /* 第一次调用活动 */
    TextView tv=null;
    @Override
    public void onCreate(Bundle savedInstanceState)
    {   // onCreate 方法
        super.onCreate(savedInstanceState);
        setContentView(R.layout.main);
        tv=(TextView)findViewById(R.id.textView2);
        tv.setText("设置选项菜单！ ");                    //设置文本
    }
    @Override
    public boolean onCreateOptionsMenu(Menu menu)
    {
        //重载 onCreateOptionsMenu 方法
        // TODO：自动存根法
        menu.add(0,1,1,"黑色");                          //添加菜单 1
```

```
            menu.add(0,2,2,"红色");                              //添加菜单2
            menu.addSubMenu(0,3,3,"字号");                        //添加子菜单
            return super.onCreateOptionsMenu(menu);
        }
        @Override
        public boolean onOptionsItemSelected(MenuItem item)
        {
            //重载 onOptionsItemSelected 方法
            if(item.getItemId()==1)                              //如果 itemId 为 1
            {
                tv.setTextColor(Color.BLUE);                     //设置字体颜色为蓝色
            }
            else if(item.getItemId()==2)                         //如果 itemId 为 2
            {
                tv.setTextColor(Color.RED);                      //设置字体颜色为红色
            }
            return super.onOptionsItemSelected(item);
        }
    }
```

运行程序，单击模拟器右边的"Menu"按钮，即可弹出如图 4-3 所示的对话框。

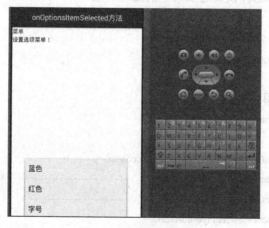

图 4-3 选项菜单处理

5. onOptionsMenuClosed 方法

onOptionsMenuClosed 方法用于设置当选项菜单关闭或退出时执行的操作，其输入参数为 Menu 对象。当用户执行了某个菜单，或按了返回按键，都将触发该方法。其语法格式为：

```
public void onOptionsMenu(Menu menu)
```

其中，参数 menu 为传入的菜单。

通过以下代码来演示怎样设置选项菜单关闭处理函数。

```java
public class MainActivity extends Activity
{
    /* 第一次调用活动 */
    TextView tv=null;
    @Override
    public void onCreate(Bundle savedInstanceState)
    {
        // onCreate 方法
        super.onCreate(savedInstanceState);
        setContentView(R.layout.main);
        tv=(TextView)findViewById(R.id.textView2);
        tv.setText("设置选项菜单！");                          //设置文本
    }
    @Override
    public boolean onCreateOptionsMenu(Menu menu)
    {
        //重载 onCreateOptionsMenu 方法
        // TODO：自动存根法
        menu.add(0,1,1,"蓝色");                              //添加菜单1
        menu.add(0,2,2,"红色");                              //添加菜单2
        menu.addSubMenu(0,3,3,"字号");                       //添加子菜单
        return super.onCreateOptionsMenu(menu);
    }
    @Override
    public boolean onOptionsItemSelected(MenuItem item)
    {
        //重载 onOptionsItemSelected 方法
        if(item.getItemId()==1)                              //如果 itemId 为 1
        {
            tv.setTextColor(Color.BLACK);                    //设置字体颜色为黑色
        }
        else if(item.getItemId()==2)                         //如果 itemId 为 2
        {
            tv.setTextColor(Color.RED);                      //设置字体颜色为红色
        }
        return super.onOptionsItemSelected(item);
    }
    @Override
    public void onOptionsMenuClosed(Menu menu)
    {
        //重载 onOptionsMenuClosed
```

```
            //显示信息
            Toast.makeText(getApplicationContext(), "关闭了选项菜单", Toast.LENGTH_LONG).show();
            super.onOptionsMenuClosed(menu);
        }
    }
```

运行程序,效果如图 4-3 所示,当调出菜单项后,单击模拟器中的"返回"按键,得到如图 4-4 所示的效果。

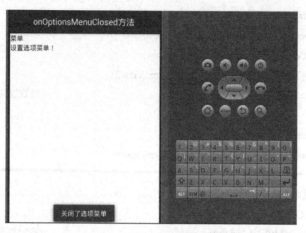

图 4-4 关闭选项菜单效果

4.1.2 Menu 菜单综合实例

创建一个将在菜单内显示的项列表,并处理用户在菜单内选定某一项时所触发的事件。

【实现步骤】

(1) 在 Eclipse 环境下,建立一个名为 E4_1_4 的工程。
(2) 打开 res/layout 目录下的 main.xml 文件,代码修改为:

```
<?xml version="1.0" encoding="utf-8"?>
<LinearLayout xmlns:android="http://schemas.android.com/apk/res/android"
    android:orientation="vertical"
    android:layout_width="fill_parent"
    android:layout_height="fill_parent"
    >
<Button android:id="@+id/btn1"
    android:layout_width="fill_parent"
    android:layout_height="wrap_content"
    android:text = "Tap and hold this for more options..." />
</LinearLayout>
```

(3) 打开 src/com.example.e4_1_4 目录下的 MainActivity.java 文件,代码修改为:

```java
package com.example.e4_1_4;

import android.app.Activity;
import android.os.Bundle;
import android.view.Menu;
import android.view.MenuItem;
import android.widget.Button;
import android.widget.Toast;
import android.view.View;
import android.view.ContextMenu;
import android.view.ContextMenu.ContextMenuInfo;
public class MainActivity extends Activity
{
    /* 第一次调用活动 */
    @Override
    public void onCreate(Bundle savedInstanceState)
    {
        super.onCreate(savedInstanceState);
        setContentView(R.layout.main);

        Button btn = (Button) findViewById(R.id.btn1);
        btn.setOnCreateContextMenuListener(this);
    }
    //只创建一次
    @Override
    public boolean onCreateOptionsMenu(Menu menu)
    {
        super.onCreateOptionsMenu(menu);
        CreateMenu(menu);
        return true;
    }
    @Override
    public boolean onOptionsItemSelected(MenuItem item)
    {
        return MenuChoice(item);
    }
        @Override
//长按按钮即可弹出菜单
    public void onCreateContextMenu(ContextMenu menu, View view,
ContextMenuInfo menuInfo)
```

```java
{
    super.onCreateContextMenu(menu, view, menuInfo);
    CreateMenu(menu);
}
//选择菜单
@Override
public boolean onContextItemSelected(MenuItem item)
{
    return MenuChoice(item);
}
private void CreateMenu(Menu menu)//使用快捷键 C 可打开菜单
{
    menu.setQwertyMode(true);
    MenuItem mnu1 = menu.add(0, 0, 0, "Item 1");
    {
        mnu1.setAlphabeticShortcut('a');
        mnu1.setIcon(R.drawable.ic_launcher);
    }
    MenuItem mnu2 = menu.add(0, 1, 1, "Item 2");
    {
        mnu2.setAlphabeticShortcut('b');
        mnu2.setIcon(R.drawable.ic_launcher);
    }
    MenuItem mnu3 = menu.add(0, 2, 2, "Item 3");
    {
        mnu3.setAlphabeticShortcut('c');
        mnu3.setIcon(R.drawable.ic_launcher);
    }
    MenuItem mnu4 = menu.add(0, 3, 3, "Item 4");
    {
        mnu4.setAlphabeticShortcut('d');
    }
    menu.add(0, 3, 3, "Item 5");
    menu.add(0, 3, 3, "Item 6");
    menu.add(0, 3, 3, "Item 7");
}
//调用 MenuChoice()方法来显示所选择的菜单项（并做任何想做的事）
private boolean MenuChoice(MenuItem item)
{
    switch (item.getItemId())
```

```
            {
                case 0:
                    Toast.makeText(this, "You clicked on Item 1",
                        Toast.LENGTH_LONG).show();
                    return true;
                case 1:
                    Toast.makeText(this, "You clicked on Item 2",
                        Toast.LENGTH_LONG).show();
                    return true;
                case 2:
                    Toast.makeText(this, "You clicked on Item 3",
                        Toast.LENGTH_LONG).show();
                    return true;
                case 3:
                    Toast.makeText(this, "You clicked on Item 4",
                        Toast.LENGTH_LONG).show();
                    return true;
                case 4:
                    Toast.makeText(this, "You clicked on Item 5",
                        Toast.LENGTH_LONG).show();
                    return true;
                case 5:
                    Toast.makeText(this, "You clicked on Item 6",
                        Toast.LENGTH_LONG).show();
                    return true;
                case 6:
                    Toast.makeText(this, "You clicked on Item 7",
                        Toast.LENGTH_LONG).show();
                    return true;
            }
            return false;
        }
    }
```

在以上代码中，运行程序，除了单击模拟器中的"Meun"按钮可弹出菜单外，还可以长按界面中的"按钮"也可弹出菜单，也可通过快捷键 C 来弹出菜单。

运行程序，通过长按界面中的按钮，弹出菜单效果如图 4-5 所示。

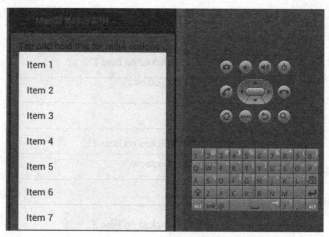

图 4-5 Menu 综合效果

4.2 MenuItem 菜单

在某些时候，如果希望所创建的菜单项是单选菜单项或多选菜单项，则可以调用 MenuItem 来实现。

4.2.1 MenuItem 菜单方法

MenuItem 类有许多重要的方法，下面分别给予介绍。

1．setIcon 方法

setIcon 方法用于设置菜单项显示的图标，可以通过指定图片 ID 或者使用 Drawable 对象来实现。通过设置图标，可以使菜单功能更明显，增强了用户体验。其语法格式为：

```
public abstract MenuItem setIcon(int iconRes)
public abstract MenuItem setIcon(Drawable icon)
```

其中，参数 iconRes 为引用图片的资源 ID；参数 icon 为图片的 drawable 对象。

2．setAlphabeticShortCut 方法

setAlphabeticShortCut 方法用于设置菜单项的字母快捷键。在显示菜单时，可以直接按相应的字母快捷键来执行菜单命令。其语法格式为：

```
public abstract MenuItem setAlphabeticShortcut(char alphaChar)
```

其中，参数 alphaChar 为字符，其值可设为 a～z。

3．setNumericShortcut 方法

setNumericShortcut 方法用于设置菜单项的数字快捷键。在显示菜单时，可以直接按相应的数字快捷键来执行菜单命令。通过设置快捷键可以增强用户体验。其语法格式为：

```
public abstract MenuItem setNumericShortcut(char numericChar)
```

其中,参数 numericChar 为数字,取值为 0~9。

应用以上三种方法建立对应的菜单,其实现代码为:

```
public class MainActivity extends Activity
{
    /* 第一次调用活动 */
    TextView tv=null;
    @Override
    public void onCreate(Bundle savedInstanceState)
    {
        // onCreate 方法
        super.onCreate(savedInstanceState);
        setContentView(R.layout.main);
        tv=(TextView)findViewById(R.id.textView2);
        tv.setText("设置选项菜单! ");                    //设置文本
    }
    @Override
    public boolean onCreateOptionsMenu(Menu menu)
    {
        //重载 onCreateOptionsMenu 方法
        // TODO Auto-generated method stub
        MenuItem menu1=menu.add(0,1,1,"黑色");          //菜单项 1
        MenuItem menu2=menu.add(0,2,2,"红色");          //菜单项 2
        menu1.setIcon(R.drawable.cn);                   //设置图标
        menu2.setIcon(R.drawable.cn);                   //设置图标
        menu1.setAlphabeticShortcut('b');               //快捷键 b
        menu2.setAlphabeticShortcut('c');               //快捷键 c
        menu1.setNumericShortcut('1');                  //快捷键 1
        menu2.setNumericShortcut('2');                  //快捷键 2
        return super.onCreateOptionsMenu(menu);
    }
}
```

4. setShortcut 方法

setShortcut 方法用于设置菜单项的字母快捷键和数字快捷键。在显示菜单时,可以直接按相应的字母或数字快捷键来执行菜单命令。其语法格式为:

```
public abstract MenuItem setShortcut(char numericChar,char alphaChar)
```

其中,参数 numericChar 为数字快捷键,可以为 0~9,主要用于 12 键数字键盘的手机中;参数 alphaChar 为字母快捷键,可以为 a~z,主要用于 QWER 全键盘的手机中。

可用以下代码完成数字及字母快捷键的选择。

```java
@Override
public boolean onCreateOptionsMenu(Menu menu)
{
    //重载 onCreateOptionsMenu 方法
    // TODO ：自动存根法
    MenuItem menu1=menu.add(0,1,1,"黑色");              //菜单项1
    MenuItem menu2=menu.add(0,2,2,"红色");              //菜单项2
    menu1.setIcon(R.drawable.icon);                     //设置图标
    menu2.setIcon(R.drawable.icon);                     //设置图标
    menu1.setShortcut('1', 'b');                        //设置快捷键
    menu2.setShortcut('2', 'r');                        //设置快捷键
    return super.onCreateOptionsMenu(menu);
}
```

5. setOnMenuItemClickListener 方法

sctOnMenuItemClickListener 方法用于设置菜单项监听器。当菜单项被单击时，将触发该监听器。用户可以在该监听器中放置菜单执行代码。其语法格式为：

public void MenuItem setOnMenuItemClickListener(MenuItem.OnMenuItemClickListener menuItemClickListener)

通过以下代码来演示怎样设置选项菜单项的监听器。

```java
public class MainActivity extends Activity
{
    /* 第一次调用活动*/
    TextView tv=null;
    @Override
    public void onCreate(Bundle savedInstanceState)
    {
        // onCreate 方法
        super.onCreate(savedInstanceState);
        setContentView(R.layout.main);
        tv=(TextView)findViewById(R.id.textView2);
        tv.setText("设置选项菜单！ ");                    //设置文本
    }
    @Override
    public boolean onCreateOptionsMenu(Menu menu)
    {
        //重载 onCreateOptionsMenu 方法
        // TODO Auto-generated method stub
        MenuItem menu1=menu.add(0,1,1,"黑色");           //菜单项1
        MenuItem menu2=menu.add(0,2,2,"红色");           //菜单项2
        menu1.setIcon(R.drawable.ic_launcher);           //设置图标
```

```java
            menu2.setIcon(R.drawable.ic_launcher);              //设置图标
            menu1.setOnMenuItemClickListener(new MenuItem.OnMenuItemClickListener()
            {
                //设置监听器
                @Override
                public boolean onMenuItemClick(MenuItem item)
                {
                    // TODO：自动存根法
                    tv.setText("单击了菜单项 1");                //重置文本
                    return false;
                }
            });
            menu2.setOnMenuItemClickListener(new enuItem.OnMenuItemClickListener()
            {
                //设置监听器
                @Override
                public boolean onMenuItemClick(MenuItem item)
                {
                    // TODO：自动存根法
                    tv.setText("单击了菜单项 2");                //重置文本
                    return false;
                }
            });
        return super.onCreateOptionsMenu(menu);
    }
}
```

代码中通过重载 onCreateOptionsMenu 方法来实现，先通过 add 方法设置菜单项，接着设置图标，然后分别设置了监听器。

运行程序，效果如图 4-6 所示。

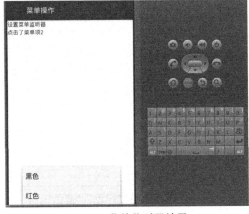

图 4-6 菜单监听器效果

4.2.2 MenuItem 菜单综合实例

下面程序演示怎样开发复选菜单和单选菜单。
【实现步骤】
（1）在 Eclipse 环境下，建立一个名为 E4_2_2 的工程。
（2）打开 res/layout 目录下的 Main.xml 文件，代码修改为：

```xml
<?xml version="1.0" encoding="utf-8"?>
<LinearLayout xmlns:android="http://schemas.android.com/apk/res/android"
    android:orientation="vertical"
    android:layout_width="fill_parent"
    android:layout_height="fill_parent">
<EditText
    android:id="@+id/txt"
    android:layout_width="fill_parent"
    android:layout_height="wrap_content"
    android:editable="false"    />
</LinearLayout>
```

（3）打开 src/example.e4_2_2 目录下的 MainActivity.java 文件，代码修改为：

```java
package com.example.e4_2_3;

import android.app.Activity;
import android.os.Bundle;
import android.view.Menu;
import android.view.MenuItem;
import android.view.SubMenu;
import android.widget.EditText;
public class MainActivity extends Activity
{
    // 定义性别菜单项的标识
    final int MALE = 0x111;
    final int FEMALE = 0x112;
    // 定义颜色菜单项的标识
    final int RED = 0x113;
    final int GREEN = 0x114;
    final int BLUE = 0x115;
    //定义3个颜色菜单项
    MenuItem[] items = new MenuItem[3];
    String[] colorNames = new String[]{"红色","绿色","蓝色"};
    private EditText edit;
```

```java
@Override
public void onCreate(Bundle savedInstanceState)
{
    super.onCreate(savedInstanceState);
    setContentView(R.layout.main);
    edit = (EditText) findViewById(R.id.txt);
}
@Override
public boolean onCreateOptionsMenu(Menu menu)
{
    //向 menu 中添加选择性别的子菜单
    SubMenu genderMenu = menu.addSubMenu("性别");
    // 设置菜单的图标
    genderMenu.setIcon(R.drawable.fn);
    // 设置菜单头的图标
    genderMenu.setHeaderIcon(R.drawable.fn);
    // 设置菜单头的标题
    genderMenu.setHeaderTitle("选择的性别");
    genderMenu.add(0, MALE,   0, "男");
    genderMenu.add(0, FEMALE, 0, "女");
    //设置 genderMenu 菜单内 0 组菜单项为单选菜单项
    genderMenu.setGroupCheckable(0 , true , true);
    //向 menu 中添加颜色的子菜单
    SubMenu colorMenu = menu.addSubMenu("喜欢的颜色");
    colorMenu.setIcon(R.drawable.cn);
    // 设置菜单头的图标
    colorMenu.setHeaderIcon(R.drawable.cn);
    // 设置菜单头的标题
    colorMenu.setHeaderTitle("选择最喜欢的颜色");
    //添加菜单项，并设置它为可勾选的菜单项
    items[0] = colorMenu.add(0, RED, 0, colorNames[0])
        .setCheckable(true);
    //添加菜单项，并设置它为可勾选的菜单项
    items[1] = colorMenu.add(0, BLUE, 0, colorNames[1])
        .setCheckable(true);
    //添加菜单项，并设置它为可勾选的菜单项
    items[2] = colorMenu.add(0, GREEN, 0, colorNames[2])
        .setCheckable(true);
    //设置快捷键
    items[2].setAlphabeticShortcut('u');
    return super.onCreateOptionsMenu(menu);
```

```java
        }
        @Override
        // 菜单项被单击后的回调方法
        public boolean onOptionsItemSelected(MenuItem mi)
        {
            //判断单击的是哪个菜单项，并有针对性地做出响应
            switch (mi.getItemId())
            {
                case MALE:
                    edit.setText("性别为：男");
                    //必须通过代码来改变勾选状态
                    mi.setChecked(true);
                    break;
                case FEMALE:
                    edit.setText("性别为：女");
                    //必须通过代码来改变勾选状态
                    mi.setChecked(true);
                    break;
                case RED:
                    //必须通过代码来改变勾选状态
                    if(mi.isChecked())
                    {
                        mi.setChecked(false);
                    }
                    else
                    {
                        mi.setChecked(true);
                    }
                    showColor();
                    break;
                case GREEN:
                    //必须通过代码来改变勾选状态
                    if(mi.isChecked())
                    {
                        mi.setChecked(false);
                    }
                    else
                    {
                        mi.setChecked(true);
                    }
                    showColor();
```

```
                break;
            case BLUE:
                //必须通过代码来改变勾选状态
                if(mi.isChecked())
                {
                    mi.setChecked(false);
                }
                else
                {
                    mi.setChecked(true);
                }
                showColor();
                break;
        }
        return true;
    }
    private void showColor()
    {
        String result = "喜欢的颜色有：";
        for (int i = 0 ; i < items.length ; i++)
        {
            if(items[i].isChecked())
            {
                result += colorNames[i] + "、";
            }
        }
        edit.setText(result);
    }
}
```

运行程序，效果如图 4-7 所示。

图 4-7　复选菜单效果

4.3 SubMenu 菜单

SubMenu 类用于设计子菜单,通过子菜单可以将相同类别的菜单命令归类,带来更好的用户体验。SubMenu 类继承于 Menu 类,每一个 SubMenu 对象代表一个子菜单。

4.3.1 SubMenu 菜单方法

SubMenu 中有很多方法与 Menu 类相类似。下面将主要介绍一些常用的方法。

1. setIcon 方法

setIcon 方法用于设置子菜单显示的图标,可通过指定图片 ID 或使用 Drawable 对象来实现。通过设置图标,可使菜单功能更明显,增强用户体验。其语法格式为:

```
public abstract SubMenu setIcon(int iconRes)
public abstract SubMenu setIcon(Drawable icon)
```

其中,参数 iconRes 为引用图片的资源 ID;参数 icon 参数为图片的 drawable 对象。

2. add 方法

add 方法用于向子菜单中添加菜单项,其返回值为 MenuItem 类型。这是设置子菜单项最主要的方法。其语法格式为:

```
public abstract MenuItem add(int groupId,int itemId,int order,CharSequence title)
public abstract MenuItem add(int groupId,int itemId,int order,int titleRes)
public abstract MenuItem add(CharSequence title)
public abstract MenuItem add(int titleRes)
```

其中,参数 groupId 为子菜单项所在的组 ID,通过对子菜单项分组,可以进行批量处理,也可以赋值 NONE 来表示不使用组 ID;参数 itemId 为子菜单项的唯一标识 ID;参数 order 为子菜单项的顺序,也可以赋值为 NONE;参数 title 为子菜单项所显示的菜单名称;参数 titleRes 为子菜单项所显示的菜单名称,通过引用字符串 ID 来表示。

通过以下代码来演示怎样设置子菜单项并为子菜单项添加图标。

```
public class MainActivity extends Activity
{
    /* 第一次调用活动 */
    TextView tv=null;
    @Override
    public void onCreate(Bundle savedInstanceState)
    {
        // onCreate 方法
        super.onCreate(savedInstanceState);
        setContentView(R.layout.main);
```

```
            tv=(TextView)findViewById(R.id.textView2);
            tv.setText("设置选项菜单！");                    //设置文本
        }
        @Override
        public boolean onCreateOptionsMenu(Menu menu)
        {
            //重载 onCreateOptionsMenu 方法
            // TODO：自动存根法
            MenuItem menu1=menu.add(0,1,1,"黑色");          //菜单项 1
            MenuItem menu2=menu.add(0,2,2,"红色");          //菜单项 2
        SubMenu menu3=menu.addSubMenu(0,3,3,"字号");        //设置子菜单
            menu1.setIcon(R.drawable.fs);                   //设置图标
            menu2.setIcon(R.drawable.fs);                   //设置图标
            menu3.setIcon(R.drawable.fs);                   //设置子菜单
            MenuItem sub1=menu3.add(1, 4, 4, "12 号");      //设置子菜单项 1
            MenuItem sub2=menu3.add(1, 5, 5, "16 号");      //设置子菜单项 2
            MenuItem sub3=menu3.add(1, 6, 6, "20 号");      //设置子菜单项 3
            MenuItem sub4=menu3.add(1, 7, 7, "25 号");      //设置子菜单项 4
        return super.onCreateOptionsMenu(menu);
        }
    }
```

运行程序，效果如图 4-8 所示。

图 4-8　子菜单项

3．setOnMenuItemClickListener 方法

setOnMenuItemClickListener 方法用于设置子菜单项监听器，当子菜单项被单击时，将触发该监听器。用户可以在该监听器中放置菜单执行代码。其语法格式为：

```
    public abstract MenuItem setOnMenuItemClickListener(MenuItem.OnMenuItemClickListener menuItemClickListener)
```

通过以下代码来演示怎样设置选项菜单项的监听器。

```
    public class MainActivity extends Activity
    {
```

```java
/* 第一次调用活动 */
TextView tv=null;
@Override
public void onCreate(Bundle savedInstanceState)
{
    // onCreate 方法
    super.onCreate(savedInstanceState);
    setContentView(R.layout.main);
    tv=(TextView)findViewById(R.id.textView2);
    tv.setText("设置选项菜单！");                        //设置文本
}
@Override
public boolean onCreateOptionsMenu(Menu menu)
{
    //重载 onCreateOptionsMenu 方法
    // TODO：自动存根法
    MenuItem menu1=menu.add(0,1,1,"黑色");             //菜单项 1
    MenuItem menu2=menu.add(0,2,2,"红色");             //菜单项 2
    SubMenu menu3=menu.addSubMenu(0,3,3,"字号");        //设置子菜单
    menu1.setIcon(R.drawable.ic_action_search);        //设置图标
    menu2.setIcon(R.drawable.ic_action_search);        //设置图标
    menu3.setIcon(R.drawable.ic_action_search);        //设置子菜单
    MenuItem sub1=menu3.add(1, 4, 4, "12 号");         //设置子菜单项 1
    MenuItem sub2=menu3.add(1, 5, 5, "16 号");         //设置子菜单项 2
    MenuItem sub3=menu3.add(1, 6, 6, "20 号");         //设置子菜单项 3
    MenuItem sub4=menu3.add(1, 7, 7, "25 号");         //设置子菜单项 4
    sub1.setOnMenuItemClickListener(new MenuItem.OnMenuItemClickListener()
    {
        //设置监听器
        @Override
        public boolean onMenuItemClick(MenuItem item)
        {
            tv.setTextSize(10);                         //设置字体大小
            return false;
        }
    });
    sub2.setOnMenuItemClickListener(new MenuItem.OnMenuItemClickListener()
    {
        //设置监听器
        @Override
        public boolean onMenuItemClick(MenuItem item)
```

```
                {
                    tv.setTextSize(15);                     //设置字体大小
                    return false;
                }
            });
        sub3.setOnMenuItemClickListener(new MenuItem.OnMenuItemClickListener()
            {
                //设置监听器
                @Override
                public boolean onMenuItemClick(MenuItem item)
                {
                    tv.setTextSize(20);                     //设置字体大小
                    return false;
                }
            });
        sub4.setOnMenuItemClickListener(new MenuItem.OnMenuItemClickListener()
            {
                //设置监听器
                @Override
                public boolean onMenuItemClick(MenuItem item)
                {
                    tv.setTextSize(25);                     //设置字体大小
                    return false;
                }
            });
        return super.onCreateOptionsMenu(menu);
    }
}
```

运行程序，效果如图 4-9 所示。

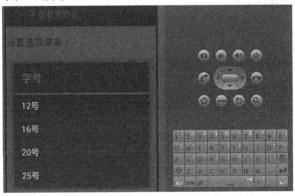

图 4-9　子菜单监听器

4.3.2　SubMenu 菜单综合实例

以下代码在 Options Menu 中添加子菜单，单击相关子菜单即进行对应的操作。
【实现步骤】
（1）在 Eclipse 环境下，建立一个名为 E4_3_3 的工程。
（2）打开 src/com.example.e4_3_3 目录下的 MainActivity.java 文件，其代码修改为：

```java
package com.example.e4_3_3;

import android.app.Activity;
import android.os.Bundle;
import android.view.Menu;
import android.view.MenuItem;
import android.view.SubMenu;
import android.widget.Toast;
public class MainActivity extends Activity
{
    public void onCreate(Bundle savedInstanceState)
    {
        super.onCreate(savedInstanceState);
        setContentView(R.layout.main);
    }
    //创建 menu
    public boolean onCreateOptionsMenu(Menu menu)
    {
    //添加两个 subMenu
    SubMenu File = menu.addSubMenu("文件").setIcon(android.R.drawable.ic_menu_add);
    SubMenu Edit = menu.addSubMenu("编辑").setIcon(android.R.drawable.ic_menu_share);
    //subMenu 里添加选项
    File.add(0,1,1,"新建");
    File.add(0,2,2,"打开");
    Edit.add(1,3,3,"编辑");
    Edit.add(1,4,4,"删除");
    return true;
    }
    //选项的单击事件
    public boolean onOptionsItemSelected(MenuItem item)
    {
    switch(item.getItemId())
    {
```

```
case 1:
    Toast.makeText(this,"好 ", Toast.LENGTH_SHORT).show() ;
    break;
case 2:
    Toast.makeText(this, "正确", Toast.LENGTH_SHORT).show();
    break;
case 3:
    Toast.makeText(this, "编辑一个新文件",Toast.LENGTH_SHORT
        ).show();
    break;
case 4:
    Toast.makeText(this, "不做任何操作", Toast.LENGTH_SHORT).show();
    break;
}
return false;
}
public boolean onPrepareOptionsMenu(Menu menu)
{
    return true;
}
}
```

运行程序,单击模拟器的"Menu"菜单,在弹出的 Menu 菜单中,选择"编辑"子菜单,效果如图 4-10 所示。

图 4-10 子菜单综合实例

4.4 ContextMenu 菜单

ContextMenu 继承自 Menu。上下文菜单不同于选项菜单,选项菜单服务于 Activity,而上下文菜单是注册到某个 View 对象上的。如果一个 View 对象注册了上下文菜单,用户可以通过长按（约 2s）该 View 对象以调出上下文菜单。

上下文菜单不支持快捷键（shortcut），其菜单选项也不能附带图标，但是可以为上下文菜单的标题指定图标。

4.4.1 ContextMenu 菜单方法

本节主要介绍 ContextMenu 类中常用的实现方法。

1. registerForContextMenu 方法

registerForContextMenu 方法用于为一个 View 对象注册上下文菜单，其输入参数为控件的 View 对象。在使用上下文菜单前，必须先使用该方法进行注册，否则将无法使用。其语法格式为：

```
public void registerForContextMenu(View view)
```

其中，参数 view 为 Activity 上的一个 View 对象，主要为控件。

2. onCreateContextMenu 方法

onCreateContextMenu 方法用于初始化上下文菜单，在应用程序每次显示上下文菜单时调用。在该方法中主要通过 add 方法设置上下文菜单。其语法格式为：

```
public void onCreateContextMenu(ContextMenu menu,View v,ContextMenuInfo menuInfo)
```

其中，参数 menu 为上下文菜单对象，通过该变量可以为应用程序添加各个菜单；参数 v 为注册上下文菜单的 View 对象；参数 menuInfo 为上下文菜单需要额外显示的信息。

通过以下代码来演示怎样设置上下文菜单。

```
public class MainActivity extends Activity
{
    /** 第一次调用活动*/
    TextView tv=null;                                    //文本框对象
    EditText et=null;                                    //编辑框对象

    @Override
    public void onCreate(Bundle savedInstanceState)
    {
        //重载 onCreate 方法
        super.onCreate(savedInstanceState);
        setContentView(R.layout.main);
        tv=(TextView)findViewById(R.id.textView1);
        et=(EditText)findViewById(R.id.editText1);
        this.registerForContextMenu(tv);     //注册上下文菜单
        this.registerForContextMenu(et);     //注册上下文菜单
    }
    public void onCreateContextMenu(ContextMenu menu, View v,ContextMenuInfo menuInfo)
```

```
        {
            //重载 onCreateContextMenu 方法
            // TODO：自动存根法
            menu.setHeaderIcon(R.drawable.sm);           //设置图标
            if(v==tv)                                    //如果是文本框
            {
                menu.add(0, 1, 1, "10 号");              //添加菜单项 1
                menu.add(0, 2, 2, "15 号");              //添加菜单项 2
                menu.add(0, 3, 3, "20 号");              //添加菜单项 3
                menu.add(0, 4, 5, "25 号");              //添加菜单项 4
            }
            else if(v==et)                               //如果是编辑框
            {
                menu.add(1, 6, 6, "被按下");              //添加菜单项 5
                menu.add(1, 7, 7, "被选择");              //添加菜单项 6
            }
            super.onCreateContextMenu(menu, v, menuInfo);
        }
    }
```

运行程序，分别长按文本框和编辑框，可看到如图 4-11 及图 4-12 所示的效果。

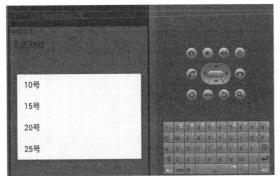

图 4-11　长按文本框效果

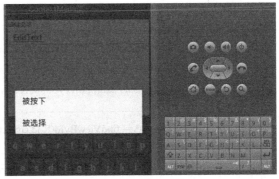

图 4-12　长按编辑框效果

3. onContextItemSelected 方法

onContextItemSelected 方法用于设置当上下文菜单中某个选项被选中时执行的操作。该方法传入的是 MenuItem 对象，通过该对象可以获知哪个菜单项被选中，这是执行菜单处理的主要方法。其语法格式为：

```
        public Boolean onContextItemSelected(MenuItem item)
```

其中，参数 item 为传入的菜单项，程序中通过该参数可以获知哪个菜单项被选中。

通过以下代码来演示怎样设置上下文菜单的处理函数。

```
        public class MainActivity extends Activity
        {
```

```java
final int MENU1 = 1;
final int MENU2 = 2;
final int MENU3 = 3;
final int MENU4 = 4;
final int MENU5 = 5;
@Override
public void onCreate(Bundle savedInstanceState)
{
    super.onCreate(savedInstanceState);
    setContentView(R.layout.main);
    //为两个文本框注册上下文菜单
    this.registerForContextMenu(findViewById(R.id.editText1));
    this.registerForContextMenu(findViewById(R.id.editText2));
}
//添加上下文菜单
@Override
public void onCreateContextMenu(ContextMenu menu, View v,ContextMenuInfo menuInfo)
{
    //此方法在每次调用上下文菜单时都会被调用一次
    menu.setHeaderIcon(R.drawable.sm);
    if (v==findViewById(R.id.editText1))
    {
        menu.add(0, MENU1, 0, R.string.mi1);
        menu.add(0, MENU2, 0, R.string.mi2);
        menu.add(0, MENU3, 0, R.string.mi3);
    }
    else if (v==findViewById(R.id.editText2))
    {
        menu.add(0, MENU4, 0, R.string.mi4);
        menu.add(0, MENU5, 0, R.string.mi5);
    }
}
//响应上下文菜单
@Override
public boolean onContextItemSelected(MenuItem item)
{
    switch (item.getItemId())
    {
    case MENU1:
    case MENU2:
    case MENU3:
```

```
            EditText et1 = (EditText)this.findViewById(R.id.editText1);
            et1.append("\n" + item.getTitle() + "被按下");
            break;
        case MENU4:
        case MENU5:
            EditText et2 = (EditText)this.findViewById(R.id.editText2);
            et2.append("\n" + item.getTitle() + "被按下");
            break;
        }
        return true;
    }
}
```

运行程序，选中编辑框，效果如图 4-13 所示。

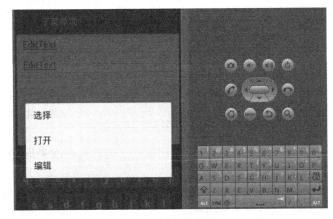

图 4-13　选择菜单处理

4．onContextMenuClosed 方法

onContextMenuClosed 方法用于设置当上下文菜单关闭或退出时执行的操作，其输入参数为 Menu 对象。当用户执行了某个菜单，或者按了返回键，都将触发该方法。其语法格式为：

```
    public void onContextMenuChosed(Menu menu)
```

其中，参数 menu 为传入的菜单。

通过以下代码来演示怎样设置上下文菜单关闭处理函数。

```
    public class MainActivity extends Activity
    {
        /* 第一次调用活动 */
        TextView tv=null;                              //文本框对象
        EditText et=null;                              //编辑框对象
        @Override
        public void onCreate(Bundle savedInstanceState)
        {
            //重载 onCreate 方法
```

```java
        super.onCreate(savedInstanceState);
        setContentView(R.layout.main);
        tv=(TextView)findViewById(R.id.textView1);
        et=(EditText)findViewById(R.id.editText1);
        this.registerForContextMenu(tv);                        //注册上下文菜单
        this.registerForContextMenu(et);                        //注册上下文菜单
    }
    @Override
    public void onCreateContextMenu(ContextMenu menu, View v,ContextMenuInfo menuInfo)
    {
        //重载 onCreateContextMenu 方法
        // TODO：自动存根法
        menu.setHeaderIcon(R.drawable.sm);                      //设置图标
        if(v==tv)                                               //如果是文本框
        {
            menu.add(0, 1, 1, "10 号");                         //添加菜单项 1
            menu.add(0, 2, 2, "15 号");                         //添加菜单项 2
            menu.add(0, 3, 3, "20 号");                         //添加菜单项 3
            menu.add(0, 4, 5, "25 号");                         //添加菜单项 4
        }
        else if(v==et)                                          //如果是编辑框
        {
            menu.add(1, 6, 6, "15 号");                         //添加菜单项 5
            menu.add(1, 7, 7, "25 号");                         //添加菜单项 6
        }
        super.onCreateContextMenu(menu, v, menuInfo);
    }
    @Override
    public boolean onContextItemSelected(MenuItem item)
    {
        //重载 onContextItemSelected 方法
        // TODO Auto-generated method stub
        switch(item.getItemId())                                //判断被单击的菜单项
        {
        case 1:
            tv.setTextSize(10);                                 //设置字体大小
            break;
        case 2:
            tv.setTextSize(30);                                 //设置字体大小
            break;
        case 3:
```

```
                    tv.setTextSize(30);                          //设置字体大小
                    break;
                case 4:
                    tv.setTextSize(40);                          //设置字体大小
                    break;
                case 5:
                    et.setTextSize(20);                          //设置字体大小
                    break;
                case 6:
                    et.setTextSize(40);                          //设置字体大小
                    break;
            }
            return super.onContextItemSelected(item);
        }
        @Override
        public void onContextMenuClosed(Menu menu)
        {
            //重载 onContextMenuClosed 方法
            // TODO：自动存根法
            Toast.makeText(getApplicationContext(),"退出了上下文菜单！",Toast.LENGTH_LONG). show();
            super.onContextMenuClosed(menu);
        }
    }
```

运行程序，单击编辑框，在弹出的子菜单项进行选择，选择后弹出如图 4-14 所示的效果。

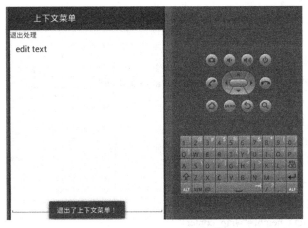

图 4-14　退出上下文菜单

4.4.2 ContextMenu 菜单综合实例

下面通过一个综合例子来说明 ContextMenu 类的用法。
【实现步骤】
(1) 在 Eclipse 环境下，建立一个名为 E4_4_4 的工程。
(2) 打开 res/values 目录下的 strings.xml，代码修改为：

```xml
<resources>
    <string name="app_name">E4_4_4</string>
    <string name="hello_world">Hello world!</string>
    <string name="menu_settings">Settings</string>
    <string name="title_activity_main">上下菜单综合实例</string>
</resources>
```

(3) 打开 res/layout 目录下的 main.xml 文件，代码修改为：

```xml
<?xml version="1.0" encoding="utf-8"?>
<LinearLayout xmlns:android="http://schemas.android.com/apk/res/android"
    android:orientation="vertical"
    android:layout_width="fill_parent"
    android:layout_height="fill_parent" >
<ListView     android:id="@android:id/list"
              android:layout_width="fill_parent"
              android:layout_height="fill_parent">
</ListView>
</LinearLayout>
```

(4) 在 res/layout 目录下新建一个名为 simple_list_item_1.xml 的文件，代码为：

```xml
<?xml version="1.0" encoding="utf-8"?>
<LinearLayout
xmlns:android="http://schemas.android.com/apk/res/android"
android:layout_width="wrap_content"
android:layout_height="wrap_content"
android:orientation="vertical">
<TextView
    android:id="@+id/label"
    android:layout_width="fill_parent"
    android:layout_height="wrap_content"
    android:textSize="30sp"></TextView>
</LinearLayout>
```

(5) 打开 src/com.example.e4_4_4 目录下的 MainActivity.java 文件，代码修改为：

```java
package com.example.e4_4_4;
```

```java
import android.app.ListActivity;
import android.os.Bundle;
import android.view.ContextMenu;
import android.view.ContextMenu.ContextMenuInfo;
import android.view.Menu;
import android.view.MenuItem;
import android.view.View;
import android.widget.ArrayAdapter;
public class MainActivity extends ListActivity
{
    public static final int EIGHT_ID = Menu.FIRST+1;
    public static final int SIXTEEN_ID = Menu.FIRST+2;
    public static final int TWENTY_FOUR_ID = Menu.FIRST+3;
    public static final int TWO_ID = Menu.FIRST+4;
    public static final int THIRTY_TWO_ID = Menu.FIRST+5;
    public static final int FORTY_ID = Menu.FIRST+6;
    public static final int ONE_ID = Menu.FIRST+7;
    String[] items={"lorem", "ipsum", "dolor", "sit", "amet",
            "consectetuer", "adipiscing", "elit", "morbi", "vel",
            "ligula", "vitae", "arcu", "aliquet", "mollis",
            "etiam", "vel", "erat", "placerat", "ante",
            "porttitor", "sodales", "pellentesque", "augue", "purus"};
    @Override
    public void onCreate(Bundle savedInstanceState)
    {
        super.onCreate(savedInstanceState);
        setContentView(R.layout.main);
        setListAdapter(new ArrayAdapter(this,R.layout.simple_list_item_1,R.id.label,items));
        //注册 ContextView 到 view 中
        registerForContextMenu(getListView());
    }
    //重写该方法, 生成 ContextMenu 菜单
    @Override
    public void onCreateContextMenu(ContextMenu menu, View v,
            ContextMenuInfo menuInfo)
    {
        this.populateMenu(menu);
        super.onCreateContextMenu(menu, v, menuInfo);
    }
    private void populateMenu(Menu menu)
    {
```

```
        menu.add(Menu.NONE, ONE_ID, Menu.NONE, "1 Pixel");
        menu.add(Menu.NONE, TWO_ID, Menu.NONE, "2 Pixels");
        menu.add(Menu.NONE, EIGHT_ID, Menu.NONE, "8 Pixels");
        menu.add(Menu.NONE, SIXTEEN_ID, Menu.NONE, "16 Pixels");
        menu.add(Menu.NONE, TWENTY_FOUR_ID, Menu.NONE, "24 Pixels");
        menu.add(Menu.NONE, THIRTY_TWO_ID, Menu.NONE, "32 Pixels");
        menu.add(Menu.NONE, FORTY_ID, Menu.NONE, "40 Pixels");
    }
}
```

运行程序,效果如图 4-15 所示。

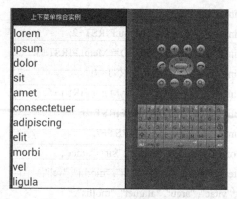

图 4-15 上下文菜单综合实例

第 5 章 Android 对话框

Activitiy 提供了一种方便管理的创建、保存、回复的对话框机制，例如，onCreateDialog(int)、onPrepareDialog(int, Dialog)、showDialog(int), dismissDialog(int)等方法，如果使用这些方法的话，Activity 将通过 getOwnerActivity()方法返回该 Activity 管理的对话框（dialog）。

下面主要来介绍 Android 常用的对话框。

5.1 AlertDialog 对话框

AlertDialog（提示）对话框是功能最丰富、实际应用最广的对话框。下面先来介绍 AlertDialog 对话框的方法。

5.1.1 创建提示对话框

在 AlertDialog 对话框中可以显示提示信息，也可以包含若干个按钮、单选按钮和复选按钮等。一般来说，提示对话框已经可以满足绝大部分的应用程序需求。AlertDialog 类可以完成提示对话框的主要功能。

下面先来介绍实现创建提示对话框的相关方法。

1．showDialog 方法

当想要显示一个对话框时，调用 showDialog(int id)方法并传递一个唯一标识这个对话框的整数。当对话框第一次被请求时，Android 从 Activity 中调用 onCreateDialog(int id)，你应该在这里初始化这个对话框 Dialog。这个回调方法与 showDialog(int id)方法有相同的 ID。当创建这个对话框后，在 Activity 的最后返回这个对象。其语法格式为：

```
public final void showDialog(int id)
```

其中，参数 id 为输入的对话框 ID。

2．onCreateDialog 方法

当使用 onCreateDialog 回调函数时，Android 系统会有效地设置这个 Activity 为每个对话框的所有者，从而自动管理每个对话框的状态并挂靠到 Activity 上。这样，每个对话框继承这个 Activity 的特定属性。例如，当一个对话框打开时，菜单键显示为这个 Activity 定义的选项菜单，音量键修改 Activity 使用的音频流。其语法格式为：

```
protected Dialog onCreateDialog(int id)
```

其中，参数 id 为输入的对话框 ID。

3. setTitle 方法

setTitle 方法用于设置一个提示对话框显示的标题，通过标题可以表明该对话框的功能。一般在 onCreateDialog 方法中使用。其语法格式为：

> public AlertDialog.Builder setTitle(CharSequence title)
> public AlertDialog.Builder setTitle(int titleId)

其中，参数 title 为标题字符串；参数 titleId 为标题字符串的资源 ID。

4. setMessage 方法

setMessage 方法用于设置一个提示对话框显示的提示信息，该信息直接呈现给用户。它一般在 onCreateDialog 方法中使用。其语法格式为：

> public AlertDialog.Builder setMessage(CharSequence message)
> public AlertDialog.Builder setMessage(int messageId)

其中，参数 message 为提示信息字符串；参数 messageId 为提示信息字符串的资源 ID。

5. create 方法

create 方法用于按照给定的参数创建一个提示对话框。在该方法调用前，必须对对话框进行必要的设置，如标题、提示信息和按钮等，否则将没有任何显示效果。其语法格式为：

> public AlertDialog create()

6. setIcon 方法

setIcon 方法用于设置一个提示对话框显示的图标。通过设置图标，可以使对话框功能更直观。其一般在 onCreateDialog 方法中使用。其语法格式为：

> public AlertDialog.Builder setIcon(int iconId)
> public AlertDialog.Builder setIcon(Drawable icon)

其中，参数 iconId 为图标的资源 ID；参数 icon 为图标的 Drawable 对象。

7. setPositiveButton 方法

setPositiveButton 方法用于为提示对话框设置一个"确定"按钮。有多个按钮时，该按钮将放置在对话框的左侧。通过该按钮可退出对话框，并返回之前的 Activicy，也可以进行一些数据的交互。为对话框设置按钮是必须的。其语法格式为：

> public AlertDialog.Builder setPositiveButton(CharSequence text,DialogInterface.OnClickListener listener)

其中，参数 text 为"确定"按钮显示的文本，也可以使用引用字符串资源 ID 的形式；参数 listener 为按钮监听器，用来设置按钮执行的动作。

8. setNegativeButton 方法

setNegativeButton 方法用于为提示对话框设置一个"取消"按钮。有多个按钮时，该按钮将放置在对话框的右侧。通过该按钮可退出对话框，并返回之前的 Activicy，也可以进行一些数据的交互，为对话框设置按钮是必须的。其语法格式为：

> public AlertDialog.Builder setNegativeButton (CharSequence text,DialogInterface.OnClickListener listener)

其中，参数 text 为"取消"按钮显示的文本，也可以使用引用字符串资源 ID 的形式；参数 listener

9. setNeutralButton 方法

setNeutralButton 方法用于为提示对话框设置一个"中间"按钮。有多个按钮时，该按钮将放置在"确定"按钮与"取消"按钮的中间。通过该按钮可退出对话框，并返回之前的 Activicy，也可以进行一些数据的交互，为对话框设置按钮是必须的。其语法格式为：

public AlertDialog.Builder setNeutralButton (CharSequence text,DialogInterface.OnClickListener listener)

其中，参数 text 为"中间"按钮显示的文本，也可以使用引用字符串资源 ID 的形式；参数 listener 为按钮监听器，用来设置按钮执行的动作。

通过以下代码来创建一个包含一个简单文本框和三个按钮的界面。

```java
public class MainActivity extends Activity
{
    @Override
    public void onCreate(Bundle savedInstanceState)
    {
        super.onCreate(savedInstanceState);
        setContentView(R.layout.main);
        Button bn = (Button)findViewById(R.id.bn01);
        //定义一个 AlertDialog.Builder 对象
        final Builder builder = new AlertDialog.Builder(this);
        //为按钮绑定事件监听器
        bn.setOnClickListener(new View.OnClickListener()
        {
            @Override
            public void onClick(View source)
            {
                // 设置对话框的图标
                builder.setIcon(R.drawable.sm);
                // 设置对话框的标题
                builder.setTitle("自定义普通对话框");
                // 设置对话框显示的内容
                builder.setMessage("一个简单的提示对话框");
                // 为对话框设置一个"确定"按钮
                builder.setPositiveButton("确定"
                    //为列表项的单击事件设置监听器
                    , new OnClickListener()
                    {
                        @Override
                        public void onClick(DialogInterface dialog, int which)
                        {
                            EditText show = (EditText) findViewById(R.id.show);
```

```
                    // 设置 EditText 内容
                    show.setText("用户单击了"确定"按钮！");
                }
            });
            // 为对话框设置一个"中间"按钮
            builder.setNeutralButton("返回"
                , new OnClickListener()
            {
                @Override
                public void onClick(DialogInterface dialog, int which)
                {
                    EditText show = (EditText) findViewById(R.id.show);
                    // 设置 EditText 内容
                    show.setText("用户单击了"返回"按钮！");
                }
            });
            // 为对话框设置一个"取消"按钮
            builder.setNegativeButton("取消"
                , new OnClickListener()
            {
                @Override
                public void onClick(DialogInterface dialog, int which)
                {
                    EditText show = (EditText) findViewById(R.id.show);
                    // 设置 EditText 内容
                    show.setText("用户单击了"取消"按钮！");
                }
            });
            //创建、并显示对话框
            builder.create().show();
        }
    });
}
```

在以上代码中，设置了对话框的图标、标题等属性，并为按钮添加了三个按钮。运行程序，单击界面中的"进入对话框"按钮，并选择对话框中的"返回"按钮，得到如图 5-1 所示的效果。

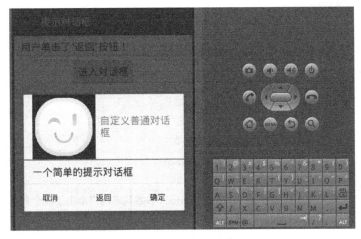

图 5-1　提示对话框

5.1.2　创建列表对话框

在 Android 中提供了 setItem 方法用于为提示对话框设置选择列表，这就构成了一种列表形式的对话框，通常用于多个项目进行选择的场合。列表对话框同样通过 AlertDialog 类来实现，只不过其传递参数不同。其语法格式为：

```
public AlertDialog.Builder setItems(CharSequence[] items,DialogInterface.OnClickListener listener)
public AlertDialog.Builder setItems(int itemsId,DialogInterface.OnClickListener listener)
```

其中，参数 items 为列表项的字符串数组；参数 itemsId 为列表项字符串数组的资源 ID；参数 listener 为监听器，设置单击某个列表项时执行的动作。

通过以下代码来实现列表对话框的创建。

```
public class MainActivity extends Activity
{
    @Override
    public void onCreate(Bundle savedInstanceState)
    {
        super.onCreate(savedInstanceState);
        setContentView(R.layout.main);
        Button bn = (Button)findViewById(R.id.bn);
        final Builder b = new AlertDialog.Builder(this);
        //为按钮绑定事件监听器
        bn.setOnClickListener(new View.OnClickListener()
        {
            @Override
            public void onClick(View source)
            {
                //设置对话框的图标
```

```
                    b.setIcon(R.drawable.sm);
                    //设置对话框的标题
                    b.setTitle("选择喜欢的颜色");
                    //为对话框设置多个列表
                    b.setItems(
                        new String[] {"红色","黄色","蓝色"}
                        //为按钮设置监听器
                        , new OnClickListener()
                        {
                            //该方法的 which 参数代表用户单击了哪个列表项
                            @Override
                            public void onClick(DialogInterface dialog
                                , int which)
                            {
                                TextView show = (TextView)findViewById(R.id.show);
                                //which 代表哪个列表项被单击了
                                switch(which)
                                {
                                    case 0:
                                        show.setBackgroundColor(Color.RED);
                                        break;
                                    case 1:
                                        show.setBackgroundColor(Color.YELLOW);
                                        break;
                                    case 2:
                                        show.setBackgroundColor(Color.BLUE);
                                        break;
                                }
                            }
                        });
                    //创建并显示对话框
                    b.create().show();
                }
            });
        }
    }
```

运行程序，单击界面中的"弹出列表框"按钮，即弹出如图 5-2 所示的效果。

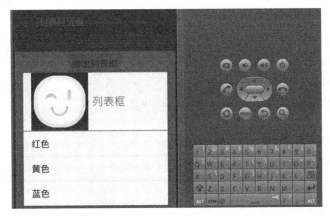

图 5-2 列表对话框

5.1.3 单选列表对话框

在 Android 中提供了 setSingleChoiceItem 方法用于为对话框设置单选按钮。这就构成了一种单选列表形式的对话框，通常用于多个项目选择其一的场合。单选列表对话框同样通过 AlertDialog 类来实现，只不过传递的参数不同。其语法格式为：

public AlertDialog.Builder setSingleChoiceItems(CharSequence[] items,int checkedItem,DialogInterface.OnClickListener listener)

其中，参数 items 为列表项的字符串数组；参数 checkedItem 为默认哪一项被选中，如果取值为-1 时，即未被选中；参数 listener 为监听器，设置单击某个列表项时执行的动作。

通过以下代码来创建单选列表对话框。

```
public class MainActivity extends Activity
{
    final int SINGLE_DIALOG = 0x113;
    @Override
    public void onCreate(Bundle savedInstanceState)
    {
        super.onCreate(savedInstanceState);
        setContentView(R.layout.main);
        Button bn = (Button)findViewById(R.id.bn);
        //为按钮绑定事件监听器
        bn.setOnClickListener(new View.OnClickListener()
        {
            @Override
            public void onClick(View source)
            {
                //显示对话框
                showDialog(SINGLE_DIALOG);
            }
```

```java
        });
}
//重写 onCreateDialog 方法创建对话框
@Override
public Dialog onCreateDialog(int id, Bundle state)
{
        //判断需要生成哪种类型的对话框
        switch (id)
        {
                case SINGLE_DIALOG:
                        Builder b = new AlertDialog.Builder(this);
                        // 设置对话框的图标
                        b.setIcon(R.drawable.sm);
                        // 设置对话框的标题
                        b.setTitle("单选列表对话框");
                        // 为对话框设置多个列表
                        b.setSingleChoiceItems(new String[]
                                { "红色", "黄色", "蓝色" }
                        // 默认选中第二项
                                , 1
                        //为列表项的单击事件设置监听器
                                , new OnClickListener()
                                {
                                        @Override
                                        public void onClick(DialogInterface dialog,
                                                int which)
                                        {
                                                TextView show = (TextView) findViewById(R.id.show);
                                                // which 代表哪个列表项被单击了
                                                switch (which)
                                                {
                                                        //修改文本框的背景色
                                                        case 0:
                                                                show.setBackgroundColor(Color.RED);
                                                                break;
                                                        case 1:
                                                                show.setBackgroundColor(Color.YELLOW);
                                                                break;
                                                        case 2:
                                                                show.setBackgroundColor(Color.BLUE);
                                                                break;
```

```
                                }
                            }
                        });
                    // 添加一个"确定"按钮，用于关闭该对话框
                    b.setPositiveButton("确定", null);
                    // 创建对话框
                    return b.create();
            }
            return null;
        }
    }
```

运行程序，在初始界面中单击"进入单选列表对话框"按钮，效果如图5-3所示。

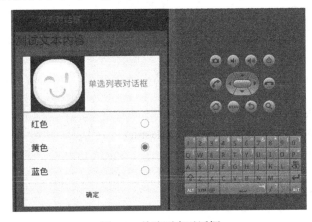

图 5-3 单选列表对话框

5.1.4 复选列表对话框

在 Android 中提供了 setMultiChoiceItems 方法用于为提示对话框设置复选框，这就构成了一种复选列表对话框，通常用于多个项目进行选择的场合。复选列表对话框同样通过 AlertDialog 类来实现，只不过传递的参数不同而已。其语法格式为：

  public AlertDialog.Builder setMultiChoiceItems(CharSequence[] items,Boolean[] checkedItems,DialogInterface.OnMultiChoiceClickListener listener)

其中，参数 items 为列表项的字符串数组；参数 checkedItems 为默认哪一项被选中，其为布尔型数组；参数 listener 为监听器，设置单击某个列表项时执行的动作。

通过以下代码来创建复选列表对话框。

```
public class MainActivity extends Activity
{
    final int SINGLE_DIALOG = 0x113;
    String[] colorNames = new String[]
    { "打球","跳远","跑步" };
```

```java
@Override
public void onCreate(Bundle savedInstanceState)
{
    super.onCreate(savedInstanceState);
    setContentView(R.layout.main);
    Button bn = (Button) findViewById(R.id.bn);
    // 为按钮绑定事件监听器
    bn.setOnClickListener(new View.OnClickListener()
    {
        @Override
        public void onClick(View source)
        {
            // 显示对话框
            showDialog(SINGLE_DIALOG);
        }
    });
}
@Override
public Dialog onCreateDialog(int id, Bundle state)
{
    // 判断需要生成哪种类型的对话框
    switch (id)
    {
        case SINGLE_DIALOG:
            Builder b = new AlertDialog.Builder(this);
            // 设置对话框的图标
            b.setIcon(R.drawable.sm);
            // 设置对话框的标题
            b.setTitle("多选列表对话框");
            final boolean[] checkStatus = new boolean[] { true, false, true };
            // 为对话框设置多个列表
            b.setMultiChoiceItems(new String[] { "打球", "跳远", "跑步" }
                // 设置默认勾选了哪些列表项
                , checkStatus
                // 为列表项的单击事件设置监听器
                , new OnMultiChoiceClickListener()
                {
                    @Override
                    public void onClick(DialogInterface dialog, int which,
                        boolean isChecked)
                    {
```

```
                        EditText show = (EditText) findViewById(R.id.show);
                        String result = "喜欢的运动是：";
                        for (int i = 0; i < checkStatus.length; i++)
                        {
                                // 如果该选项被选中
                                if (checkStatus[i])
                                {
                                        result += colorNames[i] + "、";
                                }
                        }
                        show.setText(result);
                }
            });
            // 添加一个"确定"按钮，用于关闭该对话框
            b.setPositiveButton("确定", null);
            // 创建对话框
            return b.create();
        }
        return null;
    }
}
```

运行程序，单击初始界面中的"选择喜欢的运动"按钮，效果如图 5-4 所示。

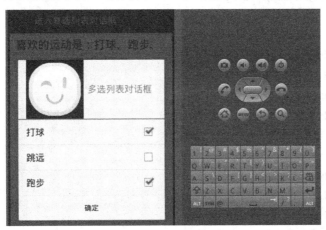

图 5-4 复选列表对话框

5.1.5 AlertDialog 对话框综合实例

在前面几小节中分别介绍了几种常用的 AlertDialog 对话框，下面综合介绍一种用户登录对话框。登录对话框中界面包含三个输入框及三个按钮。

【实现步骤】

(1) 在 Eclipse 环境下建立名为 E5_1_5 的工程。

(2) 打开 res/layout 目录下的 main.xml 文件,其代码修改为:

```xml
<?xml version="1.0" encoding="utf-8"?>
<LinearLayout xmlns:android="http://schemas.android.com/apk/res/android"
    android:orientation="vertical"
    android:layout_width="fill_parent"
    android:layout_height="fill_parent"
    android:gravity="center_horizontal">
<Button
    android:id="@+id/bn"
    android:layout_width="wrap_content"
    android:layout_height="wrap_content"
    android:text="登录对话框"/>
</LinearLayout>
```

(3) 在 res/layout 目录下新建 "dcnglu.xml" 文件。

```xml
<?xml version="1.0" encoding="utf-8"?>
<TableLayout xmlns:android="http://schemas.android.com/apk/res/android"
    android:id="@+id/loginForm"
    android:orientation="vertical"
    android:layout_width="fill_parent"
    android:layout_height="fill_parent">
<TableRow>
<TextView
    android:layout_width="fill_parent"
    android:layout_height="wrap_content"
    android:text="用户名:"
    android:textSize="10pt"     />
<!--输入用户名的文本框 -->
<EditText
    android:layout_width="fill_parent"
    android:layout_height="wrap_content"
    android:hint="请填写登录账号"
    android:selectAllOnFocus="true"  />
</TableRow>
<TableRow>
<TextView
    android:layout_width="fill_parent"
    android:layout_height="wrap_content"
    android:text="密码:"
    android:textSize="10pt"     />
```

```xml
<!-- 输入密码的文本框 -->
<EditText
    android:layout_width="fill_parent"
    android:layout_height="wrap_content"
    android:password="true"
    android:hint="******"
    android:textSize="10pt"  />
</TableRow>
<TableRow>
<TextView
    android:layout_width="fill_parent"
    android:layout_height="wrap_content"
    android:text="电话号码:"
    android:textSize="10pt"  />
<!-- 输入电话号码的文本框 -->
<EditText
    android:layout_width="fill_parent"
    android:layout_height="wrap_content"
    android:hint="请输入电话号码"
    android:selectAllOnFocus="true"
    android:phoneNumber="true"  />
</TableRow>
<Button
    android:layout_width="wrap_content"
    android:layout_height="wrap_content"
    android:text="注册"/>
</TableLayout>
```

（4）打开 src/com.example.e5_1_5 目录下的 MainActivity.java 文件，代码修改为：

```java
package com.example.e5_1_5;

import android.app.Activity;
import android.app.AlertDialog;
import android.app.AlertDialog.Builder;
import android.content.DialogInterface;
import android.content.DialogInterface.OnClickListener;
import android.os.Bundle;
import android.view.View;
import android.widget.Button;
import android.widget.EditText;
import android.widget.LinearLayout;
import android.widget.TableLayout;
```

```java
public class MainActivity extends Activity
{
    @Override
    public void onCreate(Bundle savedInstanceState)
    {
        super.onCreate(savedInstanceState);
        setContentView(R.layout.main);
        Button bn = (Button)findViewById(R.id.bn);
        //定义一个 AlertDialog.Builder 对象
        final Builder builder = new AlertDialog.Builder(this);
        //为按钮绑定事件监听器
        bn.setOnClickListener(new View.OnClickListener()
        {
            @Override
            public void onClick(View source)
            {
                // 设置对话框的图标
                builder.setIcon(R.drawable.sm);
                // 设置对话框的标题
                builder.setTitle("自定义普通对话框");
                //装载/res/layout/login.xml 界面布局
                TableLayout loginForm = (TableLayout)getLayoutInflater()
                    .inflate( R.layout.denglu, null);
                // 设置对话框显示的 View 对象
                builder.setView(loginForm);
                // 为对话框设置一个"确定"按钮
                builder.setPositiveButton("登录"
                    // 为按钮设置监听器
                    , new OnClickListener()
                    {
                        @Override
                        public void onClick(DialogInterface dialog, int which)
                        {
                            //此处可执行登录处理
                        }
                    });
                // 为对话框设置一个"取消"按钮
                builder.setNegativeButton("取消"
                    , new OnClickListener()
                    {
                        @Override
```

```
                    public void onClick(DialogInterface dialog, int which)
                    {
                        //取消登录，不做任何事情
                    }
                });
                //创建并显示对话框
                builder.create().show();
            }
        });
    }
}
```

运行程序，在登录界面中单击"登录对话框"按钮，得到如图 5-5 所示的效果。

图 5-5   登录对话框界面

5.2   DatePickerDialog 与 TimePickerDialog 对话框

5.2.1   DatePickerDialog 与 TimePickerDialog 概述

DatePickerDialog 类实现了日期选择对话框，这种对话框类似于日期选择控件，只不过是通过对话框的形式呈现给用户的。Android 在 DatePickerDialog 类中进行了很好的封装，对于开发者来说，只需要简单地调用即可使用日期选择对话框。

日期选择对话框同样需要在 onCreateDialog 方法中进行设置，并通过 showDialog 方法来呈现给用户。在此，主要使用 onDateSetListener()方法实现日期选择对话框。其语法格式为：

```
public DatePickerDialog onDateSetListener()
```

TimePickerDialog 类实现了时间选择对话框，这种对话框类似于时间选择控件，只不过是通过对话框的形式呈现给用户的。Android 在 TimePickerDialog 类中进行了很好的封装，对于开发者来说，只须简单地调用即可使用时间选择对话框。

时间选择对话框同样需要在 onCreateDialog 方法中进行设置，并通过 showDialog 方法来呈现给用户。此处主要使用 OnTimeSetListener()方法，其语法格式为：

public TimerPickerDialog OnTimeSetListener()

5.2.2 DatePickerDialog 与 TimePickerDialog 对话框综合实例

下面设计一个含有日期选择对话框和时间选择对话框的界面。
【实现步骤】
（1）在 Eclipse 环境下，创建一个名为 E5_2 的工程。
（2）打开 res/layout 目录下的 main.xml 文件，其代码修改为：

```
<?xml version="1.0" encoding="utf-8"?>
<LinearLayout xmlns:android="http://schemas.android.com/apk/res/android"
    android:orientation="vertical"
    android:layout_width="fill_parent"
    android:layout_height="fill_parent"
    android:gravity="center_horizontal"     >
<EditText
    android:id="@+id/show"
    android:layout_width="fill_parent"
    android:layout_height="wrap_content"
    android:editable="false"/>
<LinearLayout
    android:orientation="horizontal"
    android:layout_width="fill_parent"
    android:layout_height="wrap_content"
    android:gravity="center">
<Button
    android:id="@+id/dateBn"
    android:layout_width="wrap_content"
    android:layout_height="wrap_content"
    android:text="设置日期"/>
<Button
    android:id="@+id/timeBn"
    android:layout_width="wrap_content"
    android:layout_height="wrap_content"
    android:text="设置时间"/>
</LinearLayout>
<AnalogClock
    android:id="@+id/AnalogClock01"
    android:layout_width="fill_parent"
    android:layout_height="wrap_content"/>
</LinearLayout>
```

(3) 打开 src/com.example.e5_2 目录下的 MainActivity.java 文件，代码修改为：

```java
package com.example.e5_2;

import java.util.Calendar;
import android.app.Activity;
import android.app.DatePickerDialog;
import android.app.TimePickerDialog;
import android.os.Bundle;
import android.view.View;
import android.view.View.OnClickListener;
import android.widget.Button;
import android.widget.DatePicker;
import android.widget.EditText;
import android.widget.TimePicker;
public class MainActivity extends Activity
{
    @Override
    public void onCreate(Bundle savedInstanceState)
    {
        super.onCreate(savedInstanceState);
        setContentView(R.layout.main);
        Button dateBn = (Button)findViewById(R.id.dateBn);
        Button timeBn = (Button)findViewById(R.id.timeBn);
        //为"设置日期"按钮绑定监听器
        dateBn.setOnClickListener(new OnClickListener()
        {
            @Override
            public void onClick(View source)
            {
                Calendar c = Calendar.getInstance();
                // 直接创建一个 DatePickerDialog 对话框实例，并将它显示出来
                new DatePickerDialog(MainActivity.this,
                    // 绑定监听器
                    new DatePickerDialog.OnDateSetListener()
                    {
                        @Override
                        public void onDateSet(DatePicker dp, int year,
                            int month, int dayOfMonth)
                        {
                            EditText show = (EditText) findViewById(R.id.show);
                            show.setText("您选择了：" + year + "年" + month + "月"
```

```java
                                + dayOfMonth + "日");
                    }
                }
                //设置初始日期
                , c.get(Calendar.YEAR)
                , c.get(Calendar.MONTH)
                , c.get(Calendar.DAY_OF_MONTH)).show();
        }
    });
    //为"设置时间"按钮绑定监听器
    timeBn.setOnClickListener(new OnClickListener()
    {
        @Override
        public void onClick(View source)
        {
            Calendar c = Calendar.getInstance();
            // 创建一个 TimePickerDialog 实例，并把它显示出来
            new TimePickerDialog(MainActivity.this,
                    // 绑定监听器
                    new TimePickerDialog.OnTimeSetListener()
                    {
                        @Override
                        public void onTimeSet(TimePicker tp, int hourOfDay,
                            int minute)
                        {
                            EditText show = (EditText) findViewById(R.id.show);
                            show.setText("您选择了：" + hourOfDay + "时" + minute
                                + "分");
                        }
                    }
                //设置初始时间
                , c.get(Calendar.HOUR_OF_DAY)
                , c.get(Calendar.MINUTE)
                //true 表示采用 24 小时制
                , true).show();
        }
    });
}
```

运行程序，效果如图 5-6 所示。

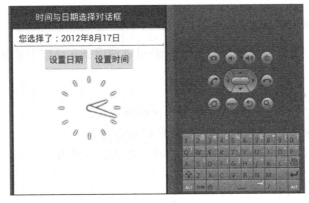

图 5-6　时间与日期选择对话框

在图 5-6 中，如果单击"设置日期"按钮即可弹出"日期对话框"设置界面，可选择对应的日期，接着单击对话框中的"Done"按钮即可在界面中显示你所设置的日期。同理，如果单击"设置时间"按钮即可弹出"时间选择对话框"，接着单击对话框中的"Done"按钮即可在界面中显示你所设置的时间。

5.3　ProgressDailog 对话框

ProgressDailog 类实现了 Android 中的进度条对话框。该对话框主要包括一个进度条，主要用于显示一些操作的进度。ProgressDailog 类似于 ProgressBar 类，很多方法都与进度条控件类似，只不过是通过对话框的形式呈现给用户的。

5.3.1　ProgressDailog 对话框方法

下面将介绍实现 ProgressDailog 对话框的相关方法。

1. setMax 方法

setMax 方法用于设置一个 ProgressDailog 进度条对话框中进度条的最大值，在默认情况下进度条的最大值为 100。该方法主要用于一些需要自定义进度条的场合。其语法格式为：

　　public void setMax(int max)

其中，参数 max 为设置的进度条最大值，此时，该进度条的取值范围为 0~max。

2. setTile 方法

setTile 方法用于设置一个 ProgressDailog 进度条对话框的标题，通过标题可以表明该对话框的功能。其语法格式为：

　　public setTitle(CharSequence title)
　　public setTitle(int titleId)

其中，参数 title 为标题字符串；参数 titleId 为标题字符串的资源 ID。

3. setProgressStyle 方法

setProgressStyle 方法用于设置一个 ProgressDailog 进度条对话框中进度条的样式。其语法格式为:

```
public void setProgressStyle(int style)
```

其中,参数 style 为进度条的样式,可以选取下面两种样式。

- ProgressDialog.STYLE_SPINNER:圆圈样式的进度条。
- ProgressDialog.STYLE_HORIZONTAL:水平样式的进度条。

4. incrementProgressBy 方法

incrementProgressBy 方法用于设置一个 ProgressDailog 进度条对话框中进度增加的步长。其语法格式为:

```
public void incrementProgressBy(int diff)
```

其中,参数 diff 为进度条进度增加的步长。

5.3.2 ProgressDailog 进度条对话框综合实例

下面通过程序来显示一个 ProgressDailog 进度条对话框,界面中只有一个简单的按钮,当用户单击按钮时系统启动进度条对话框,该进度条对话框显示一个耗时任务完成进度。

【实现步骤】

(1) 在 Eclipse 环境下,建立一个名为 E5_3 的工程。
(2) 打开 res/layout 目录下的 main.xml,其代码修改为:

```xml
<?xml version="1.0" encoding="utf-8"?>
<LinearLayout xmlns:android="http://schemas.android.com/apk/res/android"
    android:orientation="vertical"
    android:layout_width="fill_parent"
    android:layout_height="fill_parent"
    android:gravity="center_horizontal">
    <Button
        android:id="@+id/exec"
        android:layout_width="wrap_content"
        android:layout_height="wrap_content"
        android:text="进入 ProgressDialog 对话框"   />
</LinearLayout>
```

(3) 打开 src/com.example.e5_3 目录下的 MainActivity.java 文件,其代码修改为:

```
package com.example.e5_3;
import android.app.Activity;
import android.app.Dialog;
import android.app.ProgressDialog;
import android.os.Bundle;
```

```java
import android.os.Handler;
import android.os.Message;
import android.view.View;
import android.view.View.OnClickListener;
import android.widget.Button;

public class MainActivity extends Activity
{
    // 该程序模拟填充长度为 100 的数组
    private int[] data = new int[100];
    int hasData = 0;
    // 定义进度条对话框的标识
    final int PROGRESS_DIALOG = 0x112;
    // 记录进度条对话框的完成百分比
    int progressStatus = 0;
    ProgressDialog pd;
    // 定义一个负责更新进度的 Handler
    Handler handler;
    @Override
    public void onCreate(Bundle savedInstanceState)
    {
        super.onCreate(savedInstanceState);
        setContentView(R.layout.main);
        Button execBn = (Button) findViewById(R.id.exec);
        execBn.setOnClickListener(new OnClickListener()
        {
            @Override
            public void onClick(View source)
            {
                showDialog(PROGRESS_DIALOG);
            }
        });
        handler = new Handler()
        {
            @Override
            public void handleMessage(Message msg)
            {
                // 表明消息是由该程序发送的
                if (msg.what == 0x111)
                {
                    pd.setProgress(progressStatus);
```

```java
                    }
                }
            };
        }
        @Override
        public Dialog onCreateDialog(int id, Bundle status)
        {
            System.out.println("==========create==========");
            switch (id)
            {
                case PROGRESS_DIALOG:
                    // 创建进度条对话框
                    pd = new ProgressDialog(this);
                    pd.setMax(100);
                    // 设置对话框的标题
                    pd.setTitle("任务完成百分比");
                    // 设置对话框显示的内容
                    pd.setMessage("耗时任务的完成百分比");
                    // 设置对话框不能用"取消"按钮关闭
                    pd.setCancelable(false);
                    // 设置对话框的进度条风格
                    pd.setProgressStyle(ProgressDialog.STYLE_SPINNER);
                    pd.setProgressStyle(ProgressDialog.STYLE_HORIZONTAL);
                    // 设置对话框的进度条是否显示进度
                    pd.setIndeterminate(false);
                    break;
            }
            return pd;
        }
        // 该方法将在 onCreateDialog 方法调用之后被回调
        @Override
        public void onPrepareDialog(int id, Dialog dialog)
        {
            System.out.println("==========prepare==========");
            super.onPrepareDialog(id, dialog);
            switch (id)
            {
                case PROGRESS_DIALOG:
                    // 对话框进度清零
                    pd.incrementProgressBy(-pd.getProgress());
                    new Thread()
```

```java
                {
                    public void run()
                    {
                        while (progressStatus < 100)
                        {
                            // 获取耗时操作的完成百分比
                            progressStatus = doWork();
                            // 发送消息到 Handler
                            Message m = new Message();
                            m.what = 0x111;
                            // 发送消息
                            handler.sendMessage(m);
                        }
                        // 如果任务已经完成
                        if (progressStatus >= 100)
                        {
                            // 关闭对话框
                            pd.dismiss();
                        }
                    }
                }.start();
                break;
        }
    }
    // 模拟一个耗时的操作
    public int doWork()
    {
        // 为数组元素赋值
        data[hasData++] = (int) (Math.random() * 100);
        try
        {
            Thread.sleep(100);
        }
        catch (InterruptedException e)
        {
            e.printStackTrace();
        }
        return hasData;
    }
}
```

运行程序，效果如图 5-7 所示。

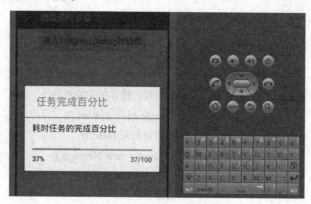

图 5-7　进度条对话框

5.4　Notification 通知

Notification 是 Android 中一种常用的通知方式，可将通知显示到屏幕的状态栏上，这时可以通过下拉状态栏读取通知，常用于手机接收到新短信、未接来电时状态栏显示的信息，或者显示后台下载的进度提示等。

5.4.1　常用的 Notification

本小节将介绍一种常见的 Notification 用法，可以为 Notification 设置显示图标、内容及声音。

1．获取 NotificationManager 对象

NotificationManager 有三个公共方法。

（1）cancel(int id)：取消以前显示的一个通知。假如是一个短暂的通知，试图将其隐藏；假如是一个持久的通知，将从状态条中移走。

（2）cancelAll()：取消以前显示的所有通知。

（3）notify(int id, Notification notification)：将持久的通知发送到状态条上。

2．初始化 Notification 对象

Notification 的属性主要有以下内容。

- audioStreamType：当声音响起时，所用的音频流的类型。
- contentIntent：当通知条目被单击，就执行这个被设置的 Intent。
- contentView：当通知被显示在状态条上的时候，同时显示这个被设置的视图。
- defaults：指定哪个值要被设置成默认的。
- deleteIntent：当用户单击"Clear All Notifications"按钮去删除所有的通知的时候，这个被设置的 Intent 被执行。
- icon：状态条所用的图片。

- iconLevel：假如状态条的图片有几个级别，就在这里设置。
- ledARGB LED：灯的颜色。
- ledOffMS LED：关闭时的闪光时间（以毫秒计算）。
- ledOnMS LED：开始时的闪光时间（以毫秒计算）。
- number：这个代表通知事件的号码。
- sound：通知的声音。
- tickerText：通知被显示在状态条时，所显示的信息。
- vibrate：振动模式。
- when：通知的时间戳。

注意：如果要使 Notification 常驻在状态栏，可以把 Notification 的 flags 属性设置为 FLAG_ONGOING_EVENT，例如：

```
1 n.flags=Notification.FLAG_ONGOING_EVENT
```

3．设置通知的显示参数

使用 PendingIntent 来包装通知 Intent，使用 Notification 的 setLatestEventInfo 来设置通知的标题、通知内容等信息。

4．发送通知

使用 NotificationManager 的 notify(int id, Notification notification)方法来发送通知。

下面的示例演示怎样通过 NotificationManager 来添加、删除 Notification。程序界面很简单，只包含两个普通按钮，分别用于添加 Notification 和删除 Notification。其具体实现操作如下：

（1）在 Eclipse 环境下，建立一个名为 E5_4_1 的工程。

（2）打开 res/values 目录下的 strings.xml 文件，代码修改为：

```
<resources>
    <string name="app_name">E5_4_1</string>
    <string name="hello_world">Hello world!</string>
    <string name="menu_settings">Settings</string>
    <string name="title_activity_main">系统通知</string>
    <string name="other_activity">Notification 启动的 Activity</string>
</resources>
```

（3）打开 res/layout 目录下 main.xml 文件，代码修改为：

```
<?xml version="1.0" encoding="utf-8"?>
<LinearLayout xmlns:android="http://schemas.android.com/apk/res/android"
    android:orientation="horizontal"
    android:layout_width="fill_parent"
    android:layout_height="fill_parent"
    android:gravity="center_horizontal">
<Button
    android:id="@+id/notify"
    android:layout_width="wrap_content"
    android:layout_height="wrap_content"
```

```
            android:text="添加 Notification"/>
    <Button
        android:id="@+id/cancel"
        android:layout_width="wrap_content"
        android:layout_height="wrap_content"
        android:text="删除 Notification"/>
</LinearLayout>
```

（4）在 res/layout 目录下新建一个名为 other.xml 的文件，代码修改为：

```
<?xml version="1.0" encoding="utf-8"?>
<LinearLayout  xmlns:android="http://schemas.android.com/apk/res/android"
    android:layout_width="fill_parent"
    android:layout_height="wrap_content"
    android:gravity="center_horizontal"
    android:orientation="vertical">
    <!-- 定义一个 ImageView -->
    <ImageView
        android:layout_width="fill_parent"
        android:layout_height="wrap_content"
        android:src="@drawable/br"
        android:layout_gravity="center_horizontal"/>
</LinearLayout>
```

（5）在 src/com.example.e5_4_1 目录下，新建一个名为 OtherActivity.java 的文件，代码修改为：

```
package com.example.e5_4_1;

import android.app.Activity;
import android.os.Bundle;
public class OtherActivity extends Activity
{
    @Override
    public void onCreate(Bundle savedInstanceState)
    {
        super.onCreate(savedInstanceState);
        //设置该 Activity 显示的页面
        setContentView(R.layout.other);
    }
}
```

（6）打开 src/com.example.e5_4_1 目录下的 MainActivity.java 文件，代码修改为：

```
package com.example.e5_4_1;

import android.app.Activity;
import android.app.Notification;
```

```java
import android.app.NotificationManager;
import android.app.PendingIntent;
import android.content.Intent;
import android.os.Bundle;
import android.view.View;
import android.view.View.OnClickListener;
import android.widget.Button;

public class MainActivity extends Activity
{
    private Button notifyBtn;
    private Button cancelBtn;
    private NotificationManager nm;
    private Notification n;
    public static final int ID = 0;
    @Override
    public void onCreate(Bundle savedInstanceState)
    {
        super.onCreate(savedInstanceState);
        setContentView(R.layout.main);
        notifyBtn = (Button) findViewById(R.id.notify);
        cancelBtn = (Button) findViewById(R.id.cancel);
        notifyBtn.setOnClickListener(new MyOnClickListener());
        cancelBtn.setOnClickListener(new MyOnClickListener());
        //获取通知管理
        nm = (NotificationManager) getSystemService(NOTIFICATION_SERVICE);
        n = new Notification();
        n.flags = Notification.FLAG_ONGOING_EVENT;
        n.icon = R.drawable.br;
        n.tickerText = "一个通知";
        n.when = System.currentTimeMillis();
    }
    class MyOnClickListener implements OnClickListener
    {
        @Override
        public void onClick(View v)
        {
            switch (v.getId())
            {
                case R.id.notify:
                    Intent intent = new Intent(MainActivity.this, OtherActivity.class);
```

```
                    PendingIntent pi = PendingIntent.getActivity(MainActivity.this, 0, intent, 0);
                    n.setLatestEventInfo(MainActivity.this, "通知标题", "通知内容", pi);
                    nm.notify(ID, n);
                    break;
                case R.id.cancel:
                    nm.cancel(ID);
                    break;
            }
        }
    }
}
```

运行程序，在界面中出现"添加 Notification"及"删除 Notification"两个按钮，单击"添加 Notification"按钮，将在手机屏幕上方出现一个 Notification，将状态栏向下拖动可看到 Notification 的详细情况，如图 5-8 所示。

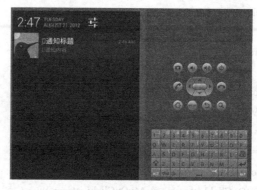

图 5-8 Notification 通知

5.4.2 带进度条的 Notification

本小节将介绍一种稍复杂的 Notification 的用法，下面以带进度条的 Notification 的实现为例，来介绍怎样自定义 Notification 的 View。

自定义 Notification 的 View 主要依赖于 Notification 的的 content View 属性。

【实现步骤】

(1) 在 Eclipse 环境下，建立一个名为 E5_4_2 的工程。
(2) 自定义的 other.xml 文件及 NotitificationActivity.java 文件与 5.4.1 节所创建的文件一致。
(3) 打开 res/layout 目录下的 main.xml 文件，代码修改为：

```xml
<?xml version="1.0" encoding="utf-8"?>
<LinearLayout
    xmlns:android="http://schemas.android.com/apk/res/android"
    android:orientation="horizontal"
    android:layout_width="fill_parent"
    android:layout_height="fill_parent"
```

```xml
        android:padding="10dp"
        android:background="#880490FF" >
<ImageView android:id="@+id/image"
        android:layout_width="wrap_content"
        android:layout_height="fill_parent" />
<ProgressBar
        android:id="@+id/pb"
        android:layout_width="180dip"
        android:layout_height="wrap_content"
        style="?android:attr/progressBarStyleHorizontal"
android:layout_gravity="center_vertical"/>
<Button
        android:id="@+id/bt"
        android:layout_width="wrap_content"
        android:layout_height="wrap_content"
        android:text="下载进度"/>
<TextView
        android:id="@+id/tv"
        android:layout_width="wrap_content"
        android:layout_height="fill_parent"
        android:textSize="16px"
        android:textColor="#FF0000"/>
</LinearLayout>
```

（4）打开 src/com.example.e5_4_2 目录下的 MainActivity.java 文件，代码修改为：

```java
package com.example.e5_4_2;

import android.app.Activity;
import android.app.Notification;
import android.app.NotificationManager;
import android.app.PendingIntent;
import android.content.Intent;
import android.os.Bundle;
import android.os.Handler;
import android.os.Message;
import android.view.View;
import android.widget.Button;
import android.widget.RemoteViews;

public class MainActivity extends Activity
{
    //当前进度条里的进度值
    private int progress=0;
```

```java
    private RemoteViews view=null;
    private Notification notification=new Notification();
    private boolean flag=true;
    private NotificationManager manager=null;
    private Intent intent=null;
    private PendingIntent pIntent=null;//更新显示
    private Handler handler=new Handler()
    {
        @Override
        public void handleMessage(Message msg)
        {
            // TODO: 自动存根法
            view.setProgressBar(R.id.pb, 100, progress, false);
            //关键部分，如果你不重新更新通知，进度条是不会更新的
            view.setTextViewText(R.id.tv, "下载"+progress+"%");
            notification.contentView=view;
            notification.contentIntent=pIntent;
            manager.notify(0, notification);
            if(progress==100)
            {
                flag=false;
            }
            super.handleMessage(msg);
        }
    };
    @Override
    public void onCreate(Bundle savedInstanceState)
    {
        super.onCreate(savedInstanceState);
        setContentView(R.layout.main);
        //获取通知管理器
        manager=(NotificationManager)getSystemService(NOTIFICATION_SERVICE);
        view=new RemoteViews(getPackageName(),R.layout.other);
        intent=new Intent(MainActivity.this,NotificationService.class);
        pIntent=PendingIntent.getService(MainActivity.this, 0, intent, 0);
        //开始进度按钮
        Button button=(Button)findViewById(R.id.bt);
        button.setOnClickListener(new Button.OnClickListener()
        {
            @Override
            public void onClick(View v)
            {
```

```
//通知的图标必须设置(其他属性为可选设置),否则通知无法显示
//启动一个线程用来更新 progress
notification.icon=R.drawable.br;
view.setImageViewResource(R.id.image, R.drawable.br);
Thread thread=new Thread(new Runnable()
{
    @Override
    public void run()
    {
        while (flag)
        {
            progress+=5;
            //更新进度条
            Message msg=handler.obtainMessage();
            msg.arg1=progress;
            msg.sendToTarget();
            try
            {
                Thread.sleep(1000);
            }catch(Exception e)
            {
                e.printStackTrace();
            }
        }
    }
});thread.start();
    }
});
```

运行程序,效果如图 5-9 所示。

图 5-9 带进度条的 Notification

第 6 章 Android 视图与动画

本章将介绍 Android 的图像、图片及动画等内容。

6.1 Android 图像

Android 中提供了许多方法用于显示图像，本节给出补充介绍。

6.1.1 ImageSwitcher 类

在第 3 章中演示了怎样将 Gallery 视图与一个 ImageView 视图一起使用来显示一系列缩略图像，以便当其中之一被选中时，使选定的图像在 ImageView 中显示。然而，有时并不想在 Gallery 视图中选择一张图像时，图像则显示得太突然（无过渡过程）。例如，用户也许希望在图像之间过渡时应用一些动画效果。这时，Gallery 视图就需要配合使用 ImageSwitcher 类。

1. ImageSwitcher 类的原理

总的来说。ImageSwitcher 是一个能动机构。如果将它称做"父对象"的话，那么就包含了两个"子对象"——ImageView。ImageSwitcher 负责的就是两个子对象的切换动画，切换动作，切换时机，以及设定两个子对象的内容等。

ImageSwitcher 类必须设置一个 ViewFactory，主要用来将显示的图片和父窗口区分开来，因此需要实现 ViewSwitcher.ViewFactory 接口，通过 makeView()方法来显示图片，这里会返回一个 ImageView 对象，而方法 setImageResource 用来指定图片资源。

2. ImageSwitcher 类的重要方法

在 ImageSwitcher 类中主要有以下 3 个重要方法：
- setImageURI(Uri uri)：设置图片地址。
- setImageResource(int resid)：设置图片资源库。
- setImageDrawable(Drawable drawable)：绘制图片。

3. ImageSwitcher 类实例

下面利用 ImageSwitcher 类实现图片的切换效果。
【实现步骤】
（1）在 Eclipse 环境下，建立一个名为 E6_1_1 的工程。
（2）把图片存放在 res/drawable-hipi 目录下。
（3）打开 res/layout 目录下的 main.xml 文件，代码修改为：

```xml
<?xml version="1.0" encoding="utf-8"?>
<LinearLayout xmlns:android="http://schemas.android.com/apk/res/android"
    android:orientation="vertical" android:layout_width="fill_parent"
    android:layout_height="fill_parent">
    <LinearLayout android:layout_width="fill_parent"
        android:layout_height="wrap_content" >
        <ImageSwitcher android:id="@+id/is_imageswitch"
            android:layout_width="wrap_content" android:layout_height="300px">
        </ImageSwitcher>;
    </LinearLayout>
    <LinearLayout android:layout_width="fill_parent"
        android:layout_height="wrap_content">
        <Button android:id="@+id/btn_last" android:layout_width="150px"
            android:layout_height="wrap_content" android:text="上一张"
            android:onClick="onClickLast" >
        </Button>
        <Button android:id="@+id/btn_next" android:layout_width="150px"
            android:layout_height="wrap_content" android:text="下一张"
            android:onClick="onClickNext" >
        </Button>
    </LinearLayout>
</LinearLayout>
```

（4）打开 src/com.example.e6_1_1 目录下的 MainActivity.java 文件，代码修改为：

```java
package com.example.e6_1_1;
import android.app.Activity;
import android.content.Context;
import android.os.Bundle;
import android.view.View;
import android.widget.Button;
import android.widget.ImageSwitcher;
import android.widget.ImageView;
import android.widget.ViewSwitcher.ViewFactory;

public class MainActivity extends Activity
{
    private ImageSwitcher is_imageSwitcher;
    //存放图片 ID 的 int 数组
    private int[] images={R.drawable.fj1,
        R.drawable.fj2,
        R.drawable.fj3,
        R.drawable.fj4,
```

```java
            R.drawable.fj5,
            R.drawable.fj6};
    //下一张按钮和上一张按钮
    private Button btn_next;
    private Button btn_last;
    private int index=0;
    @Override
    public void onCreate(Bundle savedInstanceState)
    {
        super.onCreate(savedInstanceState);
        setContentView(R.layout.main);
        is_imageSwitcher=(ImageSwitcher)findViewById(R.id.is_imageswitch);
        btn_last=(Button)findViewById(R.id.btn_last);
        btn_next=(Button)findViewById(R.id.btn_next);
        //imageSwticher 必须设置一个 viewfactory 后才可以查看图片
        is_imageSwitcher.setFactory(new ImageViewFactory(this));
        //设置图片资源 id
        is_imageSwitcher.setBackgroundResource(images[index]);
    }
    //重写了的 viewFactory
    class ImageViewFactory implements ViewFactory
    {
        private Context context;
        public ImageViewFactory(Context context)
        {
            this.context = context;
        }
        @Override
        public View makeView()
        {
            // TODO：自动存根法
            //定义每个图像的显示大小
            ImageView iv = new ImageView(this.context);
            iv.setLayoutParams(new ImageSwitcher.LayoutParams(320, 320));
            return iv;
        }
    }
    //上一张的按钮事件
    public void onClickLast(View v)
    {
        if(index>=0&&index<images.length-1)
```

```
            {
                index++;
                is_imageSwitcher.setBackgroundResource(images[index]);
            }else
            {
                index=images.length-1;
            }
    }
        //下一张的按钮事件
        public void onClickNext(View v)
        {
            if(index>0&&index<images.length)
            {
                index--;
                is_imageSwitcher.setBackgroundResource(images[index]);
            }else
            {
                index=images.length-1;
            }
        }
    }
```

运行程序，效果如图 6-1 所示。

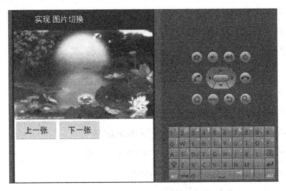

图 6-1　实现图片切换

6.1.2　ScrollView 类

ScrollView 类是一种可供用户滚动的层次结构布局容器，允许显示比实际多的内容。ScrollView 是一种 FrameLayout，意味着需要在其上放置有自己滚动内容的子元素。子元素可以是一个复杂对象的布局管理器。通常用的子元素是垂直方向的 LinearLayout，可显示在最上层的垂直方向供用户滚动的箭头。ScrollView 只支持垂直方向的滚动。

其语法格式为：

```
public ScrollView (Context context)                          //创建一个默认属性的 ScrollView 实例
public ScrollView (Context context, AttributeSet attrs)      //创建一个带有 attrs 属性的 ScrollView 实例
public ScrollView (Context context, AttributeSet attrs, int defStyle)  //创建一个带有 attrs 属性，并且指定
                                                             //其默认样式的 ScrollView 实例
```

下面通过两种方式来实现 ScrollView 类。

1. 仅通过布局文件来演示

下面只通过修改布局文件代码即利用 ScrollView 类完成图片的纵向拖动。其代码为：

```xml
<?xml version="1.0" encoding="utf-8"?>
<ScrollView xmlns:android="http://schemas.android.com/apk/res/android"
    android:layout_width="fill_parent"
    android:layout_height="fill_parent"
    android:scrollbars="vertical">
    <LinearLayout android:orientation="vertical"
        android:layout_width="fill_parent"
        android:layout_height="fill_parent">
        <ImageView android:layout_width="wrap_content"
            android:layout_height="wrap_content"
            android:src="@drawable/p1"
            android:layout_gravity="center_horizontal"/>
        <ImageView android:layout_width="wrap_content"
            android:layout_height="wrap_content"
            android:src="@drawable/p1"
            android:layout_gravity="center_horizontal"/>
        <ImageView android:layout_width="wrap_content"
            android:layout_height="wrap_content"
            android:src="@drawable/p1"
            android:layout_gravity="center_horizontal"/>
        <ImageView android:layout_width="wrap_content"
            android:layout_height="wrap_content"
            android:src="@drawable/p1"
            android:layout_gravity="center_horizontal"/>
        <ImageView android:layout_width="wrap_content"
            android:layout_height="wrap_content"
            android:src="@drawable/p1"
            android:layout_gravity="center_horizontal"/>
        <ImageView android:layout_width="wrap_content"
            android:layout_height="wrap_content"
            android:src="@drawable/p1"
            android:layout_gravity="center_horizontal"/>
```

```xml
        <ImageView android:layout_width="wrap_content"
            android:layout_height="wrap_content"
            android:src="@drawable/p1"
            android:layout_gravity="center_horizontal"/>
        <ImageView android:layout_width="wrap_content"
            android:layout_height="wrap_content"
            android:src="@drawable/p1"
            android:layout_gravity="center_horizontal"/>
        <ImageView android:layout_width="wrap_content"
            android:layout_height="wrap_content"
            android:src="@drawable/p1"
            android:layout_gravity="center_horizontal"/>
    </LinearLayout>
</ScrollView>
```

运行程序，效果如图 6-2 所示。

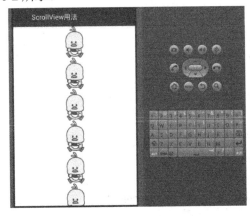

图 6-2　通过布局文件来演示 ScrollView 类

2．通过 Activity 类活动来演示

ScrollView（卷轴视图）是在拥有很多内容以至一屏显示不完时，需要通过滚动条来显示的视图。

【实现步骤】

（1）在 Eclipse 环境下，建立一个名为 E6_1_2 的工程。

（2）打开 res/layout 目录下的 main.xml 文件，代码修改为：

```xml
<?xml version="1.0" encoding="utf-8"?>
<ScrollView xmlns:android="http://schemas.android.com/apk/res/android"
    android:id="@+id/ScrollView" android:layout_width="fill_parent"
    android:layout_height="wrap_content" android:scrollbars="vertical">
    <LinearLayout android:id="@+id/LinearLayout"
        android:orientation="vertical" android:layout_width="fill_parent"
        android:layout_height="wrap_content">
```

```xml
        <TextView android:id="@+id/TestView" android:layout_width="fill_parent"
            android:layout_height="wrap_content" android:text="TestView0" />
        <Button android:id="@+id/Button"
            android:text="Button0" android:layout_width="fill_parent"
            android:layout_height="wrap_content"></Button>
    </LinearLayout>
</ScrollView>
```

（3）打开 src/com.example.e6_1_2 目录下的 MainActivity.java 文件，代码修改为：

```java
package com.example.e6_1_2;

import android.app.Activity;
import android.os.Bundle;
import android.os.Handler;
import android.view.KeyEvent;
import android.view.View;
import android.widget.Button;
import android.widget.LinearLayout;
import android.widget.ScrollView;
import android.widget.TextView;
public class MainActivity extends Activity
{
    /* 第一次调用活动*/
    private LinearLayout mLayout;
    private ScrollView sView;
    private final Handler mHandler = new Handler();
    @Override
    public void onCreate(Bundle savedInstanceState)
    {
        super.onCreate(savedInstanceState);
        setContentView(R.layout.main);
        // 创建一个线性布局
        mLayout = (LinearLayout) this.findViewById(R.id.LinearLayout);
        // 创建一个 ScrollView 对象
        sView = (ScrollView) this.findViewById(R.id.ScrollView);
        Button mBtn = (Button) this.findViewById(R.id.Button);
        mBtn.setOnClickListener(mClickListener);        // 添加单击事件监听
    }
    public boolean onKeyDown(int keyCode, KeyEvent event)
    {
        Button b = (Button) this.getCurrentFocus();
        int count = mLayout.getChildCount();
```

```java
            Button bm = (Button) mLayout.getChildAt(count-1);
            if(keyCode==KeyEvent.KEYCODE_DPAD_UP && b.getId()==R.id.Button)
            {
                bm.requestFocus();
                return true;
            }else if(keyCode==KeyEvent.KEYCODE_DPAD_DOWN && b.getId()==bm.getId())
            {
                this.findViewById(R.id.Button).requestFocus();
                return true;
            }
            return false;
}
// Button 事件监听,当单击第一个按钮时增加一个 button 和一个 textview
private Button.OnClickListener mClickListener = new Button.OnClickListener()
{
    private int index = 1;
    @Override
    public void onClick(View v)
    {
        TextView tView = new TextView(MainActivity.this);//定义一个 TextView
        tView.setText("TextView" + index);//设置 TextView 的文本信息
        //设置线性布局的属性
        LinearLayout.LayoutParams params = new LinearLayout.LayoutParams(
                LinearLayout.LayoutParams.FILL_PARENT,
                LinearLayout.LayoutParams.WRAP_CONTENT);
        mLayout.addView(tView, params);//添加一个 TextView 控件
        Button button = new Button(MainActivity.this);//定义一个 Button
        button.setText("Button" + index);//设置 Button 的文本信息
        button.setId(index++);
        mLayout.addView(button, params);//添加一个 Button 控件
        mHandler.post(mScrollToButton);//传递一个消息进行滚动
    }
};
private Runnable mScrollToButton = new Runnable()
{
    @Override
    public void run()
    {
        int off = mLayout.getMeasuredHeight() - sView.getHeight();
        if (off > 0) {
            sView.scrollTo(0, off);//改变滚动条的位置
```

            }
         }
      };
   }
```

运行程序,单击界面的"Button"按钮,效果如图 6-3 所示。

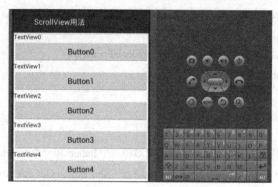

图 6-3  ScrollView 演示效果

## 6.1.3  GridView 类

GridView(网格视图)是按照行列的方式来显示内容的,一般用于显示图片等内容,例如,实现九宫格图用 GridView 是首选,也是最简单的。GridView 类主要用于设置 Adapter。

GridView 常用的 XML 属性有以下内容。
- android:columnWidth:设置列的宽度。
- android:gravity:设置此组件中的内容在组件中的位置,可选的值有:top、bottom、left、right、center_vertical、fill_vertical、center_horizontal、fill_horizontal、center、fill、clip_vertical。可以多选,其间用"|"分开。
- android:horizontalSpacing:两列之间的间距。
- android:numColumns:设置列数。
- android:stretchMode:缩放模式。
- android:verticalSpacing:两行之间的间距。

以下程序通过使用 GridView 类实现显示自定义的图片文字。

【实现步骤】

(1)在 Eclipse 环境下,建立一个名为 E6_1_3 的工程。
(2)打开 res/layout 目录下的 main.xml 文件,代码修改为:

```xml
<?xml version="1.0" encoding="utf-8"?>
<GridView xmlns:android="http://schemas.android.com/apk/res/android"
    android:id="@+id/gridview"
    android:layout_width="fill_parent"
    android:layout_height="fill_parent"
    android:columnWidth="90dp"
```
```

```
        android:numColumns="auto_fit"
        android:verticalSpacing="10dp"
        android:horizontalSpacing="10dp"
        android:stretchMode="columnWidth"
        android:gravity="center"/>
```

（3）在打开的 res/layout 目录下新建一个名为 picture_item.xml 的文件，代码修改为：

```
<?xml version="1.0" encoding="utf-8"?>
<LinearLayout
    xmlns:android="http://schemas.android.com/apk/res/android"
    android:id="@+id/root"
    android:orientation="vertical"
    android:layout_width="wrap_content"
    android:layout_height="wrap_content"
    android:layout_marginTop="5dp">
    <ImageView
        android:id="@+id/image"
        android:layout_width="100dp"
        android:layout_height="150dp"
        android:layout_gravity="center"
        android:scaleType="fitXY"
        android:padding="4dp"/>
    <TextView
        android:id="@+id/title"
        android:layout_width="wrap_content"
        android:layout_height="wrap_content"
        android:layout_gravity="center"
        android:gravity="center_horizontal"/>
</LinearLayout>
```

（4）打开 src/com.example.e6_1_3 目录下的 MainActivity.java 文件，代码修改为：

```
package com.example.e6_1_3;
import java.util.ArrayList;
import java.util.List;
import android.app.Activity;
import android.content.Context;
import android.os.Bundle;
import android.view.LayoutInflater;
import android.view.View;
import android.view.ViewGroup;
import android.widget.AdapterView;
import android.widget.BaseAdapter;
import android.widget.GridView;
```

```java
import android.widget.ImageView;
import android.widget.TextView;
import android.widget.Toast;
import android.widget.AdapterView.OnItemClickListener;
public class MainActivity extends Activity
{
    private GridView gridView;
    //图片的文字标题
    private String[] titles = new String[]
    { "fj1", "fj2", "fj3", "fj4", "fj5", "fj6"};
    //图片 ID 数组
    private int[] images = new int[]{
            R.drawable.fj1, R.drawable.fj2, R.drawable.fj3,
            R.drawable.fj4, R.drawable.fj5, R.drawable.fj6
    };
    @Override
    public void onCreate(Bundle savedInstanceState)
    {
        super.onCreate(savedInstanceState);
        setContentView(R.layout.activity_main);
        gridView = (GridView) findViewById(R.id.gridview);
        PictureAdapter adapter = new PictureAdapter(titles, images, this);
        gridView.setAdapter(adapter);

        gridView.setOnItemClickListener(new OnItemClickListener()
        {
            public void onItemClick(AdapterView<?> parent, View v, int position, long id)
            {
                Toast.makeText(MainActivity.this, "pic" + (position+1), Toast.LENGTH_SHORT).show();
            }
        });
    }
}
//自定义适配器
class PictureAdapter extends BaseAdapter
{
    private LayoutInflater inflater;
    private List<Picture> pictures;
    public PictureAdapter(String[] titles, int[] images, Context context)
    {
```

```java
        super();
        pictures = new ArrayList<Picture>();
        inflater = LayoutInflater.from(context);
        for (int i = 0; i < images.length; i++)
        {
            Picture picture = new Picture(titles[i], images[i]);
            pictures.add(picture);
        }
    }
    @Override
    public int getCount()
    {
        if (null != pictures)
        {
            return pictures.size();
        } else
        {
            return 0;
        }
    }
    @Override
    public Object getItem(int position)
    {
        return pictures.get(position);
    }
    @Override
    public long getItemId(int position)
    {
        return position;
    }
    @Override
    public View getView(int position, View convertView, ViewGroup parent)
    {
        ViewHolder viewHolder;
        if (convertView == null)
        {
            convertView = inflater.inflate(R.layout.picture_item, null);
            viewHolder = new ViewHolder();
            viewHolder.title = (TextView) convertView.findViewById(R.id.title);
            viewHolder.image = (ImageView) convertView.findViewById(R.id.image);
            convertView.setTag(viewHolder);
```

```java
            } else
            {
                viewHolder = (ViewHolder) convertView.getTag();
            }
            viewHolder.title.setText(pictures.get(position).getTitle());
            viewHolder.image.setImageResource(pictures.get(position).getImageId());
            return convertView;
        }
    }
    class ViewHolder
    {
        public TextView title;
        public ImageView image;
    }
    class Picture
    {
        private String title;
        private int imageId;
        public Picture()
        {
            super();
        }
        public Picture(String title, int imageId)
        {
            super();
            this.title = title;
            this.imageId = imageId;
        }
        public String getTitle()
        {
            return title;
        }
        public void setTitle(String title)
        {
            this.title = title;
        }
        public int getImageId()
        {
            return imageId;
        }
        public void setImageId(int imageId)
```

```
                {
                    this.imageId = imageId;
                }
        }
```

运行程序，效果如图 6-4 所示。

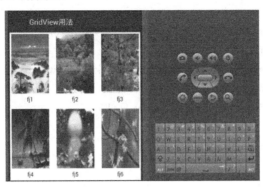

图 6-4　利用 GridView 类自定义显示图片效果

6.1.4　WebView 类

WebView 类类似于常用的浏览器，在 Android 手机中内置了一款高性能的 webkit 内核浏览器，WebView 组件就是由 webkit 封装而来的，可以用它来显示一个 Web 页面。

下面展示怎样以编程方式加载一个 Web 页面的内容并在活动中显示出来。

【实现步骤】

（1）在 Eclipse 环境下，建立一个名为 E6_1_4 的工程。

（2）打开 res/layout 目录下的 main.xml 文件，代码修改为：

```xml
<?xml version="1.0" encoding="utf-8"?>
<LinearLayout xmlns:android="http://schemas.android.com/apk/res/android"
    android:orientation="vertical"
    android:layout_width="fill_parent"
    android:layout_height="fill_parent" >
<WebView android:id="@+id/webview1"
    android:layout_width="wrap_content"
    android:layout_height="wrap_content" />
</LinearLayout>
```

（3）打开 src/com.example.e6_1_4 目录下的 MainActivity.java 文件，代码修改为：

```java
package com.example.e6_1_4;
import android.app.Activity;
import android.os.Bundle;
import android.webkit.WebSettings;
import android.webkit.WebView;
import android.webkit.WebViewClient;
```

```
public class MainActivity extends Activity
{
    /*第一次调用活动 */
    @Override
    public void onCreate(Bundle savedInstanceState)
    {
        super.onCreate(savedInstanceState);
        setContentView(R.layout.main);
        WebView wv = (WebView) findViewById(R.id.webview1);
        //利用 WebView 加载一个 Web 页面
        wv.loadUrl("file:///android_asset/Index.html");
    }
    private class Callback extends WebViewClient
    {
        @Override
        //通过 shouldOverrideUrlLoading 方法拦截 URL
        public boolean shouldOverrideUrlLoading(WebView view, String url)
        {
            return(false);
        }
    }
}
```

运行程序，效果如图 6-5 所示。

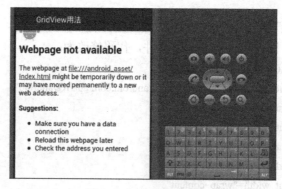

图 6-5 加载 Web 页面

6.2 Android 绘图

除了使用已有的图片之外，Android 应用常常要动态地生成图片，比如一个手机游戏，游戏界面看上去丰富多彩，而且可以随着用户的动作而动态改变，这就需要借助于 Android 的绘图支持了。

6.2.1 Paint 类

Canvas 提供的方法还涉及一个 API——Paint。Paint 为 Canvas 的画笔，因此 Paint 主要用于设置绘制风格，包括画笔的颜色、画笔触粗细、填充风格等。

1．Paint 类的方法

Paint 类提供了如下方法。

（1）setColor 方法：用于设置画笔的颜色，可以通过 Color 类中的预定义颜色来设置，也可以通过指定 RGB 值来设置。该方法是设置颜色的主要方法，通过改变画笔颜色，可以绘制出色彩缤纷的图形。其语法格式为：

```
public void setColor(int color)
```

其中，参数 color 为颜色值，也可以直接使用系统中 Color 类定义的颜色。

- Color.BLACK：黑色。
- Color.BLUE：蓝色。
- Color.CYAN：青绿色。
- Color.DKGRAY：灰黑色。
- Color.YELLOW：黄色。
- Color.GRAY：灰色。
- Color.GREEN：绿色。
- Color.LTGRAY：浅灰色。
- Color.MAGENTA：红紫色。
- Color.RED：红色。
- Color.TRANSPARENT：透明。

（2）setAlpha 方法：用于设置画笔的透明度，直观上表现为颜色变淡，具有一定的透明效果。该方法经常用于一些图片重叠或者特效显示的场合。其语法格式为：

```
public void setAlpha(int a)
```

其中，参数 a 为透明度，其取值范围为 0~255，数值越小越透明。

（3）setStyle 方法：用于设置画笔的风格，可以指定是实心还是空心。该方法在矩形、圆形等图形上有明显的效果。其语法格式为：

```
public void setStyle(Paint.Style style)
```

其中，参数 style 为画笔的风格，为 Paint.Style 类型，有下面几种取值。

- Style.FILL：实心。
- Style.FILL_AND_STROKE：同时显示实心和空心，该参数在某些场合会带来不可预期的显示效果。
- Style.STROKE：空心。

（4）setStrokeWidth 方法：用于设置画笔的空心线宽。该方法在矩形、圆形等图形上有明显的效果。其语法格式为：

```
public void setStrokeWidth(float width)
```

其中，参数 width 为线宽，浮点型数据。

（5）setTextSize 方法：用于设置画笔的字体大小，该方法主要用在绘制字符串的场合，对于一些图形则没有效果。其语法格式为：

```
public void setTextSize(float textSize)
```

其中，参数 textSize 为字体大小，浮点型数据。

（6）setTypeface 方法：用于设置画笔的字体样式，可以指定系统自带的字段，也可以使用自定义的字体。其语法格式为：

```
public Typeface setTypeface(Typeface typeface)
```

其中，参数 typeface 为字体样式，有下面几种取值。

- Typeface.DEFAULT：默认字体。
- Typeface.DEFAULT_BOLD：加粗字体。
- Typeface.MONOSPACE：monospace 字体。
- Typeface.SANS_SERIF：sans 字体。
- Typeface.SERIF：serif 字体。

（7）setTextScaleX 方法：用于设置画笔字体的比例因子，默认为 1。当大于 1 时表示横向拉伸，当小于 1 时表示横向压缩。其语法格式为：

```
public void setTextScaleX(float scaleX)
```

其中，参数 scaleX 为字体比例因子，当大于 1 时表示横向拉伸，当小于 1 时表示横向压缩。

（8）setARGB 方法：用于设置画笔的颜色和透明度，其中颜色采用 RGB 数值的方式来指定。该方法的功能相当于 setColor 方法与 setAlpha 方法，使用非常方便、灵活。其语法格式为：

```
public void setARGB(int a,int r,int g,int b)
```

其中，参数 a 为透明度，取值范围为 0～255，数值越小越透明；参数 r 为红色的颜色值，取值范围为 0～255；参数 g 为绿色的颜色值，取值范围为 0～255；参数 b 为蓝色的颜色值，取值范围为 0～255。

（9）setUnderlineText 方法：用于设置画笔的下画线。该方法主要用于绘制字符串的场合，对于其他一些图形则没有效果。其语法格式为：

```
public void setUnderlineText(Boolean underlineText)
```

其中，参数 underlineText 表示是否显示下画线，当取值为 true 时表示显示下画线，取值为 false 时表示不显示下画线。

（10）setTextSkewX 方法：用于设置画笔的倾斜因子，默认为 0。当取值为正数时表示向左倾斜，取值为负数时表示向右倾斜。setTextSkewX 方法主要用于绘制字符串的场合，对于其他一些图形则没有效果。其语法格式为：

```
public void setTextSkewX(float skewX)
```

其中，参数 skewX 为倾向因子，正数表示向左倾斜，负数表示向右倾斜。

2. Paint 类实例

以下代码用于演示 Paint 类方法的使用。

【实现步骤】

（1）在 Eclipse 环境下，建立一个名为 E6_2_1 工程。

（2）打开 src/com.example.e6_2_1 目录下的 MainActivity.java 文件，代码修改为：

```
public class MainActivity extends Activity
{
```

```java
        private PaintTest paintTest=null;                    //声明自定义 View 对象
        /* 第一次调用活动*/
        @Override
        public void onCreate(Bundle savedInstanceState)
        {
            //重载 onCreate 方法
            super.onCreate(savedInstanceState);
            this.paintTest=new PaintTest(this);               //创建自定义 View 对象
            setContentView(paintTest);                        //设置显示自定义 View
        }
    }
```

（3）在 src/com.example.e6_2_1 目录下新建一个名为 PaintTest.java 的文件，代码修改为：

```java
    public class PaintTest extends View implements Runnable
    {
        //自定义 View
        private Paint paint=null;                             //声明画笔对象
        public PaintTest(Context context)
        {
            super(context);
            // TODO：自动存根法
            paint=new Paint();                                //构建对象
            new Thread(this).start();                         //开启线程
        }
        protected void onDraw(Canvas canvas)
        {
            //重载 onDraw 方法
            // TODO：自动存根法
            super.onDraw(canvas);
            paint.setColor(Color.RED);                        //设置画笔颜色
            canvas.drawColor(Color.BLUE);
            paint.setTextSkewX((float) -1.5);                 //设置倾斜因子
            paint.setARGB(255, 0, 0, 0);                      //设置字体颜色
            paint.setTextSize(20);                            //设置画笔字体的大小
            canvas.drawText("Hello Android!", 10, 50, paint);
            paint.setARGB(255, 255, 0, 0);                    //设置字体颜色
            paint.setTextSize(30);                            //设置画笔字体的大小
            canvas.drawText("Hello Android!", 10, 150, paint);
            paint.setARGB(255, 0, 255, 0);                    //设置字体颜色
            paint.setTextSize(40);                            //设置画笔字体的大小
            canvas.drawText("Hello Android!", 10, 250, paint);
            paint.setARGB(255, 0, 0, 255);                    //设置字体颜色
```

```
        paint.setTextSize(50);                                //设置画笔字体的大小
        canvas.drawText("Hello Android!", 10, 350, paint);
    }
    @Override
    public void run()
    {                                                         //重载 run 方法
        // TODO：自动存根法
        while(!Thread.currentThread().isInterrupted())
        {
            try
            {
                Thread.sleep(100);
            }
            catch(InterruptedException e)
            {
                Thread.currentThread().interrupt();
            }
            postInvalidate();                                 //更新界面
        }
    }
}
```

运行程序，效果如图 6-6 所示。

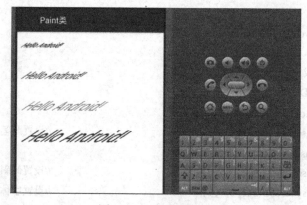

图 6-6　Paint 类的方法

6.2.2　Canvas 类

Canvas 类像是一块画布，可以在上面画想画的东西。当然，还可以设置画布的属性，如画布的颜色、尺寸等。

1. Canvas 类方法

Canvas 提供了如下一些方法。

（1）drawColor 方法：用于设置画布的背景颜色，可以通过 Color 类中的预定义颜色来设置，也可以通过指定 RGB 值来设置。其语法格式为：

    public void drawColor(int color)

其中，参数 color 为颜色值。也可以直接使用系统 Color 类中定义的颜色。

（2）drawLine 方法：用于在画布上绘制直线，通过指定直线的两个端点坐标来绘制。该方法只能绘制单条直线；如果需要同时绘制多条直线，则可以使用 drawLines 方法。其语法格式为：

    public void drawLine(float startX,float startY,float stopX,float,stopY,Paint paint)

其中，参数 startX 为起始端点的 X 坐标；参数 startY 为起始端点的 Y 坐标；参数 stopX 为终止端点的 X 坐标；stopY 参数为终止端点的 Y 坐标；参数 paint 为绘制直线所使用的画笔。

（3）drawLines 方法：用于在画布上绘制多条直线，通过指定直线的端点坐标数组来绘制。该方法可以绘制多条直线，非常灵活，其语法格式为：

    public void drawLines(float[] pts,Paint paint)

其中，参数 pts 为绘制直线的端点数组，每条直线占用 4 个数据；参数 paint 为绘制直线所使用的画笔。

（4）drawPoint 方法：用于在画布上绘制一个点，通过指定端点坐标来绘制。该方法只能绘制单个点，如果需要同时绘制多个点，则可以使用 drawPoints 方法。其语法格式为：

    public void drawPoint(float x,float y,Paint paint)

其中，参数 x 为绘制点的 X 坐标；参数 y 为绘制点的 Y 坐标；参数 paint 为绘制点所使用的画笔。

（5）drawPoints 方法：用于在画布上绘制多个点，通过指定端点坐标数组来绘制。该方法可以绘制多个点，同时也可以指定哪些点绘制，而哪些点不绘制，非常灵活。其语法格式为：

    public void drawPoints(float[] pts,Paint paint)
    public void drawPoints(float[] pts,int offset,int count,Paint paint)

其中，参数 pts 为绘制点的数组，每个端点占用两个数据；参数 offset 为跳过的数据个数，这些数据将不参与绘制过程；参数 count 为实际参与绘制的数据个数；参数 paint 为绘制时所使用的画笔。

（6）drawRect 方法：用于在画布上绘制矩形，可以通过指定矩形的四条边来实现，也可以通过指定 Rect 对象来实现。同时还可以通过设置画笔的空心效果来绘制空心的矩形。其语法格式为：

    public void drawRect(Rect r,Paint paint);
    public void drawRect(RectF rect,Paint paint);
    public void drawRect(float left,float top,float right,float below,Paint paint)

其中，参数 r 为 Rect 对象；参数 rect 为 RectF 对象；参数 left 为矩形的左边位置；参数 top 为矩形的上边位置；参数 right 为矩形的右边位置；参数 below 为矩形的下边位置；参数 paint 为绘制矩形时所使用的画笔。

（7）drawRoundRect 方法：用于在画面上绘制圆角矩形，通过指定 RectF 对象及圆角半径来实现。该方法是绘制圆角矩形的主要方法，同时也可以通过设置画笔的空心效果来绘制空心的圆角矩形。其语法格式为：

```
public void drawRoundRect(RectF rect,float rx,float ry,Paint paint)
```
其中，参数 rect 为 RectF 对象；参数 rx 为 X 方向上的圆角半径；参数 ry 为 Y 方向上的圆角半径；参数 paint 为绘制时所使用的画笔。

（8）drawCircle 方法：用于在画布上绘制圆形，通过指定圆形圆心的坐标和半径来实现。该方法是绘制圆形的主要方法，同时也可以通过设置画笔的空心效果来绘制空心的圆形。其语法格式为：

```
public void drawCircle(float cx,float cy,float radius,Paint paint)
```
其中，参数 cx 为圆心的 X 坐标；参数 cy 为圆心的 Y 坐标；参数 radius 为圆的半径；参数 paint 为绘制时所使用的画笔。

（9）drawOval 方法：用于在画布上绘制椭圆形，通过指定椭圆外切矩形的 RectF 对象来实现。该方法为绘制椭圆形的主要方法，同时也可以通过设置画笔的空心效果来绘制空心的椭圆形。其语法格式为：

```
public void drawOval(RectF oval,Paint paint)
```
其中，参数 oval 为椭圆外切矩形的 RectF 对象；参数 paint 为绘制时所使用的画笔。

（10）drawPath 方法：用于在画布上绘制任意多边形，通过指定 Path 对象来实现。在 Path 对象中规划了多边形的路径信息。其语法格式为：

```
public void drawPath(Path path,Paint paint)
```
其中，参数 path 为包含路径信息的 Path 对象；参数 paint 为绘制时所使用的画笔。

（11）drawArc 方法：用于在画布上绘制圆弧，通过指定圆弧所在的椭圆对象、起始角度、终止角度来实现。该方法是绘制圆弧的主要方法。其语法格式为：

```
public void drawArc(RectF oval,float startAngle,float sweepAngle,Boolean useCenter,Paint paint)
```
其中，参数 oval 为圆弧所在的椭圆对象；参数 startAngle 为圆弧的起始角度；参数 sweepAngle 为圆弧的角度；参数 useCenter 表示是否显示半径连线，当取值为 true 时表示显示圆弧与圆心的半径连线，取值为 false 时表示不显示；参数 paint 为绘制时所使用的画笔。

（12）drawText 方法：用于在画布上绘制字符串，通过指定字符串的内容和显示的位置来实现。在画布上绘制字符串是常用的操作，Android 系统提供了非常灵活的绘制字符串的方法，可以根据不同的需要调用不同的方法来实现。字体的大小、样式等信息都需要在 Paint 画笔中指定。其语法格式为：

```
public void drawText(String text,float x,float y,Paint pain)
public void drawText(char[] text,int index,int count,float x,float y,Paint paint)
public void drawText(CharSequence text,int start,int end,float x,float y,Paint paint)
public void drawText(String text,int start,int end,float x,float y,Paint paint)
```
其中，参数 text 为字符串内容，可以采用 String 格式，也可以采用 char 字符数组形式；参数 x 为显示位置的 X 坐标；参数 y 为显示位置的 Y 坐标；参数 index 为显示的起始字符位置；参数 count 为显示字符的个数；参数 start 为显示的起始字符位置；参数 end 为显示的终止字符位置；参数 paint 为绘制时所使用的画笔。

（13）drawBitmap 方法：用于在画布上绘制图像，通过指定 Bitmap 对象来实现。前面的各种方法都是自己绘制各图形，但我们的应用程序往往需要直接引用一些图片资源。这时即可使用 drawBitmap 方法在画布上直接显示图像。其语法格式为：

```
public void drawBitmap(Bitmap bitmap,float left,float top,Paint paint)
```

其中,参数 bitmap 为 Bitmap 对象,代表图像资源;参数 left 为图像显示的左边位置;参数 top 为图像显示的上边位置;参数 paint 为绘制时所使用的画笔。

(14) save 方法:用于锁定画布,这种方法主要用于锁定画布中的某一个或某几个对象,用于锁定对象操作的场合。使用 save 方法锁定画布并完成操作之后,需要使用 restore 方法解除锁定。其语法格式为:

> public int save()

(15) restore 方法:用于解除锁定的画布,这种方法主要用在 save 方法之后,使用 save 方法锁定画布并完成操作后,需要使用 restore 方法解除锁定。

(16) clipRect 方法:用于裁剪画布,即设置画布的显示区域。在使用时,可以使用 Rect 对象来指定裁剪区,也可以通过指定矩形的 4 条边来指定裁剪区。该方法主要用于部分显示以及对画布中的部分对象进行操作的场合。其语法格式为:

> public Boolean clipRect(Rect rect)
> public Boolean clipRect(float left,float top,float right,float bottom)
> public Boolean clipRect(int left,int top,int right,int bottom)

其中,参数 rect 为 Rect 对象,用于定义裁剪区的范围;参数 left 为矩形裁剪区的左边位置,可以为浮点型或整型;参数 top 为矩形裁剪区的上边位置,可以为浮点型或整型;参数 right 为矩形裁剪区的右边位置,可以为浮点型或整型;参数 bottom 为矩形裁剪区的下边位置,可以为浮点型或整型。

(17) rotate 方法:用于旋转画布。通过旋转画布,可以将画布上绘制的对象旋转。在使用这个方法时,将会把画布上的所有对象都旋转。为了只对某一个对象进行旋转,则可以通过 save 方法锁定画布,然后执行旋转操作,最后通过 restore 方法解锁,此后再绘制其他图形。其语法格式为:

> public void rotate(float degrees)
> public final void rotate(float degrees,float px,float py)

其中,参数 degrees 为旋转角度,正数为顺时针方向,负数为逆时针方向;参数 px 为旋转点的 X 坐标;参数 py 为旋转点的 Y 坐标。

2. Canvas 类实例

通过以下代码演示怎样在画布上绘制圆形及显示图像。

【实现步骤】

(1) 在 Eclipse 环境下,建立一个名为 E6_2_2 的工程。

(2) 打开 src/com.example.e6_2_2 目录下的 MainActivity.java 文件,代码修改为:

```
public class MainActivity extends Activity
{
    private CanvasTest canvasTest=null;           //声明自定义 View 对象
    /* 第一次调用活动 */
    @Override
    public void onCreate(Bundle savedInstanceState)
    {
        //重载 onCreate 方法
```

```
            super.onCreate(savedInstanceState);
            this.canvasTest=new CanvasTest(this);              //创建自定义 View 对象
            setContentView(canvasTest);                        //设置显示自定义 View
        }
    }
```

(3) 在 src/com.example.e6_2_2 目录下新建一个名为 CanvasTest.java 的文件，代码修改为：

```
    public class CanvasTest extends View implements Runnable
    {
        //自定义 View
        private Paint paint=null;                              //声明画笔对象
        public CanvasTest (Context context)
        {
            super(context);
            // TODO：自动存根法
            paint=new Paint();                                 //构建对象
            new Thread(this).start();                          //开启线程
        }
        protected void onDraw(Canvas canvas)
        {
            // TODO ：自动存根法
            super.onDraw(canvas);
            paint.setAntiAlias(true);                          //设置画笔为无锯齿
            paint.setColor(Color.BLACK);                       //设置画笔颜色
            canvas.drawColor(Color.WHITE);                     //白色背景
            paint.setStrokeWidth((float) 3.0);                 //线宽
            paint.setStyle(Style.STROKE);
            RectF oval=new RectF();                            //RectF 对象
            oval.left=100;                                     //左边
            oval.top=100;                                      //上边
            oval.right=400;                                    //右边
            oval.bottom=300;                                   //下边
            canvas.drawOval(oval, paint);                      //绘制椭圆
            oval.left=150;                                     //左边
            oval.top=400;                                      //上边
            oval.right=350;                                    //右边
            oval.bottom=700;                                   //下边
            canvas.drawOval(oval, paint);                      //绘制椭圆
            Bitmap bitmap=null;                                //Bitmap 对象
```

```
bitmap=((BitmapDrawable)getResources().getDrawable(R.drawable.ch)).getBitmap();
        canvas.drawBitmap(bitmap, 50, 50, null);              //绘制图像
bitmap=((BitmapDrawable)getResources().getDrawable(R.drawable.ra)).getBitmap();
        canvas.drawBitmap(bitmap, 50, 150, null);             //绘制图像
bitmap=((BitmapDrawable)getResources().getDrawable(R.drawable.bc)).getBitmap();
        canvas.drawBitmap(bitmap, 50, 450, null);             //绘制图像
    }
    @Override
    public void run()
    {                                                          //重载 run 方法
        // TODO：自动存根法
        while(!Thread.currentThread().isInterrupted())
        {
            try
            {
                Thread.sleep(100);
            }
            catch(InterruptedException e)
            {
                Thread.currentThread().interrupt();
            }
            postInvalidate();                                  //更新界面
        }
    }
}
```

运行程序，效果如图 6-7 所示。

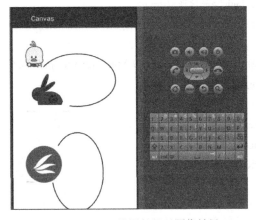

图 6-7　Canvas 绘图与显示图像效果

6.2.3 Canvas 与 Paint 类综合实例

下面的代码演示了怎样在 Android 中绘制基本的集合图形。程序的关键在于一个自定义的 View 组件，其重写了 onDraw(Canvas)方法，接下来在该 Canvas 上绘制大量几何图形。

【实现步骤】

（1）在 Eclipse 环境下，建立一个名为 E6_2_1 的工程。

（2）打开 res/layout 目录下的 main.xml 文件，代码修改为：

```xml
<?xml version="1.0" encoding="utf-8"?>
<LinearLayout xmlns:android="http://schemas.android.com/apk/res/android"
    android:orientation="vertical"
    android:layout_width="fill_parent"
    android:layout_height="fill_parent">
<com.example.e6_2_1.ViewTest
    android:layout_width="wrap_content"
    android:layout_height="wrap_content"/>
</LinearLayout>
```

（3）打开 res/values 目录下的 strings.xml 文件，代码修改为：

```xml
<resources>
    <string name="app_name">E6_2_1</string>
    <string name="hello_world">Hello world!</string>
    <string name="menu_settings">Settings</string>
    <string name="title_activity_main">Canvas 画布</string>
    <string name="circle">圆形</string>
    <string name="square">正方形</string>
    <string name="rect">长方形</string>
    <string name="round_rect">圆角矩形</string>
    <string name="oval">椭圆形</string>
    <string name="triangle">三角形</string>
    <string name="pentagon">五角形</string>
</resources>
```

（4）打开 src/com.example.e6_2_1 目录下的 MainActivity.java 文件，代码修改为：

```java
package com.example.e6_2_1;

import android.app.Activity;
import android.os.Bundle;
public class MainActivity extends Activity
{
    @Override
    public void onCreate(Bundle savedInstanceState)
    {
```

```java
        super.onCreate(savedInstanceState);
        setContentView(R.layout.main);
    }
}
```

(5) 在 src/com.example.e6_2_1 目录下新建一个名为 ViewTest.java 的文件，代码为：

```java
package com.example.e6_2_1;

import android.content.Context;
import android.graphics.Canvas;
import android.graphics.Bitmap;
import android.graphics.Color;
import android.graphics.LinearGradient;
import android.graphics.Paint;
import android.graphics.Path;
import android.graphics.RectF;
import android.graphics.Shader;
import android.graphics.drawable.BitmapDrawable;
import android.util.AttributeSet;
import android.view.View;
public class ViewTest extends View
{
    public ViewTest(Context context, AttributeSet set)
    {
        super(context, set);
    }
    @Override
    // 重写该方法，进行绘图
    protected void onDraw(Canvas canvas)
    {
        super.onDraw(canvas);
        // 把整张画布绘制成白色
        canvas.drawColor(Color.WHITE);
        Paint paint = new Paint();
        // 去锯齿
        paint.setAntiAlias(true);
        paint.setColor(Color.BLUE);
        paint.setStyle(Paint.Style.STROKE);
        paint.setStrokeWidth(3);
        // 绘制圆形
        canvas.drawCircle(40, 40, 30, paint);
```

```
// 绘制正方形
canvas.drawRect(10, 80, 70, 140, paint);
// 绘制矩形
canvas.drawRect(10, 150, 70, 190, paint);
RectF re1 = new RectF(10, 200, 70, 230);
// 绘制圆角矩形
canvas.drawRoundRect(re1, 15, 15, paint);
RectF re11 = new RectF(10, 240, 70, 270);
// 绘制椭圆
canvas.drawOval(re11, paint);
// 定义一个 Path 对象，封闭成一个三角形
Path path1 = new Path();
path1.moveTo(10, 340);
path1.lineTo(70, 340);
path1.lineTo(40, 290);
path1.close();
// 根据 Path 进行绘制，绘制三角形
canvas.drawPath(path1, paint);
// 定义一个 Path 对象，封闭成一个五角形
Path path2 = new Path();
path2.moveTo(26, 360);
path2.lineTo(54, 360);
path2.lineTo(70, 392);
path2.lineTo(40, 420);
path2.lineTo(10, 392);
path2.close();
// 根据 Path 进行绘制，绘制五角形
canvas.drawPath(path2, paint);
/*
 * 设置填充风格后绘制
 */
paint.setStyle(Paint.Style.FILL);
paint.setColor(Color.RED);
canvas.drawCircle(120, 40, 30, paint);
//绘制正方形
canvas.drawRect(90, 80, 150, 140, paint);
//绘制矩形
canvas.drawRect(90, 150, 150, 190, paint);
RectF re2 = new RectF(90, 200, 150, 230);
//绘制圆角矩形
```

```
canvas.drawRoundRect(re2, 15, 15, paint);
RectF re21 = new RectF(90, 240, 150, 270);
// 绘制椭圆
canvas.drawOval(re21, paint);
Path path3 = new Path();
path3.moveTo(90, 340);
path3.lineTo(150, 340);
path3.lineTo(120, 290);
path3.close();
//绘制三角形
canvas.drawPath(path3, paint);
Path path4 = new Path();
path4.moveTo(106, 360);
path4.lineTo(134, 360);
path4.lineTo(150, 392);
path4.lineTo(120, 420);
path4.lineTo(90, 392);
path4.close();
//绘制五角形
canvas.drawPath(path4, paint);
/*
 * 设置渐变器后绘制
 */
// 为 Paint 设置渐变器
Shader mShader = new LinearGradient(0, 0, 40, 60
    , new int[] {
        Color.RED, Color.GREEN, Color.BLUE, Color.YELLOW }
    , null , Shader.TileMode.REPEAT);
paint.setShader(mShader);
//设置阴影
paint.setShadowLayer(45 , 10 , 10 , Color.GRAY);
// 绘制圆形
canvas.drawCircle(200, 40, 30, paint);
// 绘制正方形
canvas.drawRect(170, 80, 230, 140, paint);
// 绘制矩形
canvas.drawRect(170, 150, 230, 190, paint);
RectF re3 = new RectF(170, 200, 230, 230);
// 绘制圆角矩形
canvas.drawRoundRect(re3, 15, 15, paint);
```

```
RectF re31 = new RectF(170, 240, 230, 270);
// 绘制椭圆
canvas.drawOval(re31, paint);
Path path5 = new Path();
path5.moveTo(170, 340);
path5.lineTo(230, 340);
path5.lineTo(200, 290);
path5.close();
// 根据 Path 进行绘制，绘制三角形
canvas.drawPath(path5, paint);
Path path6 = new Path();
path6.moveTo(186, 360);
path6.lineTo(214, 360);
path6.lineTo(230, 392);
path6.lineTo(200, 420);
path6.lineTo(170, 392);
path6.close();
// 根据 Path 进行绘制，绘制五角形
canvas.drawPath(path6, paint);
/*
*设置字符大小后绘制
*/
paint.setTextSize(24);
paint.setShader(null);
Bitmap bitmap=null;
// 绘制 7 个字符串
canvas.drawText(getResources().getString(R.string.circle), 240, 50, paint);
canvas.drawText(getResources().getString(R.string.square), 240, 120,paint);
canvas.drawText(getResources().getString(R.string.rect), 240, 175,paint);
canvas.drawText(getResources().getString(R.string.round_rect), 230,220, paint);
canvas.drawText(getResources().getString(R.string.oval), 240,260, paint);
canvas.drawText(getResources().getString(R.string.triangle), 240, 325,paint);
canvas.drawText(getResources().getString(R.string.pentagon), 240, 390,paint);
//显示图形
bitmap=((BitmapDrawable)getResources().getDrawable(R.drawable.ra)).getBitmap();
canvas.drawBitmap(bitmap, 50, 450, null);              //绘制图像
    }
}
```

运行程序，效果如图 6-8 所示。

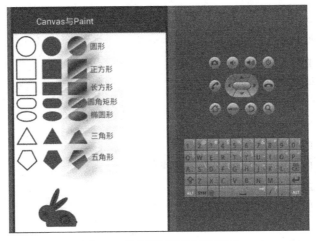

图 6-8　基本图形的绘制及显示

6.2.4　Path 类

从 6.2.3 节综合例子中的程序代码可看出，Android 提供的 Path 为一个非常有用的类，其可以预先在 View 上将 N 个点连成一条 "路径"，然后调用 Canvas 的 drawPath(path,paint)即可沿着路径绘制图形。实际上，Android 还为路径绘制提供了 PathEffect 来定义绘制效果，PathEffect 包含了如下子类（每个子类代表一种绘制效果）：

- ComposePathEffect
- CornerPathEffect
- DashPathEffect
- DiscretePathEffect
- PathDashPathEffect
- SumPathEffect

下面通过一个实例来让读者了解这些绘制效果。以下代码绘制了 7 条路径，分别演示了不使用路径效果和使用上面 6 种子类的效果。

```
public class MainActivity extends Activity
{
    @Override
    protected void onCreate(Bundle savedInstanceState)
    {
        super.onCreate(savedInstanceState);
        setContentView(new MyView(this));
    }
    class MyView extends View
    {
        float phase;
        PathEffect[] effects = new PathEffect[7];
```

```java
int[] colors;
private Paint paint;
Path path;
public MyView(Context context)
{
    super(context);
    paint = new Paint();
    paint.setStyle(Paint.Style.STROKE);
    paint.setStrokeWidth(4);
    //创建并初始化 Path
    path = new Path();
    path.moveTo(0, 0);
    for (int i = 1; i <= 15; i++)
    {
        //生成 15 个点,随机生成它们的 Y 坐标,并将它们连成一条 Path
        path.lineTo(i * 30, (float) Math.random() * 80);
    }
    //初始化 7 个颜色
    colors = new int[] {Color.BLACK, Color.BLUE, Color.CYAN
        , Color.GREEN, Color.MAGENTA, Color.RED , Color.BLUE};
}
@Override
protected void onDraw(Canvas canvas)
{
    //将背景填充成白色
    canvas.drawColor(Color.WHITE);
    /*
    * 下面开始初始化 7 种路径效果
    */
    //不使用路径效果
    effects[0] = null;
    //使用 CornerPathEffect 路径效果
    effects[1] = new CornerPathEffect(10);
    //初始化 DiscretePathEffect
    effects[2] = new DiscretePathEffect(3.0f , 5.0f);
    //初始化 DashPathEffect
    effects[3] = new DashPathEffect(new float[]
        { 20, 10, 5, 10 }, phase);
    //初始化 PathDashPathEffect
```

```java
Path p = new Path();
p.addRect(0 , 0, 8, 8, Path.Direction.CCW);
effects[4] = new PathDashPathEffect(p, 12, phase,
    PathDashPathEffect.Style.ROTATE);
//初始化 PathDashPathEffect
effects[5] = new ComposePathEffect(effects[2], effects[4]);
effects[6] = new SumPathEffect(effects[4], effects[3]);
//将画布移动到(8, 8)处开始绘制
canvas.translate(8, 8);
//依次使用7种不同路径效果、7种不同的颜色来绘制路径
for (int i = 0; i < effects.length; i++)
{
    paint.setPathEffect(effects[i]);
    paint.setColor(colors[i]);
    canvas.drawPath(path, paint);
    canvas.translate(0, 60);
}
//改变 phase 值，形成动画效果
phase += 1;
invalidate();
    }
  }
}
```

运行程序，效果如图 6-9 所示。

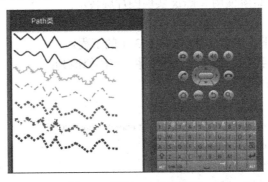

图 6-9　设置 Path 路径效果

6.3　Android 图形特效处理

在 Android 中还提供了一些额外的更高级的图形特效支持,这些图形特效支持可以让开发者开发出更"绚丽"的 UI 界面。

6.3.1 Matrix 控制变换

Matrix 的操作总共分为 translate（平移）、rotate（旋转）、scale（缩放）和 skew（倾斜）四种，每一种变换在 Android 的 API 里都提供了 set、post 和 pre 三种操作方式，除了 translate，其他三种操作都可以指定中心点。

set 是直接设置 Matrix 的值，每 set 一次，整个 Matrix 的数组都会变。

post 为后乘，即当前的矩阵乘以参数得到的矩阵。可以连续多次使用 post 来完成所需的整个变换。例如，要将一个图片旋转 30°，然后平移到(100,100)的位置，则可以这样做：

```
Matrix m = new Matrix();
m.postRotate(30);
m.postTranslate(100, 100);
```

pre 是前乘，即参数乘以当前矩阵得到的矩阵。所以操作是在当前矩阵的最前面发生的。例如上面的例子，如果用 pre 的话，即为：

```
Matrix m = new Matrix();
m.setTranslate(100, 100);
m.preRotate(30);
```

旋转、缩放和倾斜都可以围绕一个中心点来进行，如果不指定，默认情况下是围绕(0,0)点来进行。

1. 图像旋转

下面介绍在 Android 中利用 Matrix 实现图像的旋转的语法。

setRotate(float degrees)：控制 Matrix 进行旋转，参数 degrees 控制旋转的角度。

setRoate(float degrees,float px,float py)：设置以 px、py 为轴心进行旋转，参数 degrees 控制旋转的角度。

下面通过一个例子来讲解使用 Matrix 实现图片旋转。

【实现步骤】

（1）在 Eclipse 环境下，建立一个名为 E6_3_1 的工程。

（2）打开 res/layout 目录下的 main.xml 文件，代码修改为：

```xml
<?xml version="1.0" encoding="utf-8"?>
<LinearLayout
    xmlns:android="http://schemas.android.com/apk/res/android"
    android:background="@drawable/white"
    android:orientation="vertical"
    android:layout_width="fill_parent"
    android:layout_height="fill_parent">
    <TextView
        android:id="@+id/myTextView1"
        android:layout_width="fill_parent"
        android:layout_height="wrap_content"
        android:text="@string/app_name"/>
```

```xml
<LinearLayout
    android:orientation="horizontal"
    android:layout_width="wrap_content"
    android:layout_height="wrap_content">
    <Button
        android:id="@+id/myButton1"
        android:layout_width="wrap_content"
        android:layout_height="wrap_content"
        android:text="@string/str_button1" />
    <ImageView
        android:id="@+id/myImageView1"
        android:layout_width="wrap_content"
        android:layout_height="wrap_content"
        android:layout_gravity="center" />
    <Button
        android:id="@+id/myButton2"
        android:layout_width="wrap_content"
        android:layout_height="wrap_content"
        android:text="@string/str_button2" />
</LinearLayout>
</LinearLayout>
```

（3）打开 res/values 目录下的 strings.xml 文件，代码修改为：

```xml
<resources>
    <string name="app_name">E6_3_1</string>
    <string name="hello_world">Hello world!</string>
    <string name="menu_settings">Settings</string>
    <string name="title_activity_main">图片旋转</string>
    <string name="str_button1">向左旋转</string>
    <string name="str_button2">向右旋转</string>
    <string name="str_button3">倾斜</string>
</resources>
```

（4）打开 src/com.example.e6_3_1 目录下的 MainActivity.java 文件，代码修改为：

```java
public class MainActivity extends Activity
{
    private Button mButton1;
    private Button mButton2;
    private TextView mTextView1;
    private ImageView mImageView1;
    private int ScaleTimes;
    private int ScaleAngle;
    /* 第一次调用活动*/
```

```java
@Override
public void onCreate(Bundle savedInstanceState)
{
    super.onCreate(savedInstanceState);
    setContentView(R.layout.main);
    mButton1 =(Button) findViewById(R.id.myButton1);
    mButton2 =(Button) findViewById(R.id.myButton2);
    mTextView1 = (TextView) findViewById(R.id.myTextView1);
    mImageView1 = (ImageView) findViewById(R.id.myImageView1);
    ScaleTimes = 1;
    ScaleAngle = 1;
    final Bitmap mySourceBmp =
    BitmapFactory.decodeResource(getResources(), R.drawable.ch);
    final int widthOrig = mySourceBmp.getWidth();
    final int heightOrig = mySourceBmp.getHeight();
    //程序刚运行,加载默认的 Drawable
    mImageView1.setImageBitmap(mySourceBmp);
    //向左旋转按钮
    mButton1.setOnClickListener(new Button.OnClickListener()
    {
        @Override
        public void onClick(View v)
        {
            // TODO: 自动存根法
            ScaleAngle--;
            if(ScaleAngle<-5)
            {
                ScaleAngle = -5;
            }
            //ScaleTimes=1,维持 1:1 的宽高比例
            int newWidth = widthOrig * ScaleTimes;
            int newHeight = heightOrig * ScaleTimes;
            float scaleWidth = ((float) newWidth) / widthOrig;
            float scaleHeight = ((float) newHeight) / heightOrig;
            Matrix matrix = new Matrix();
            // 使用 Matrix.postScale 设置维度
            matrix.postScale(scaleWidth, scaleHeight);
            //使用 Matrix.postRotate 方法旋转 Bitmap
            //matrix.postRotate(5*ScaleAngle);
            matrix.setRotate(5*ScaleAngle);
            // 创建新的 Bitmap 对象
```

```java
                        Bitmap resizedBitmap =
                        Bitmap.createBitmap
                        (mySourceBmp, 0, 0, widthOrig, heightOrig, matrix, true);
                        BitmapDrawable myNewBitmapDrawable = new BitmapDrawable(resizedBitmap);
                        mImageView1.setImageDrawable(myNewBitmapDrawable);
                        mTextView1.setText(Integer.toString(5*ScaleAngle));
                }
        });
        /* 向右旋转按钮 */
        mButton2.setOnClickListener(new Button.OnClickListener()
        {
                @Override
                public void onClick(View v)
                {
                        // TODO Auto-generated method stub
                        ScaleAngle++;
                        if(ScaleAngle>5)
                        {
                                ScaleAngle = 5;
                        }
                        /* ScaleTimes=1,维持 1:1 的宽高比例*/
                        int newWidth = widthOrig * ScaleTimes;
                        int newHeight = heightOrig * ScaleTimes;
                        /* 计算旋转的 Matrix 比例 */
                        float scaleWidth = ((float) newWidth) / widthOrig;
                        float scaleHeight = ((float) newHeight) / heightOrig;
                        Matrix matrix = new Matrix();
                        /* 使用 Matrix.postScale 设置维度 */
                        matrix.postScale(scaleWidth, scaleHeight);
                        /* 使用 Matrix.postRotate 方法旋转 Bitmap*/
                        matrix.setRotate(5*ScaleAngle);
                        /* 创建新的 Bitmap 对象 */
                        Bitmap resizedBitmap =
                        Bitmap.createBitmap
                        (mySourceBmp, 0, 0, widthOrig, heightOrig, matrix, true);
                        BitmapDrawable myNewBitmapDrawable =new BitmapDrawable(resizedBitmap);
                        mImageView1.setImageDrawable(myNewBitmapDrawable);
                        mTextView1.setText(Integer.toString(5*ScaleAngle));
                }
        });
    }
}
```

运行程序，效果如图 6-10 所示。

图 6-10　图像的旋转处理

2. 图像缩放

下面介绍在 Android 中利用 Matrix 实现图像缩放的语法。

setScale(float sx,float sy)：设置 Matrix 进行缩放，sx、sy 控制 X、Y 方向上的缩放比例。

setScale(float sx,float sy,float px,float py)：设置 Matrix 以 px、py 为轴心进行缩放，sx、sy 控制 X、Y 方向上的缩放比例。

下面通过一个例子来讲解如何使用 Matrix 实现图片缩放。

【实现步骤】

（1）在 Eclipse 环境下，建立一个名为 E6_3_2 的工程。

（2）打开 res/layout 目录下的 main.xml 文件，代码修改为：

```xml
<?xml version="1.0" encoding="utf-8"?>
<AbsoluteLayout
    android:id="@+id/layout1"
    android:layout_width="fill_parent"
    android:layout_height="fill_parent"
    xmlns:android="http://schemas.android.com/apk/res/android">
    <ImageView
        android:id="@+id/myImageView"
        android:layout_width="200px"
        android:layout_height="150px"
        android:src="@drawable/ch"
        android:layout_x="0px"
        android:layout_y="0px"/>
    <Button
        android:id="@+id/myButton1"
        android:layout_width="90px"
        android:layout_height="60px"
        android:text="@string/str_button1"
        android:textSize="18sp"
        android:layout_x="20px"
        android:layout_y="372px"/>
    <Button
```

```
                    android:id="@+id/myButton2"
                    android:layout_width="90px"
                    android:layout_height="60px"
                    android:text="@string/str_button2"
                    android:textSize="18sp"
                    android:layout_x="210px"
                    android:layout_y="372px" />
</AbsoluteLayout>
```
(3) 打开 src/com.example.e6_3_2 目录下的 MainActivity.java 文件,代码修改为:
```
public class MainActivity extends Activity
{
    /* 相关变量声明 */
    private ImageView mImageView;
    private Button mButton01;
    private Button mButton02;
    private AbsoluteLayout layout1;
    private Bitmap bmp;
    private int id=0;
    private int displayWidth;
    private int displayHeight;
    private float scaleWidth=1;
    private float scaleHeight=1;
    //第一次调用活动
    @Override
    public void onCreate(Bundle savedInstanceState)
    {
        super.onCreate(savedInstanceState);
        //载入 main.xml Layout
        setContentView(R.layout.main);
        // 取得屏幕分辨率大小
        DisplayMetrics dm=new DisplayMetrics();
        getWindowManager().getDefaultDisplay().getMetrics(dm);
        displayWidth=dm.widthPixels;
        //屏幕高度须扣除下方 Button 的高度
        displayHeight=dm.heightPixels-80;
        //初始化相关变量
        bmp=BitmapFactory.decodeResource(getResources(),R.drawable.ch);
        mImageView = (ImageView)findViewById(R.id.myImageView);
        layout1 = (AbsoluteLayout)findViewById(R.id.layout1);
        mButton01 = (Button)findViewById(R.id.myButton1);
        mButton02 = (Button)findViewById(R.id.myButton2);
```

```java
// 缩小按钮 onClickListener
mButton01.setOnClickListener(new Button.OnClickListener()
{
    @Override
    public void onClick(View v)
    {
        small();
    }
});
// 放大按钮 onClickListener
mButton02.setOnClickListener(new Button.OnClickListener()
{
    @Override
    public void onClick(View v)
    {
        big();
    }
});
}
// 图片缩小的 method
private void small()
{
    int bmpWidth=bmp.getWidth();
    int bmpHeight=bmp.getHeight();
    // 设置图片缩小的比例
    double scale=0.8;
    // 计算出这次要缩小的比例
    scaleWidth=(float) (scaleWidth*scale);
    scaleHeight=(float) (scaleHeight*scale);
    // 产生 reSize 后的 Bitmap 对象
    Matrix matrix = new Matrix();
    matrix.postScale(scaleWidth, scaleHeight);
    Bitmap resizeBmp = Bitmap.createBitmap(bmp,0,0,bmpWidth, ,matrix,true);
    if(id==0)
    {
        //如果是第一次按，就删除原来默认的 ImageView
        layout1.removeView(mImageView);
    }
    else
    {
        // 如果不是第一次按，就删除上次放大缩小所产生的 ImageView
```

```java
            layout1.removeView((ImageView)findViewById(id));
        }
        //产生新的 ImageView，放入 reSize 的 Bitmap 对象，再放入 Layout 中
        id++;
        ImageView imageView = new ImageView(MainActivity.this);
        imageView.setId(id);
        imageView.setImageBitmap(resizeBmp);
        layout1.addView(imageView);
        setContentView(layout1);
        // 因为图片放到最大时放大按钮会 disable，所以在缩小时把它重设为 enable mButton02.
        // setEnabled(true);
    }
    // 图片放大的 method
    private void big()
    {
        int bmpWidth=bmp.getWidth();
        int bmpHeight=bmp.getHeight();
        //设置图片放大的比例
        double scale=1.25;
        // 计算这次要放大的比例
        scaleWidth=(float)(scaleWidth*scale);
        scaleHeight=(float)(scaleHeight*scale);
        // 产生 reSize 后的 Bitmap 对象
        Matrix matrix = new Matrix();
        matrix.postScale(scaleWidth, scaleHeight);
        Bitmap resizeBmp = Bitmap.createBitmap(bmp,0,0,bmpWidth, bmpHeight,matrix,true);

        if(id==0)
        {
            //如果是第一次按，就删除原来设置的 ImageView
            layout1.removeView(mImageView);
        }
        else
        {
            //如果不是第一次按，就删除上次放大/缩小所产生的 ImageView
            layout1.removeView((ImageView)findViewById(id));
        }
        // 产生新的 ImageView，放入 reSize 的 Bitmap 对象，再放入 Layout 中
        id++;
        ImageView imageView = new ImageView(MainActivity.this);
        imageView.setId(id);
```

```
imageView.setImageBitmap(resizeBmp);
layout1.addView(imageView);
setContentView(layout1);
//如果再放大会超过屏幕大小，就把 Button disable
if(scaleWidth*scale*bmpWidth>displayWidth||
    scaleHeight*scale*bmpHeight>displayHeight)
{
    mButton02.setEnabled(false);
}
}
}
```

运行程序，效果如图 6-11 所示。

图 6-11　图像的缩放效果

6.3.2　drawBitmapMesh 扭曲图像

Canvas 的 drawBitmapMesh 定义如下：

```
    public void drawBitmapMesh(Bitmap bitmap, int meshWidth, int meshHeight, float[] verts, int vertOffset,
int[] colors, int colorOffset, Paint paint)
```

drawBitmapMesh 方法用于表示将图像绘制在网格上，简言之，可以将画板想象成一张格子布，在这张布上绘制图像。对于一个网格端点均匀分布的网格来说，横向有 meshWidth+1 个顶点，纵向有 meshHeight+1 个端点。顶点数组 verts 是以行优先的数组（二维数组以一维数组表示，先行后列）。网格可以不均匀分布，用到的参数主要有以下几种。

- Bitmap：需要绘制在网格上的图像。
- meshWidth：网格的宽度方向的数目（列数），为 0 时不绘制图像。
- meshHeight：网格的高度方向的数目（行数），为 0 时不绘制图像。
- verts：为(x,y)对的数组，表示网格顶点的坐标，至少需要有(meshWidth+1) × (meshHeight+1) × 2 + meshOffset 个(x,y)坐标。
- vertOffset：用于控制 verts 数组中开始跳过的(x,y)对的数目。
- colors：可以为空，不为空为每个顶点定义对应的颜色值，至少需要有(meshWidth+1) ×

(meshHeight+1) × 2 + meshOffset 个(x,y)坐标。
- colorOffset：colors 数组中开始跳过的(x,y)对的数目。
- paint：可以为空。

值得注意的是，当程序希望调用 drawBitmapMesh 方法对位图进行扭曲时，关键是计算 verts 数组的值——该数组的值记录了扭曲后的位图上各"顶点"的坐标。

下面的示例将通过 drawBitmapMesh 方法来控制图片的扭曲，当用户"触摸"图片的指定点时，该图片会在这个点被用户"按"下去——就像将这张图片铺在"极软的床上"一样。

为了实现这个效果，代码要在用户触摸图片的指定点时动态地改变 verts 数组里每个元素的位置（控制扭曲后每个顶点的坐标）——这种改变也简单：程序计算图片上每个顶点与触摸点的距离，顶点与触摸点的距离越大，该顶点向触摸点移动的距离越大。

【实现步骤】

（1）在 Eclipse 环境下，建立一个名为 E6_3_3 的工程。

（2）打开 res/layout 目录下的 main.xml 文件，代码修改为：

```xml
<?xml version="1.0" encoding="utf-8"?>
<LinearLayout xmlns:android="http://schemas.android.com/apk/res/android"
    android:orientation="vertical"
    android:layout_width="fill_parent"
    android:layout_height="fill_parent">
<TextView
    android:layout_width="fill_parent"
    android:layout_height="wrap_content"
    android:text="@string/hello_world"/>
</LinearLayout>
```

（3）打开 src/com.example.e6_3_3 目录下的 MainActivity.java 文件，代码修改为：

```java
public class MainActivity extends Activity
{
    private Bitmap bitmap;
    @Override
    public void onCreate(Bundle savedInstanceState)
    {
        super.onCreate(savedInstanceState);
        setContentView(new MyView(this , R.drawable.ch));
    }
    private class MyView extends View
    {
        //定义两个常量，这两个常量指定该图片横向、纵向上都被划分为 50 格
        private final int WIDTH = 50;
        private final int HEIGHT = 50;
        //记录该图片上包含 441 个顶点
        private final int COUNT = (WIDTH + 1) * (HEIGHT + 1);
```

```java
//定义一个数组,保存 Bitmap 上的 21×21 个点的坐标
private final float[] verts = new float[COUNT * 2];
//定义一个数组,记录 Bitmap 上的 21×21 个点经过扭曲后的坐标
//对图片进行扭曲的关键就是修改该数组里元素的值
private final float[] orig = new float[COUNT * 2];
public MyView(Context context , int drawableId)
{
    super(context);
    setFocusable(true);
    //根据指定资源加载图片
    bitmap = BitmapFactory.decodeResource(getResources(),
        drawableId);
    //获取图片宽度、高度
    float bitmapWidth = bitmap.getWidth();
    float bitmapHeight = bitmap.getHeight();
    int index = 0;
    for (int y = 0; y <= HEIGHT; y++)
    {
        float fy = bitmapHeight * y / HEIGHT;
        for (int x = 0; x <= WIDTH; x++)
        {
            float fx = bitmapWidth * x / WIDTH;
            /*
             * 初始化 orig、verts 数组
             * 初始化后,orig、verts 两个数组均匀地保存了 21×21 个点的 x,y 坐标
             */
            orig[index * 2 + 0] = verts[index * 2 + 0] = fx;
            orig[index * 2 + 1] = verts[index * 2 + 1] = fy;
            index += 1;
        }
    }
    //设置背景色
    setBackgroundColor(Color.WHITE);
}
@Override
protected void onDraw(Canvas canvas)
{
    /* 对 bitmap 按 verts 数组进行扭曲
     * 从第一个点(由第 5 个参数 0 控制)开始扭曲
     */
    canvas.drawBitmapMesh(bitmap, WIDTH, HEIGHT, verts 0, null, 0, null);
```

```
}
//工具方法，用于根据触摸事件的位置计算 verts 数组里各元素的值
private void warp(float cx, float cy)
{
    for (int i = 0; i < COUNT * 2; i += 2)
    {
        float dx = cx - orig[i + 0];
        float dy = cy - orig[i + 1];
        float dd = dx * dx + dy * dy;
        //计算每个坐标点与当前点（cx、cy）之间的距离
        float d = (float)Math.sqrt(dd);
        //计算扭曲度，距离当前点（cx、cy）越远，扭曲度越小
        float pull = 80000 / ((float) (dd * d));
        //对 verts 数组（保存 bitmap 上 21×21 个点经过扭曲后的坐标）重新赋值
        if (pull >= 1)
        {
            verts[i + 0] = cx;
            verts[i + 1] = cy;
        }
        else
        {
            //控制各顶点向触摸事件发生点偏移
            verts[i + 0] = orig[i + 0] + dx * pull;
            verts[i + 1] = orig[i + 1] + dy * pull;
        }
    }
    //通知 View 组件重绘
    invalidate();
}
@Override
public boolean onTouchEvent(MotionEvent event)
{
    //调用 warp 方法，根据触摸屏事件的坐标点来扭曲 verts 数组
    warp(event.getX(), event.getY());
    return true;
}
}
```

运行程序，单击界面中显示的图片，效果如图 6-12 所示。

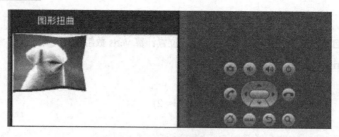

图 6-12 图像扭曲效果

6.3.3 渲染效果

Android 中提供了 Shader 类专门用来渲染图像以及一些几何图形，Shader 下面包括几个直接子类，分别是 BitmapShader、ComposeShader、LinearGradient、RadialGradient、SweepGradient。BitmapShader 主要用来渲染图像，LinearGradient 用来进行梯度渲染，RadialGradient 用来进行环形渲染，SweepGradient 用来进行梯度渲染，ComposeShader 则是一个混合渲染，可以和其他几个子类组合起来使用。

使用 Shader 类时，需要先构建 Shader 对象，然后通过 Paint 的 setShader 方法设置渲染对象，最后在绘制时使用这个 Paint 对象即可。当然，用不同的渲染时需要构建不同的对象。

【实现步骤】

（1）在 Eclipse 环境下，建立一个名为 E6_3_4 的工程。

（2）打开 src/com.example.e6_3_4 目录下的 MainActivity.java 文件，代码修改为：

```java
public class MainActivity extends Activity
{
    private GameView mGameView = null;
    @Override
    public void onCreate(Bundle savedInstanceState)
    {
        super.onCreate(savedInstanceState);
        mGameView = new GameView(this);
        setContentView(mGameView);
    }
    public boolean onKeyUp(int keyCode, KeyEvent event)
    {
        super.onKeyUp(keyCode, event);
        return true;
    }
    public boolean onKeyDown(int keyCode, KeyEvent event)
    {
        if (mGameView == null)
        {
            return false;
```

```
            }
            if (keyCode == KeyEvent.KEYCODE_BACK)
            {
                this.finish();
                return true;
            }
            return mGameView.onKeyDown(keyCode, event);
        }
    }
```

(3) 在 src/com.example.e6_3_4 目录下新建一个名为 GameView.java 的文件，代码修改为：

```
package com.example.e6_3_4;
import android.content.Context;
import android.graphics.Bitmap;
import android.graphics.BitmapShader;
import android.graphics.Canvas;
import android.graphics.Color;
import android.graphics.ComposeShader;
import android.graphics.LinearGradient;
import android.graphics.Paint;
import android.graphics.PorterDuff;
import android.graphics.RadialGradient;
import android.graphics.Shader;
import android.graphics.SweepGradient;
import android.graphics.drawable.BitmapDrawable;
import android.graphics.drawable.ShapeDrawable;
import android.graphics.drawable.shapes.OvalShape;
import android.view.KeyEvent;
import android.view.MotionEvent;
import android.view.View;
public class GameView extends View implements Runnable
{
    /* 声明 Bitmap 对象 */
    Bitmap    mBitQQ    = null;
    int       BitQQwidth  = 0;
    int       BitQQheight = 0;
    Paint     mPaint = null;
    /* Bitmap 渲染 */
    Shader mBitmapShader = null;
    /* 线性渐变渲染 */
    Shader mLinearGradient = null;
```

```java
/* 混合渲染 */
Shader mComposeShader = null;
/* 唤醒渐变渲染 */
Shader mRadialGradient = null;
/* 梯度渲染 */
Shader mSweepGradient = null;
ShapeDrawable mShapeDrawableQQ = null;
public GameView(Context context)
{
    super(context);
    /* 装载资源 */
    mBitQQ = ((BitmapDrawable) getResources().getDrawable(R.drawable.ch)).getBitmap();
    /* 得到图片的宽度和高度 */
    BitQQwidth = mBitQQ.getWidth();
    BitQQheight = mBitQQ.getHeight();
    /* 创建 BitmapShader 对象 */
    mBitmapShader = new BitmapShader(mBitQQ,Shader.TileMode.REPEAT,Shader.TileMode.MIRROR);
    /* 创建 LinearGradient 并设置渐变的颜色数组，说明一下这几个参数
     * 第一个  起始的 x 坐标
     * 第二个  起始的 y 坐标
     * 第三个  结束的 x 坐标
     * 第四个  结束的 y 坐标
     * 第五个  颜色数组
     * 第六个  这个也是一个数组，用来指定颜色数组的相对位置，如果为 null 就
     * 沿坡度线均匀分布
     * 第七个  渲染模式
     **/
    mLinearGradient = new LinearGradient(0,0,100,100,  new int[]{Color.RED,
                Color.GREEN,Color.BLUE,Color.WHITE},null,Shader.TileMode.REPEAT);
    /* 这里理解为混合渲染*/
    mComposeShader = new ComposeShader(mBitmapShader,mLinearGradient, PorterDuff.Mode.DARKEN);
    /* 构建 RadialGradient 对象，设置半径的属性 */
    //这里使用了 BitmapShader 和 LinearGradient 进行混合
    //当然也可以使用其他的组合
    //混合渲染的模式很多，可以根据自己的需要来选择
    mRadialGradient = new RadialGradient(50,200,50, new int[]{Color.GREEN,Color.RED, Color.BLUE,Color.WHITE},null,Shader.TileMode.REPEAT);
    /* 构建 SweepGradient 对象 */
```

```java
            mSweepGradient = new SweepGradient(30,30,newint[]{Color.GREEN,Color.RED, Color.BLUE, Color.WHITE},null);
            mPaint = new Paint();
            /* 开启线程 */
            new Thread(this).start();
        }
        public void onDraw(Canvas canvas)
        {
            super.onDraw(canvas);
            //将图片裁剪为椭圆形
            /* 构建 ShapeDrawable 对象并定义形状为椭圆 */
            mShapeDrawableQQ = new ShapeDrawable(new OvalShape());
            /* 设置要绘制的椭圆形状为 ShapeDrawable 图片 */
            mShapeDrawableQQ.getPaint().setShader(mBitmapShader);
            /* 设置显示区域 */
            mShapeDrawableQQ.setBounds(0,0, BitQQwidth, BitQQheight);
            /* 绘制 ShapeDrawableQQ */
            mShapeDrawableQQ.draw(canvas);
            //绘制渐变的矩形
            mPaint.setShader(mLinearGradient);
            canvas.drawRect(BitQQwidth, 0, 320, 156, mPaint);
            //显示混合渲染效果
            mPaint.setShader(mComposeShader);
            canvas.drawRect(0, 300, BitQQwidth, 300+BitQQheight, mPaint);
            //绘制环形渐变
            mPaint.setShader(mRadialGradient);
            canvas.drawCircle(50, 200, 50, mPaint);
            //绘制梯度渐变
            mPaint.setShader(mSweepGradient);
            canvas.drawRect(150, 160, 300, 300, mPaint);
        }
        // 触笔事件
        public boolean onTouchEvent(MotionEvent event)
        {
            return true;
        }
        // 按键按下事件
        public boolean onKeyDown(int keyCode, KeyEvent event)
        {
            return true;
```

```java
}
// 按键弹起事件
public boolean onKeyUp(int keyCode, KeyEvent event)
{
    return false;
}
public boolean onKeyMultiple(int keyCode, int repeatCount, KeyEvent event)
{
    return true;
}
//线程处理
public void run()
{
    while (!Thread.currentThread().isInterrupted())
    {
        try
        {
            Thread.sleep(100);
        }
        catch (InterruptedException e)
        {
            Thread.currentThread().interrupt();
        }
        //使用 postInvalidate 可以直接在线程中更新界面
        postInvalidate();
    }
}
```

运行程序，效果如图 6-13 所示。

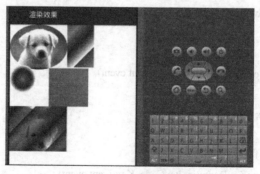

图 6-13　图像的渲染效果

6.4 Android 动画

Android 系统提供了两种实现动画的方式，一种为补间动画（Tween Animation），另一种为帧动画（Frame Animation）。

6.4.1 Animation 类

Animation 类为 Android 系统的一个动画抽象类，所有其他一些动画类都要继承该类中的实现方法。Animation 类主要用于补间动画效果，提供了动画启动、停止、重复、持续时间等方法。下面将介绍 Animation 类的常用方法。

（1）setDuration 方法：用于设置动画的持续时间，以毫秒为单位。其语法格式为：

    public void setDuration(long durationMillis)

其中，参数 durationMillis 为动画的持续时间，单位为毫秒（ms）。

（2）startNow 方法：用于启动执行一个动画。该方法是启动执行动画的主要方法，使用时需要先通过 setAnimation 方法为某一个 View 对象设置动画。另外，用户在程序中也可以使用 View 组件的 setAnimation 方法来启动执行动画。其语法格式为：

    public void startNow()

（3）start 方法：用于启动执行一个动画。该方法为启动执行动画的另一个主要方法，使用时需要先通过 setAnimation 方法为某一个 View 对象设置动画。Start 方法与 startNow 方法的区别在于，start 方法可以用于在 getTransformation 方法被调用时启动动画。其语法格式为：

    public void start()

（4）cancel 方法：用于取消一个动画的执行。该方法是取得一个正在执行中的动画的主要方法。cancel 方法和 startNow 方法相结合可实现对动画执行过程的控制。需要注意的是，通过 cancel 方法取消的动画，必须使用 reset 方法或者 setAnimation 方法重新设置，才可以再次实现动画执行过程。其语法格式为：

    public void cancel()

（5）setRepeatCount 方法：用于设置一个动画效果重复执行的次数。Android 系统默认每个动画仅执行一次，通过该方法可以设置动画执行多次。其语法格式为：

    public void setRep eatCount()

（6）setFileEnable 方法：用于使能填充效果。当该方法设置为 true 时，将执行 setFillBefore 和 setFillAfter 方法，否则将忽略 setFillBefore 和 setFillAfter 方法。其语法格式为：

    public void setFillEnabled(Boolean fillEnabled)

（7）setFillBefore 方法：用于设置一个动画效果执行完毕后，View 对象返回到起始的位置。该方法的效果是系统默认的效果。该方法的执行需要首先通过 setFillEnabled 方法使能填充效果，否则设置无效。其语法格式为：

    public void setFillBefore(Boolean fillBefore)

其中，参数 fillBefore 为是否执行起始填充效果，当其取值为 true 时表示使能该效果，当其取值为 false 时表示禁用该效果。

（8）setFillAfter 方法：用于设置一个动画效果执行完毕后，View 对象保留在终止的位置。

该方法的执行需要首先通过 setFillEnabled 方法使能填充效果,否则设置无效。其语法格式为:

      public void setFillAfter(Boolean fillAfter)

其中,参数 fillAfter 表示是否执行终止填充效果,当其取值为 true 时表示使能该效果,当取值为 false 时表示禁用该效果。

  (9)setRepeatMode 方法:用于设置一个动画效果执行的重复模式。Android 系统中提供了几种重复模式,其中最主要的即为 RESTART 模式和 REVERSE 模式。其语法格式为:

      public void setRepeatMode(int repeatMode)

其中,参数 repeatMode 为动画效果的重复模式,其取值有以下两种。

- RESTART:重新从头开始执行。
- REVERSE:反方向执行。

  (10)setStartOffset 方法:用于设置一个动画执行的启动时间,单位为毫秒。系统默认当执行 start 方法后立即执行动画,当使用该方法设置后,将延迟一定的时间启动动画。其语法格式为:

      public void setStartOffset(long startOffset)

其中,参数 startOffset 为动画的启动时间,单位为毫秒(ms)。

6.4.2 Tween 动画

Tween(补间)动画有四个主要的实现。

(1)AlphaAnimation:渐变动画,主要控制透明度变化动画类,其语法格式为 AlphaAnimation (float fromAlpha, float toAlpha);其中,

- fromAlpha:动画开始时的透明度(取值范围为 0.0~1.0)。
- toAlpha:动画结束时的透明度。

(2)ScaleAnimation:主要控制尺度变化的动画类,其语法格式为 ScaleAnimation(float fromX, float toX, float fromY, float toY, int pivotXType, float pivotXValue, int pivotYType, float pivotYValue);其中,

- fromX:动画开始 X 坐标上的伸缩尺度。
- toX:动画结束 X 坐标上的伸缩尺度。
- fromY:动画开始 Y 坐标上的伸缩尺度。
- toY:动画结束 Y 坐标上的伸缩尺度。
- pivotXType:X 坐标上的伸缩模式,取值有 Animation.ABSOLUTE、Animation.RELATIVE_TO_SELF、Animation.RELATIVE_TO_PARENT。
- pivotXValue:X 坐标上的伸缩值。
- pivotYType:Y 坐标上的伸缩模式,取值有 Animation.ABSOLUTE、Animation.RELATIVE_TO_SELF、Animation.RELATIVE_TO_PARENT。

pivotYValue:Y 坐标上的伸缩值。

(3)TranslateAnimation:主要控制位置变换的动画实现类,其语法格式为 TranslateAnimation (float fromXDelta, float toXDelta, float fromYDelta, float toYDelta);其中,

- fromXDelta:动画开始的 X 坐标。
- toXDelta:动画结束的 X 坐标。

- fromYDelta：动画开始的 Y 坐标。
- toYDelta：动画结束的 Y 坐标。

（4）RotateAnimation：主要控制旋转的动画实现类，其语法格式为 RotateAnimation(float fromDegrees, float toDegrees, int pivotXType, float pivotXValue, int pivotYType, float pivotYValue)；其中，

- fromDegrees：旋转开始角度。
- toDegrees：旋转结束角度。
- pivotXType, pivotXValue, pivotYType, pivotYValue 与尺度变化动画 ScaleAnimation 类似。

下面通过一个实例来实现补间动画。

【实现步骤】

（1）在 Eclipse 环境下，建立一个名为 E6_4_2 的工程。

（2）打开 res/layout 目录下的 main.xml 文件，代码修改为：

```
<!-- 动画 MM -->
<ImageView
    apk:id="@+id/TweenMM"
    apk:src="@drawable/hai"
    apk:layout_width="wrap_content"
    apk:layout_height="wrap_content"/>
<!-- 动画控制按钮 -->
<LinearLayout
    apk:layout_weight="1"
    apk:orientation="horizontal"
    apk:layout_width="fill_parent"
    apk:layout_height="wrap_content">
    <Button
        apk:text="改变大小"
        apk:layout_weight="1"
        apk:layout_width="fill_parent"
        apk:layout_height="wrap_content"
        apk:onClick="onBtnScaleAnimClick"/>
    <Button
        apk:text="淡入淡出"
        apk:layout_weight="1"
        apk:layout_width="fill_parent"
        apk:layout_height="wrap_content"
        apk:onClick="onBtnAlphaAnimClick"/>
</LinearLayout>
<LinearLayout
    apk:layout_weight="1"
    apk:orientation="horizontal"
```

```xml
        apk:layout_width="fill_parent"
        apk:layout_height="wrap_content">
    <Button
        apk:text="位置移动"
        apk:layout_weight="1"
        apk:layout_width="wrap_content"
        apk:layout_height="wrap_content"
        apk:onClick="onBtnTranslateAnimClick"/>
    <Button apk:text="旋转"
        apk:layout_weight="1"
        apk:layout_width="wrap_content"
        apk:layout_height="wrap_content"
        apk:onClick="onBtnRotateAnimClick"/>
    </LinearLayout>
</LinearLayout>
```

（3）在 res/layout 目录下，分别建立名为 main_rotate.xml、main_scale.xml、main_translate.xml、main_alpha.xml 的文件，代码分别为：

```xml
//main_alpha.xml 代码为：
<?xml version="1.0" encoding="utf-8"?>
<set xmlns:android="http://schemas.android.com/apk/res/android">
    <alpha
        android:fromAlpha="0.1"
        android:toAlpha="1.0"
        android:duration="2000"  />
</set>

//main_scale.xml 代码为：
<?xml version="1.0" encoding="utf-8"?>
<set xmlns:android="http://schemas.android.com/apk/res/android">
    <scale
        android:interpolator="@android:anim/accelerate_decelerate_interpolator"
        android:fromXScale="0.0"
        android:toXScale="1.0"
        android:fromYScale="0.0"
        android:toYScale="1.0"
        android:pivotX="80%"
        android:pivotY="50"
        android:fillAfter="true"
        android:duration="1000" />
</set>
//main_translate.xml 代码为：
```

```xml
<?xml version="1.0" encoding="utf-8"?>
<set xmlns:android="http://schemas.android.com/apk/res/android">
    <translate
        android:fromXDelta = "10"
        android:toXDelta = "100"
        android:fromYDelta = "10"
        android:toYDelta = "100"
        android:duration = "1000" />
</set>
```

//main_rotate.xml 代码为:

```xml
<?xml version="1.0" encoding="utf-8"?>
<set xmlns:android="http://schemas.android.com/apk/res/android">
    <rotate
        android:interpolator="@android:anim/accelerate_decelerate_interpolator"
        android:fromDegrees="0"
        android:toDegrees="360"
        android:pivotX="0"
        android:pivotY="50%"
        android:duration="1000" />
</set>
```

（4）打开 src/com.example.e6_4_2 目录下的 MainActivity.java 文件，代码修改为：

```java
package com.example.e6_4_2;

import android.app.Activity;
import android.os.Bundle;
import android.view.View;
import android.view.animation.Animation;
import android.view.animation.AnimationUtils;
import android.widget.ImageView;
import com.example.e6_4_2.R;
/*
 * 通过 XML 配置文件的方式实现 Tween 动画
 */
public class MainActivity extends Activity
{
    public static final String TAG = "TweenActivity";
    // 动画图片
    private ImageView    tweenMM;
    /**
     * @see android.app.Activity#onCreate(android.os.Bundle)
```

```java
        */
        public void onCreate(Bundle cycle)
        {
            super.onCreate(cycle);
            super.setContentView(R.layout.main);
            // 取得动画图片
            this.tweenMM = (ImageView) super.findViewById(R.id.TweenMM);
        }
        /**
         * 按钮：尺寸变化动画
         */
        public void onBtnScaleAnimClick(View view)
        {
            // 动画开始
            this.doStartAnimation(R.layout.main_scale);
        }
        // 按钮：渐变动画
        public void onBtnAlphaAnimClick(View view)
        {
            // 动画开始
            this.doStartAnimation(R.layout.main_alpha);
        }
        // 按钮：位置变化动画
        public void onBtnTranslateAnimClick(View view)
        {
            // 动画开始
            this.doStartAnimation(R.layout.main_translate);
        }
        // 按钮：旋转动画
        public void onBtnRotateAnimClick(View view)
        {
            // 动画开始
            this.doStartAnimation(R.layout.main_rotate);
        }
        // 开始动画
        private void doStartAnimation(int animId)
        {
            // 加载动画
            Animation animation = AnimationUtils.loadAnimation(this, animId);
            // 动画开始
            this.tweenMM.startAnimation(animation);
```

		}
	}

运行程序，效果如图 6-14 所示。

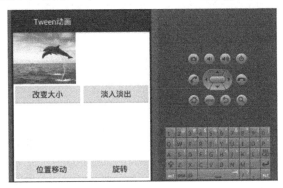

图 6-14 Tween 动画效果

6.4.3 Frame 动画

Frame（帧）动画主要是通过 AnimationDrawable 类来实现的，它有 start()和 stop()两个重要的方法，分别用来启动和停止动画。Frame 动画一般通过 XML 文件配置，在工程的 res/anim 目录下创建一个 XML 配置文件，该配置文件有一个<animation-list>根元素和若干个<item>子元素。定义帧动画的语法格式为：

```
[html] view plaincopyprint?<?xml version="1.0" encoding="utf-8"?>
<animation-list xmlns:android="http://schemas.android.com/apk/res/android"
    android:oneshot=["true" | "false"] >
    <item
        android:drawable="@[package:]drawable/drawable_resource_name"
        android:duration="integer" />
</animation-list>
```

值得注意的是：

<animation-list>元素是必需的，并且必须要作为根元素，可以包含一个或多个<item>元素；android:onshot 如果定义为 true 的话，此动画只会执行一次，如果为 false 则一直循环。

<item>元素代表一帧动画，android:drawable 指定此帧动画所对应的图片资源，android:druation 代表此帧持续的时间，整数，单位为毫秒。

下面实例实现一个卡通"滚动"的帧（Frame）动画。

【实现步骤】

（1）在 Eclipse 环境下，建立一个名为 E6_4_3 的工程。

（2）把 j1、j2、j3、j4、j5 这 5 张图片放到 res/drawable 目录下。

（3）在 res 根目录下新建一个 anim 子目录文件，实现操作为：选择 res 并右击，在弹出的快捷菜单中选择"新建"选项下的"文件夹"命令，弹出如图 6-15 所示的窗口，在"文件夹"右侧文本框中输入"anim"，单击"完成"按钮即可。

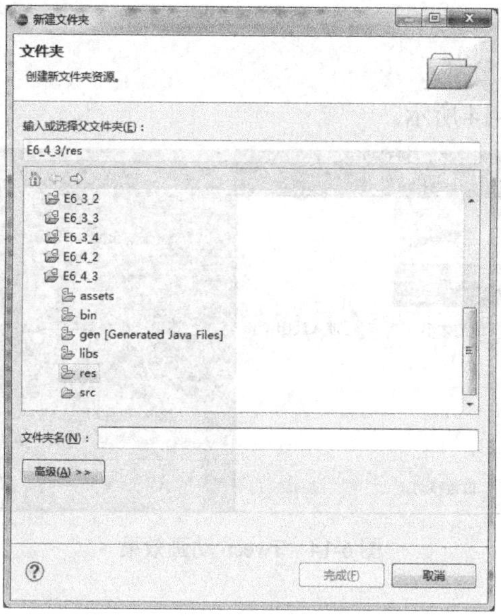

图 6-15　创建 anim 目录

（4）在 res/anim 目录下创建一个 face.xml 文件，其代码为：

```xml
<?xml version="1.0" encoding="utf-8"?>
<animation-list
xmlns:apk="http://schemas.android.com/apk/res/android"
apk:oneshot="false">
    <item apk:drawable="@drawable/j1"
    apk:duration="500" />
    <item apk:drawable="@drawable/j2"
    apk:duration="500" />
    <item apk:drawable="@drawable/j3"
    apk:duration="500" />
    <item apk:drawable="@drawable/j4"
    apk:duration="500" />
    <item apk:drawable="@drawable/j5"
    apk:duration="500" />
</animation-list>
```

（5）打开 res/layout 目录下的 main.xml 文件，代码修改为：

```xml
<?xml version="1.0" encoding="utf-8"?>
<LinearLayout
xmlns:apk="http://schemas.android.com/apk/res/android"
apk:orientation="vertical"
apk:layout_width="fill_parent"
apk:layout_height="fill_parent">
<!-- Frame 动画图片 -->
<ImageView
apk:id="@+id/ImgDance"
```

```
    apk:layout_width="wrap_content"
    apk:layout_height="wrap_content"
    apk:background="@anim/face" />
<!-- 动画控制按钮 -->
<LinearLayout
    apk:layout_width="fill_parent"
    apk:layout_height="wrap_content"
    apk:orientation="horizontal">
    <Button
    apk:text="开始"
    apk:layout_width="wrap_content"
    apk:layout_height="wrap_content"
    apk:onClick="onStartDance" />
    <Button
    apk:text="结束"
    apk:layout_width="wrap_content"
    apk:layout_height="wrap_content"
    apk:onClick="onStopDance" />
</LinearLayout> </LinearLayout>
```

（6）打开 src/com.example.e6_4_3 目录下的 MainActivity.java 文件，代码修改为：

```
package com.example.e6_4_3;

import android.app.Activity;
import android.graphics.drawable.AnimationDrawable;
import android.os.Bundle;
import android.view.View;
import android.widget.ImageView;

import com.example.e6_4_3.R;
/**
 * Frame 动画
 */
public class MainActivity extends Activity
{
    public static final String TAG = "MainActivity";
    // 显示动画的组件
    private ImageView imgDance;
    // Frame 动画
    private AnimationDrawable animDance;
    public void onCreate(Bundle cycle)
    {
```

```
        super.onCreate(cycle);
        super.setContentView(R.layout.main);
        // 实例化组件
        this.imgDance = (ImageView) super.findViewById(R.id.ImgDance);
        // 获得背景（5 个图片形成的动画）
        this.animDance = (AnimationDrawable) this.imgDance.getBackground();
    }
    //按钮：开始"滚动"动画
    public void onStartDance(View view)
    {
        this.animDance.start();
    }
    //按钮：停止"滚动"动画
    public void onStopDance(View view)
    {
        this.animDance.stop();
    }
}
```

运行程序，效果如图 6-16 所示。

图 6-16　Frame 动画效果

6.4.4　Frame 动画与 Tween 动画综合实例

很多实际的动画往往同时运行两个动画，比如在此要实现一个小游戏，需要让用户控制游戏中的主角移动——当主角移动时，不仅要控制它的位置改变，还应该在它移动时播放 Frame 动画来让用户感觉更"逼真"。

下面为一个利用 Tween 动画与 Frame 动画开发的一个"老鹰飞翔"的效果，在这个实例中，老鹰飞翔时的振翅效果为 Frame 动画，而老鹰飞翔时的位置为 Tween 动画。

【实现步骤】

（1）在 Eclipse 环境下，建立一个名为 E6_4_4 的工程。

（2）在 res 根目录下新建一个 anim 子目录，在该子目录下新建一个名为 laoyangfly.xml 的文件，代码为：

```xml
<?xml version="1.0" encoding="utf-8"?>
<!-- 定义动画循环播放 -->
<animation-list xmlns:android="http://schemas.android.com/apk/res/android"
    android:oneshot="false">
    <item
        android:drawable="@drawable/ly1" android:duration="120" />
    <item
        android:drawable="@drawable/ly2" android:duration="120" />
    <item android:drawable="@drawable/ly3" android:duration="120" />
    <item
        android:drawable="@drawable/ly4" android:duration="120" />
    <item
        android:drawable="@drawable/ly5" android:duration="120" />
    <item
        android:drawable="@drawable/ly6" android:duration="120" />
</animation-list>
```

(3) 打开 res/layout 目录下的 main.java 文件，代码修改为：

```xml
<?xml version="1.0" encoding="utf-8"?>
<LinearLayout
    xmlns:apk="http://schemas.android.com/apk/res/android"
    apk:orientation="vertical"
    apk:layout_width="fill_parent"
    apk:layout_height="fill_parent">
    <ImageView
    apk:id="@+id/laoyangfly"
    apk:layout_width="wrap_content"
    apk:layout_height="wrap_content"
    apk:background="@anim/laoyangfly" />
</LinearLayout>
```

(4) 打开 src/com.example.e6_4_4 目录下的 MainActivity.java 文件，代码修改为：

```java
package com.example.e6_4_4;
import java.util.Timer;
import java.util.TimerTask;
import android.app.Activity;
import android.graphics.drawable.AnimationDrawable;
import android.os.Bundle;
import android.os.Handler;
import android.os.Message;
import android.view.View;
```

```java
import android.view.View.OnClickListener;
import android.view.animation.TranslateAnimation;
import android.widget.ImageView;
public class MainActivity extends Activity
{
    //记录老鹰 ImageView 当前的位置
    private float curX = 0;
    private float curY = 30;
    //记录老鹰 ImageView 下一个位置的坐标
    private AnimationDrawable animDance;
    float nextX = 0;
    float nextY = 0;
    @Override
    public void onCreate(Bundle savedInstanceState)
    {
        super.onCreate(savedInstanceState);
        setContentView(R.layout.main);
        //获取显示老鹰的 ImageView 组件
        final ImageView imageView = (ImageView)findViewById(R.id.laoyangfly);
        final Handler handler = new Handler()
        {
            @Override
            public void handleMessage(Message msg)
            {
                if (msg.what == 0x123)
                {
                    //横向上一直向右飞
                    if(nextX > 320)
                    {
                        curX = nextX = 0;
                    }
                    else
                    {
                        nextX += 8;
                    }
                    //纵向上可以随机上下
                    nextY = curY + (float)(Math.random() * 10 - 5);
                    //设置显示老鹰的 ImageView 发生位移改变
                    TranslateAnimation anim
```

```
                    = new TranslateAnimation(curX , nextX , curY , nextY);
                curX = nextX;
                curY = nextY;
                anim.setDuration(200);
                //开始位移动画
                imageView.startAnimation(anim);
            }
        }
    };
    final AnimationDrawable butterfly = (AnimationDrawable)imageView
        .getBackground();
    imageView.setOnClickListener(new OnClickListener()
    {
        @Override
        public void onClick(View v)
        {
            //开始播放老鹰振翅的逐帧动画
            butterfly.start();
            //通过定制器控制每 0.2s 运行一次 TranslateAnimation 动画
            new Timer().schedule(new TimerTask()
            {
                @Override
                public void run()
                {
                    handler.sendEmptyMessage(0x123);
                }
            }, 0, 50);
        }
    });
}
```

运行程序，单击图片，效果如图 6-17 所示。

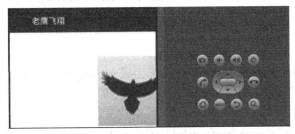

图 6-17　老鹰飞翔效果

6.4.5 SurfaceView 类

1. SurfaceView 类定义

SurfaceView 由于可以直接从内存或者 DMA 等硬件接口取得图像数据,因此是一个非常重要的绘图容器。其他的特性是:可以在主线程之外的线程中向屏幕绘图,从而可以避免画图任务繁重时造成主线程阻塞,提高了程序的反应速度。在游戏开发中多用到 SurfaceView,游戏中的背景、人物、动画等尽量在画布 Canvas 中画出。

SurfaceView 的核心在于提供了两个线程:UI 线程和渲染线程。

注意:

(1)所有 SurfaceView 和 SurfaceHolder.Callback 的方法都应该在 UI 线程里调用,一般来说就是应用程序主线程。渲染线程所要访问的各种变量应该做同步处理。

(2)由于 surface 可能被销毁,它只在 SurfaceHolder.Callback.surfaceCreated()和 SurfaceHolder.Callback.surfaceDestroyed()之间有效,所以要确保渲染线程访问的是合法有效的 surface。

2. SurfaceView 类实现

首先继承 SurfaceView 并实现 SurfaceHolder.Callback 接口。

使用接口的原因:因为使用 SurfaceView 有一个原则,所有的绘图工作必须得在 Surface 被创建之后才能开始(Surface——表面,这个概念在图形编程中常常被提到。基本上可以把它当做显存的一个映射,写入到 Surface 的内容可以被直接复制到显存从而显示出来,这使得显示速度会非常快),而在 Surface 被销毁之前必须结束。所以 Callback 中的 surfaceCreated 和 surfaceDestroyed 就成了绘图处理代码的边界。

需要重写的方法有以下几种。

(1) public void surfaceChanged(SurfaceHolder holder,int format,int width,int height){}:在 surface 的大小发生改变时激发。

(2) public void surfaceCreated(SurfaceHolder holder){}:在创建时激发,一般在这里调用画图的线程。

(3) public void surfaceDestroyed(SurfaceHolder holder) {}:销毁时激发,一般在这里将画图的线程停止、释放。

SurfaceView 类实现过程为:首先继承 SurfaceView 并实现 SurfaceHolder.Callback 接口;接着 SurfaceView.getHolder()获得 SurfaceHolder 对象;然后为 SurfaceHolder.addCallback(callback)添加回调函数;再接着用 SurfaceHolder.lockCanvas()获得 Canvas 对象并锁定画布;然后用 Canvas 绘画;最后应用 SurfaceHolder.unlockCanvasAndPost(Canvas canvas)结束锁定画图,并提交改变,显示图形。

3. SurfaceHolder

这里用到了一个类 SurfaceHolder,可以把它当成 surface 的控制器,用来操纵 surface。处理其 Canvas 上画的效果和动画,控制表面、大小、像素等。

下面介绍几个方法。

(1) abstract void addCallback(SurfaceHolder.Callback callback):给 SurfaceView 当前的持有者

一个回调对象。

（2）abstract Canvas lockCanvas()：锁定画布，一般在锁定后就可以通过其返回的画布对象 Canvas，在其上面进行画图等操作了。

（3）abstract Canvas lockCanvas(Rect dirty)：锁定画布的某个区域进行画图等，因为画完图后，会调用下面的 unlockCanvasAndPost 来改变显示内容。如果相对部分内存要求比较高的游戏来说，可以不用重画 dirty 外的其他区域的像素，以此来提高速度。

（4）abstract void unlockCanvasAndPost(Canvas canvas)：结束锁定画图，并提交改变。

下面利用 SurfaceView 开发一个绘图程序，如果用户单击界面上的按钮即会自动绘制对应的图形。

【实现步骤】

（1）在 Eclipse 环境下，建立一个名为 E6_4_5 的工程。

（2）打开 res/layout 目录下的 main.xml 文件，代码修改为：

```xml
<?xml version="1.0" encoding="utf-8"?>
<LinearLayout
xmlns:android="http://schemas.android.com/apk/res/android"
 android:layout_width="fill_parent"
 android:layout_height="fill_parent"
 android:orientation="vertical">
<LinearLayout android:id="@+id/LinearLayout01"
     android:layout_width="wrap_content"
     android:layout_height="wrap_content">
<Button android:id="@+id/Button01"
android:layout_width="wrap_content"
android:layout_height="wrap_content"
android:text="简单绘画">
</Button>
<Button android:id="@+id/Button02"
android:layout_width="wrap_content"
android:layout_height="wrap_content"
android:text="正弦曲线">
</Button>
</LinearLayout>
  <SurfaceView android:id="@+id/SurfaceView01"
      android:layout_width="fill_parent"
      android:layout_height="fill_parent">
  </SurfaceView>
</LinearLayout>
```

（3）打开 src/com.example.e6_4_5 目录下的 MainActivity.java 文件，代码修改为：

```java
public class MainActivity extends Activity
{
```

```java
/* 第一次调用活动*/
Button btnSimpleDraw, btnTimerDraw;
SurfaceView sfv;
SurfaceHolder sfh;
private Timer mTimer;
 private MyTimerTask mTimerTask;
int Y_axis[],//保存正弦波的Y轴上的点
 centerY,//中心线
 oldX,oldY,//上一个XY点
 currentX;//当前绘制到的X轴上的点
@Override
public void onCreate(Bundle savedInstanceState)
{
    super.onCreate(savedInstanceState);
    setContentView(R.layout.main);
    btnSimpleDraw = (Button) this.findViewById(R.id.Button01);
    btnTimerDraw = (Button) this.findViewById(R.id.Button02);
    btnSimpleDraw.setOnClickListener(new ClickEvent());
    btnTimerDraw.setOnClickListener(new ClickEvent());
    sfv = (SurfaceView) this.findViewById(R.id.SurfaceView01);
    sfh = sfv.getHolder();
    //动态绘制正弦波的定时器
    mTimer = new Timer();
    mTimerTask = new MyTimerTask();
    // 初始化Y轴数据
    centerY = (getWindowManager().getDefaultDisplay().getHeight() - sfv.getTop())/ 2;
    Y_axis = new int[getWindowManager().getDefaultDisplay().getWidth()];
    for (int i = 1; i < Y_axis.length; i++)
    {
        // 计算正弦波
        Y_axis[i - 1] = centerY - (int) (100 * Math.sin(i * 2 * Math.PI / 180));
    }
}
class ClickEvent implements View.OnClickListener
{
    @Override
    public void onClick(View v)
    {
        if (v == btnSimpleDraw)
        {
```

```java
                    SimpleDraw(Y_axis.length-1);              //直接绘制正弦波
                } else if (v == btnTimerDraw)
                {
                    oldY = centerY;
                    mTimer.schedule(mTimerTask, 0, 5);        //动态绘制正弦波
                }
            }
        }
        class MyTimerTask extends TimerTask
        {
            @Override
            public void run()
            {
                SimpleDraw(currentX);
                currentX++;//往前进
                if (currentX == Y_axis.length - 1)
                {
                    //如果到了终点，则清屏重来
                    ClearDraw();
                    currentX = 0;
                    oldY = centerY;
                }
            }
        }
        //绘制指定区域
        void SimpleDraw(int length)
        {
            if (length == 0)
                oldX = 0;
            // 关键:获取画布
            Canvas canvas = sfh.lockCanvas(new Rect(oldX, 0, oldX + length, getWindowManager().getDefaultDisplay().getHeight()));
            Log.i("Canvas:", String.valueOf(oldX) + "," + String.valueOf(oldX + length));
            Paint mPaint = new Paint();
            mPaint.setColor(Color.GREEN);                // 画笔为绿色
            mPaint.setStrokeWidth(2);                    // 设置画笔粗细
            int y;
            for (int i = oldX + 1; i < length; i++)
            {
                // 绘画正弦波
```

```
                    y = Y_axis[i - 1];
                    canvas.drawLine(oldX, oldY, i, y, mPaint);
                    oldX = i;
                    oldY = y;
                }
            sfh.unlockCanvasAndPost(canvas);            // 解锁画布，提交画好的图像
        }
        void ClearDraw()
        {
            Canvas canvas = sfh.lockCanvas(null);
            canvas.drawColor(Color.BLACK);              // 清除画布
            sfh.unlockCanvasAndPost(canvas);
        }
    }
```

运行程序，单击界面中的"正弦曲线"按钮，效果如图 6-18 所示。

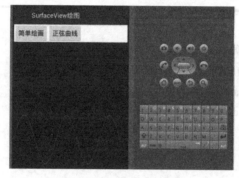

图 6-18　绘制正弦曲线

6.4.6　动画组件（ViewAnimator）

在实际开发中可能更需要统一管理需要实现某一类动画效果的 View，此时就需要一个容器，本节将介绍在 Android 中常用的动画组件，通过这些组件能方便地实现一组 View 的动画。

ViewAnimator 的基类是 FrameLayout，作用是为 FrameLayout 里面的 View 切换提供动画效果。

一般不直接使用 ViewAnimator，而是使用它的子类 ViewFlipper 和 ViewSwitcher，其中 ViewSwitcher 的子类又包含 ImageSwitcher 和 TextSwitcher。ViewFlipper 和 ViewSwitcher 的主要区别在于 ViewSwitcher 最多能有两个子 View，而 ViewFlipper 可以有多个。

1. ViewSwitcher 类

ImageSwitcher 是一个控制图片切换显示的组件，可添加图片切换动画，效果很好，很适合制作相册或动态展示图片。在 Android 中还有一个比较类似的组件就是 TextSwitcher，它们的用法基本相同。

使用 ImageSwitcher 或者 TextSwitcher 必须设置一个 ViewFactory，主要用来在 ViewSwitcher

中创建 View，因此需要实现 ViewSwitcher.ViewFactory 接口，最后通过 makeView()方法创建相应的 View，即 ImageSwitcher 对应 ImageView，TextSwitcher 对应 TextView。

2．ViewFlipper 类

ViewFlipper 可以添加两个或两个以上的 View，并且在一个时间只有一个 View 会显示。如果需要，每个 View 能在固定的间隔切换。

3．ViewSwitcher 类与 ViewFlipper 类综合实例

本实例使用 ViewSwitcher 类与 ViewFlipper 类在不同布局中切换。

【实现步骤】

（1）在 Eclipse 环境下，建立一个名为 E6_4_6 的工程。

（2）在 res 目录下建立一个子目录 anim，并在其中建立名为 slider_in_top.xml 及 slider_in_bottom.xml 的两个文件。

slider_in_top.xml 的代码为：

```xml
<?xml version="1.0" encoding="utf-8"?>
<set xmlns:android="http://schemas.android.com/apk/res/android" >
    <translate
        android:duration="2000"
        android:fromYDelta="-100%p"
        android:toYDelta="0" />
</set>
```

slider_in_bottom.xml 的代码为：

```xml
<?xml version="1.0" encoding="utf-8"?>
<set xmlns:android="http://schemas.android.com/apk/res/android" >
    <translate
        android:duration="2000"
        android:fromYDelta="0"
        android:toYDelta="100%p" />
</set>
```

（3）打开 res/layout 目录下的 main.xml 文件，代码修改为：

```xml
<?xml version="1.0" encoding="utf-8"?>
<LinearLayout xmlns:android="http://schemas.android.com/apk/res/android"
    android:layout_width="fill_parent"
    android:layout_height="fill_parent"
    android:orientation="vertical" >
    <ViewSwitcher
        android:id="@+id/switcher"
        android:layout_width="wrap_content"
        android:layout_height="wrap_content"
        android:inAnimation="@anim/slide_in_top"
```

```xml
                android:outAnimation="@anim/slide_in_bottom" >
                <RelativeLayout
                    android:layout_width="wrap_content"
                    android:layout_height="wrap_content" >
                    <TextView
                        android:layout_width="fill_parent"
                        android:layout_height="100dp"
                        android:background="#F2345F"
                        android:text="ViweFlipper 切换"
                        android:textSize="32sp" >
                    </TextView>
                </RelativeLayout>
                <RelativeLayout
                    android:layout_width="wrap_content"
                    android:layout_height="wrap_content" >
                    <TextView
                        android:layout_width="fill_parent"
                        android:layout_height="100dp"
                        android:background="#FF0FF0FF"
                        android:text="ViewSwitcher 切换"
                        android:textSize="32sp" >
                    </TextView>
                </RelativeLayout>
            </ViewSwitcher>
            <Button
                android:id="@+id/prev"
                android:layout_width="wrap_content"
                android:layout_height="wrap_content"
                android:text="前一个" />
            <Button
                android:id="@+id/next"
                android:layout_width="wrap_content"
                android:layout_height="wrap_content"
                android:text="下一个" />
</LinearLayout>
```

（4）打开 src/com.example.e6_4_6 目录下的 MainActivity.java 文件，代码修改为：

```java
package com.example.e6_4_6;

import android.app.Activity;
```

```java
import android.os.Bundle;
import android.view.View;
import android.widget.Button;
import android.widget.ViewSwitcher;
public class MainActivity extends Activity
{
    /*第一次调用活动 */
    private ViewSwitcher mSwitcher;
    private Button btn_prev, btn_next;
    @Override
    public void onCreate(Bundle savedInstanceState)
    {
        super.onCreate(savedInstanceState);
        setContentView(R.layout.main);
        mSwitcher=(ViewSwitcher) findViewById(R.id.switcher);
        mSwitcher.setDisplayedChild(0);
        btn_next =(Button) findViewById(R.id.next);
        btn_prev = (Button) findViewById(R.id.prev);
        btn_next.setOnClickListener(new View.OnClickListener()
        {
            @Override
            public void onClick(View v)
            {
                mSwitcher.showNext();
            }
        });
        btn_prev.setOnClickListener(new View.OnClickListener()
        {
            @Override
            public void onClick(View v)
            {
                mSwitcher.showPrevious();
            }
        });
    }
}
```

运行程序，效果如图 6-19 所示。

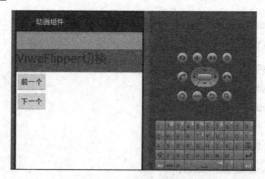

图 6-19 动画组件切换效果

注意：ViewSwitcher 只可以在两种布局中切换，ViewFlipper 可以应用多种布局，用法基本一样，但后者另外还有下面几种方法。

- isFlipping：用来判断 View 切换是否正在进行。
- setFilpInterval：设置 View 之间切换的时间间隔。
- startFlipping：使用上面设置的时间间隔来开始切换所有的 View，切换会循环进行。
- stopFlipping：停止 View 切换。

第 7 章　Android 数据存储

在实际应用中，我们总是希望应用程序能够保存一部分数据，以便程序下次启动时读取。例如，可以保存用户登录信息，方便下次快捷登录；也可以保存用户对程序的配置信息。Android 同样也提供了一种标准方式供应用软件将私有数据开放给其他应用软件。本节将会描述一个应用软件存储和获取数据、开放数据给其他应用软件、从其他应用软件请求数据并开放它们的多种方式。

在 Android 中，可供选择的存储方式有如下 5 种：
- SharedPreferences 存储数据；
- 文件存储数据；
- SQLite 数据库存储数据；
- 使用 ContentProvider 存储数据；
- 网络存储数据。

7.1　SharedPreferences 存储

SharedPreferences 是 Android 平台上一个轻量级的存储类，主要是保存一些常用的配置，如窗口状态，一般在 Activity 中由 onSaveInstanceState 完成 SharedPreferences 的窗口重载保存，它提供了 Android 平台常规的 Long（长整型）、Int（整型）、String（字符串型）的保存。

它是什么样的处理方式呢？SharedPreferences 类似过去 Windows 系统上的 ini 配置文件，但是它分为多种权限，可以全局共享访问，Android123 提示最终是以 xml 方式来保存的，整体效率来看不太高，对于常规的轻量级而言比 SQLite 要好不少，如果存储量不大则可以考虑自己定义文件格式。xml 处理时 Dalvik 会通过自带底层的本地 XML Parser 解析，如 XMLpull 方式，这样对于内存资源占用比较好。

它的本质是基于 XML 文件存储 key-value 键值对数据，通常用来存储一些简单的配置信息，其存储位置在/data/data/<文件名>/shared_prefs 目录下。

SharedPreferences 对象本身只能获取数据而不支持存储和修改，存储和修改是通过 Editor 对象实现的。

实现 SharedPreferences 存储的步骤如下：
（1）根据 Context 获取 SharedPreferences 对象；
（2）利用 edit()方法获取 Editor 对象；
（3）通过 Editor 对象存储 key-value 键值对数据；
（4）通过 commit()方法提交数据。

7.1.1 SharedPreferences 存储类效率分析

两个 Activity 之间的数据传递除了可以通过 Intent 来传递外，还可以使用 SharedPreferences 来共享数据。SharedPreferences 用法很简单，如在 A 中设置如下代码。

```
Editor sharedate=getSharedPreferences("data",0).edit();
sharedata.putString("item","hello getSharedPreferences");
sharedata.commit();
```

然后即可在 B 中编写如下代码。

```
SharedPreferences sharedata=getSharedPreferences("data",0);
String data=sharedata.getString("item",null);
Log.v("cola","data="+data);
```

通过以下 Java 代码将数据显示出来。

```
<SPAN class=hilite1>SharedPreferences</SPAN>sharedata=getSharedPreferences("data",0);
String data=sharedata.getString("item",null);
Log.v("cola","data="+data);
```

SharedPreferences 以一种简单、透明的方式来保存一些用户个性化设置的字体、颜色、位置等参数信息。一般的应用程序都会提供"设置"或者"首选项"这样的界面，那么这些设置最后就可以通过 Preferences 来保存，而程序员不需要知道它到底以什么形式保存，以及保存在什么地方。当然，如果希望获取它的保存信息，也没有什么限制。

在 Android 系统中，这些信息以 XML 文件的形式保存在 /data/data/PACKAGE_NAME/shared_prefs 目录下。

1. 数据读取

可以通过下面的代码实现数据读取。

```
String PREFS_NAME="Note.sample.roiding.com";
SharedPreferences settings=getSharedPreferences(PREFS_NAME,0);
boolean silent=settings.getBoolean("silentMode",false);
String hello=settings.getString("hello","Hi");
```

然后通过以下代码显示出来。

```
String PREFS_NAME="Note.sample.roiding.com";
<SPAN class=hilite1>SharedPreferences</SPAN>
settings=getSharedPreferences(PREFS_NAME,0);
boolean silent=settings.getBoolean("silentMode",false);
String hello=settings.getString("hello","Hi");
```

在以上代码中，"SharedPreferences settings=getSharedPreferences(PREFS_NAME,0);"通过名称，得到一个 SharedPreferences，顾名思义，这个 Preferences 是共享的，共享的数据在同一个 Package 中，这里所说的 Package 和 Java 里面的那个 Package 不同，这里的 Package 是指在 AndroidManifest.xml 文件中的：

```
<manifest xmlns:android="http://schemas.android.com/apk/res/android"
package="com.roiding.sample.note"
```

```
android:versionCode="1"
android:versionName="1.0.0">
```

对于"boolean silent=settings.getBoolean("silentMode",falde);",获得一个 boolean 值,在这里就会看到用 Preferences 的好处了:可以提供一个默认值。也就是说,如果 Preference 中不存在这个值的话,那么就用后面的值作为返回值,这样就省去了 if 为空的判断。

2. 数据写入

可以通过以下代码实现数据写入。

```
String PREFS_NAME="Note.sample.roiding.com";
SharedPreferences.settings=getSharedPreferences(PREFS_NAME,0);
SharedPreferences.Editor editor=settings.edit();
editor.putBoolean("silentMode",true);
editor.putString("hello","Hello~");
editor.commit();
```

注意:这组访问接口用于修改数据,其中 Shared Preferences 对象由函数 getSharedPreferences (String,int)返回。为了统一设置参数,有一个单例类供所有的客户端共享。修改参数必须通过一个 SharedPreferences.Editor 对象,在存储它们时,以确保参数值有统一的状态和控制,当前此类不支持多线程。

7.1.2 SharedPreferences 类实例

本节通过一个实例的实现过程来讲解 SharedPreferences 的使用过程。

【实现步骤】

(1) 在 Eclipse 环境下,建立一个名为 E7_1_2 的工程。

(2) 打开 res/layout 目录下的 main.xml 文件,代码修改为:

```
<?xml version="1.0" encoding="utf-8"?>
<LinearLayout xmlns:android="http://schemas.android.com/apk/res/android"
    android:orientation="vertical"
    android:layout_width="fill_parent"
    android:layout_height="fill_parent">
<TextView
    android:layout_width="fill_parent"
    android:layout_height="wrap_content"
    android:text="@string/hello_world"/>
</LinearLayout>
```

(3) 打开 src/com.example.e7_1_2 目录下的 MainActivity.java 文件,代码修改为:

```
package com.example.e7_1_2;
import android.app.Activity;
import android.os.Bundle;
import android.widget.TextView;
```

```java
public class MainActivity extends Activity
{
    public final static String COLUMN_NAME ="name";
    public final static String COLUMN_MOBILE ="mobile";
    sharedP sp;
    /* 第一次调用活动 */
    @Override
    public void onCreate(Bundle savedInstanceState)
    {
        super.onCreate(savedInstanceState);
        sp = new sharedP(this, "contacts");

        //1. 存储一些值
        sp.putValue(COLUMN_NAME, "July");
        sp.putValue(COLUMN_MOBILE, "SharedPreferences 存储");
        //2. 获取值
        String name = sp.getValue(COLUMN_NAME);
        String mobile = sp.getValue(COLUMN_MOBILE);
        TextView tv = new TextView(this);
        tv.setText("NAME:"+ name + "\n" + "MOBILE:" + mobile);
        setContentView(tv);
    }
}
```

（4）在 src/com.example.e7_1_2 目录下建立一个名为 sharedP.java 的文件，其代码为：

```java
package com.example.e7_1_2;
import android.content.Context;
import android.content.SharedPreferences;
public class sharedP
{
    SharedPreferences sp;
    SharedPreferences.Editor editor;
    Context context;
    public sharedP(Context c,String name)
    {
        context = c;
        sp = context.getSharedPreferences(name, 0);
        editor = sp.edit();
    }
    public void putValue(String key, String value)
    {
```

```
            editor = sp.edit();
            editor.putString(key, value);
            editor.commit();
        }
        public String getValue(String key)
        {
            return sp.getString(key, null);
        }
    }
```
运行程序，效果如图 7-1 所示。

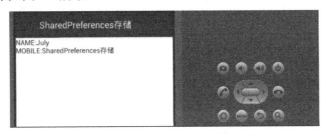

图 7-1　执行 SharedPreferences 存储效果

这样，"NAME"和"MOBILE"就是在 SharedPreferences 中存储的。

注意：因为上面例子的 pack_name 为 package com.example _SharedPreferences，所以存放数据的路径为 data/data/com.example.SharedPreferences/share_prefs/contacts.xml，contacts.xml 的代码为：

```xml
<?xml version="1.0" encoding="utf-8" standalone='yes'?>
<map>
<string name="mobile">987654321</string>
<string name="name">Grayphone</string>
</map>
```

7.2　文件存储数据

文件存储方式是一种较常用的方法，在 Android 中读取/写入文件的方法与 Java 中实现 I/O 的程序是完全一样的，Android 提供了 openFileInput() 和 openFileOutput() 方法来读取设备上的文件。

至于文件的存储方式，在 Android 中有以下几种不同的用法：
- 直接使用 java.io 包提供的方法实现文件的增、删、读、写。
- 使用 Android 提供的 openFileInput(String name) 和 openFileOutput(String name,int mode) 方法读写数据。
- 从应用目录 res 中的 raw 文件夹中获取文件并读取数据。
- 从应用目录 asset 中获取文件并读取数据。

7.2.1 java.io 包的方法

java.io 包提供了一些操作文件的方法,包括读写、创建、删除等,通过这些方法可以方便地实现文件的常用操作,下面列举常用的方法。

(1) 创建文件,代码为:

```java
/**
 * 创建新文件
 * @param fill 文件
 * @return boolean 是否成功
 */
public boolean newFile(File f)
{
    try
    {
        f.createNewFile();
    }
    catch (Exception e)
    {
        return false;
    }
    return true;
}
```

(2) 复制文件,代码为:

```java
/**
 * 复制单个文件
 * @param oldPath    原文件路径  如:/xx
 * @param newPath    复制后路径  如:/xx/ss
 * @return boolean 是否成功
 */
public boolean copyFile(String oldPath,String newPath)
{
    try
    {
        int bytesum=0;
        int byteread=0;
        String f_new="";
        File f_old=new File(oldPath);
        if (newPath.endsWith(File.separator))
        {
            f_new=newPath+f_old.getName();
```

```
            }else{
                f_new=newPath+File.separator+f_old.getName();
            }
            //如果文件夹不存在,则建立新文件夹
            new File(newPath).mkdirs();
            //如果文件不存在,则建立新文件
            new File(f_new).createNewFile();
            //文件存在时
            if (f_old.exists())
            {
                //读入原文件
                InputStream inStream=new FileInputStream(oldPath);
                FileOutputStream fs=new FileOutputStream(f_new);
                byte[] buffer=new byte[1444];
                while((byteread=inStream.read(buffer))!=-1)
                {
                    //文件总字节数
                    bytesum+=byteread;
                    fs.write(buffer,0,byteread);
                }
                inStream.close();
            }
    }
    catch (Exception e)
    {
        e.printStackTrace();
        return false;
        return true;
    }
/**
* 复制文件夹
* @param oldPath   原文件路径  如:/xx/yy
* @param newPath 复制后路径  如://aa/dd
* @return boolean 是否成功
*/
public boolean copyDir(String oldPath,String newPath)
{
    try
    {
        //要复制的文件夹  /xx/yy
        File f_old=new File(oldPath);
```

```
                        String d_old="";
                        //新文件夹路径
                        String d_new=newPath+File.separator+f_old.getName();
                        //如果文件夹不存在,则建立新文件夹
                        new File(d_new).mkdirs();
                        File [] files=f_old.listFiles();
                        for (int i=0;i<files.length;i++)
                        {
                            //要复制的文件夹下的文件
                            d_old=oldPath+File.separator+files[i].getName();
                            if (files[i].isFile())
                            {
                                copyFile(d_old,d_new);
                            }
                            else{
                                copyFile(d_old,d_new);
                            }
                        }
                    }
                catch (Exception e)
                {
                    return true;
                }
            }
        }
```

(3) 移动文件,代码为:

```
    /*
     * 移动文件到指定目录
     * @param oldPath    如: /aa.txt
     * @param newPath    如: /xx/aa.txt
     * @return boolean 是否成功
     */
    public boolean moveFile(String oldPath,String newPath)
    {
        try
        {
            if(copyFile(oldPath,newPath))
            {
                new File(oldPath).delete();
            }
```

```java
        }
        catch (Exception e)
        {
            return true;
        }
        /*
         * 移动文件夹到指定目录
         * @param oldPath  如：/aa
         * @param newPath  如：//aa//bb
         * @param boolean 是否成功
         */
        public boolean moveDir(String oldPath,String newPath)
        {
            try
            {
                if(copyDir(oldPath,newPath))
                {
                    delDir(new File(oldPath));
                }
            }
            catch (Exception e)
            {
                return false;
            }
            return true;
        }
    }
```

（4）删除文件，代码为：

```java
    /*
     * 移除单个文件
     * @param file 文件
     * @return boolean 是否成功
     */
    public boolean delFile(File f)
    {
        try
        {
            if(f.exists())
            {
                f.delete();
```

```java
            }
        }
        catch (Exception e)
        {
            return false;
        }
        return true;
    }
    /*
    * 删除文件夹
    * @param file 文件
    * @return boolean 是否成功
    */
    public boolean delDir(File f)
    {
        boolean ret=false;
        try
        {
            if(f.exists())
            {
                File[] files=f.listFiles();
                for(int i=0;i<files.length;i++)
                {
                    if(files[i].isDirectory())
                    {
                        if(!delDir(files[i]))
                        {
                            return false;
                        }
                    }
                    else{
                        files[i].delete();
                    }
                }
                //删除空文件夹
                f.delete();
                Sret=true;
            }
        }
```

```
            catch (Exception e)
            {
                return false;
            }
            return ret;
        }
```

注意：读写 SD 卡上的文件夹需要增加如下权限：

```
<!--在 SDCard 中创建与删除文件权限-->
<user-permission android:name="android.permission.MOUNT_UNMOUNT_FILESYSTEMS"/>
<!--往 SDCard 写入数据权限-->
<user-permission android:name="android.permission.WRITE_EXTERNAL_STORAGE"/>
```

7.2.2 openFileInput 和 openFileOutput

关于文件存储，Activity 提供了 openFileOutput()方法可以用于把数据输出到文件中，具体的实现过程与在 J2SE 环境中保存数据到文件中是一样的。文件可用来存放大量数据，如文本、图片、音频等。默认位置：/data/data/<package name>/files/***.***。

示例代码为：

```
        public void save()
        {
            try {
                FileOutputStream outStream = this.openFileOutput("a.txt",Context.MODE_WORLD_READABLE);
                outStream.write(text.getText().toString().getBytes());
                outStream.close();
                Toast.makeText(MyActivity.this,"Saved",Toast.LENGTH_LONG).show();
            } catch (FileNotFoundException e)
            {
                return;
            }
            catch (IOException e)
            {
                return ;
            }
        }
```

openFileOutput()方法的第一参数用于指定文件名称，不能包含路径分隔符"/"，如果文件不存在，Android 会自动创建它。创建的文件保存在/data/data/<package name>/files 目录，例如，/data/data/cn.itcast.action/files/itcast.txt，通过单击 Eclipse 菜单栏中的"窗口"菜单下的"显示视图"子菜单下的"其他"选项，在弹出的对话窗口中展开 android 文件夹，选择下面的"File Explorer"视图，然后在"File Explorer"视图中展开/data/data/<package name>/files 目录就可以看到该文件。

openFileOutput()方法的第二参数用于指定操作模式，有四种模式，分别为：
- Context.MODE_PRIVATE = 0
- Context.MODE_APPEND = 32768
- Context.MODE_WORLD_READABLE = 1
- Context.MODE_WORLD_WRITEABLE = 2

Context.MODE_PRIVATE：默认操作模式，代表该文件是私有数据，只能被应用本身访问，在该模式下，写入的内容会覆盖原文件的内容，如果想把新写入的内容追加到原文件中。可以使用 Context.MODE_APPEND。

Context.MODE_APPEND：模式会检查文件是否存在，存在就在文件中追加内容，否则就创建新文件。

Context.MODE_WORLD_READABLE 和 Context.MODE_WORLD_WRITEABLE 用来控制其他应用是否有权限读写该文件。

MODE_WORLD_READABLE：表示当前文件可以被其他应用读取。

MODE_WORLD_WRITEABLE：表示当前文件可以被其他应用写入。

如果希望文件被其他应用读和写，可以传入：openFileOutput("itcast.txt", Context.MODE_WORLD_READABLE + Context.MODE_WORLD_WRITEABLE); Android 有一套自己的安全模型，当应用程序(.apk)在安装时系统就会分配一个 userid，当该应用要去访问其他资源如文件的时候，就需要 userid 匹配。默认情况下，任何应用创建的文件、SharedPreferences、数据库都应该是私有的（位于/data/data/<package name>/files），其他程序无法访问。

除非在创建时指定了 Context.MODE_WORLD_READABLE 或者 Context.MODE_WORLD_WRITEABLE，只有这样其他程序才能正确访问。

读取文件的代码为：

```
public void load()
{
    try
    {
        FileInputStream inStream=this.openFileInput("a.txt");
        ByteArrayOutputStream stream=new ByteArrayOutputStream();
        byte[] buffer=new byte[1024];
        int length=-1;
        while((length=inStream.read(buffer))!=-1)
        {
            stream.write(buffer,0,length);
        }
        stream.close();
        inStream.close();
        text.setText(stream.toString());
        Toast.makeText(MyActivity.this,"Loaded",Toast.LENGTH_LONG).show();
    } catch (FileNotFoundException e)
```

```
            {
                e.printStackTrace();
            }
        catch (IOException e)
            {
                return ;
            }
    }
```

对于私有文件只能被创建该文件的应用访问，如果希望文件能被其他应用读和写，可以在创建文件时，指定Context.MODE_WORLD_READABLE和Context.MODE_WORLD_WRITEABLE权限。

7.2.3 从 resource 中的 raw 文件夹中读取文件

在 Android 中允许用户将程序用到文件放到 res 中的 raw 文件夹中，随程序一起打包成.apk 文件发布，可以通过 ID 读取，代码示例为：

```
/*
 * 从 resource 中的 raw 文件夹中获取文件并读取数据
 * @param fileName 文件路径
 * @return 读取的字符串
 */
public String getFromRaw(String fileName)
{
    String result="";
    try
    {
        //从 Resources 中的 raw 中的文件获取输入值
        InputStream in=getResources().openRawResource(R.raw.test);
        //获取文件的字节数
        int length=in.available();
        //创建 byte 数组
        byte [] buffer=new byte[length];
        //将文件中的数据读取到 byte 数组中
        in.read(buffer);
        //将 byte 数组转换成指定格式的字符串
        result=EncodingUtils.getString(buffer,"UTF-8");
        //关闭文件输入流
        in.close();
    }
    catch (Exception e)
```

```
        {
            //捕获异常并打印
            e.printStackTrace();
        }
        return result;
    }
```

7.2.4 从 asset 中读取文件

在 asset 目录下的文件在被打包成.apk 文件时是不会被压缩的，而 res 下的文件在打包成.apk 文件时会被压缩。

Android 中提供了 AssetManager 类来处理应用对 asset 文件的访问。通过 open(String fileName) 方法，返回一个 InputStream 对象，代码示例为：

```
/*
 * 从 asset 中获取文件并读取数据/asset 目录
 * @param fileName 文件路径
 * @return 读取的字符串
 */
public String getFromAsset(String fileName)
{
    String result="";
    try
    {
        //从 Assets 中的文件获取输入流
        InputStream in=getResources().getAssets().open(fileName);
        //获取文件的字节数
        int length=in.available();
        //创建 byte 数组
        byte[] buffer=new byte[length];
        //将文件中的数据读取到 byte 数组中
        in.read(buffer);
        //将 byte 数组转换成指定格式的字符串
        result=EncodingUtils.getString(buffer,"UTF-8");
    }
    catch (Exception e)
    {
        //捕获异常并打印
        e.printStackTrace();
    }
    return result;
}
```

7.3 SQLite 数据库存储

7.3.1 SQLite 数据库存储概述

SQLite 是轻量级嵌入式数据库引擎，它支持 SQL 语言，并且只利用很少的内存就有很好的性能。此外它还是开源的，任何人都可以使用它。许多开源项目（Mozilla, PHP, Python）都使用了 SQLite.SQLite，由以下几个组件组成：SQL 编译器、内核、后端及附件。SQLite 通过利用虚拟机和虚拟数据库引擎（VDBE），使调试、修改和扩展 SQLite 的内核变得更加方便。

SQLite 数据库存储具有如下特点：
- 面向资源有限的设备；
- 没有服务器进程；
- 所有数据存放在同一文件中跨平台；
- 可自由复制。

SQLite 内部结构如图 7-2 所示。

SQLite 基本上符合 SQL—92 标准，它的优点就是高效，Android 运行时环境包含了完整的 SQLite。

SQLite 和其他数据库最大的不同就是对数据类型的支持，创建一个表时，可以在 CREATE TABLE 语句中指定某列的数据类型，但是你可以把任何数据类型放入任何列中。当某个值插入数据库时，SQLite 将检查它的类型。如果该类型与关联的列不匹配，则 SQLite 会尝试将该值转换成该列的类型。如果不能转换，则该值将作为其本身具有的类型存储，例如，可以把一个字符串（String）放入 INTEGER 列。SQLite 称这为"弱类型"（manifest typing.）。此外，SQLite 不支持一些标准的 SQL 功能，特别是外键约束（FOREIGN KEY constrains），嵌套 transcaction 和 RIGHT OUTER JOIN 和 FULL OUTER JOIN，还有一些 ALTER TABLE 功能。除了上述功能外，SQLite 是一个完整的 SQL 系统，拥有完整的触发器、交易，等等。

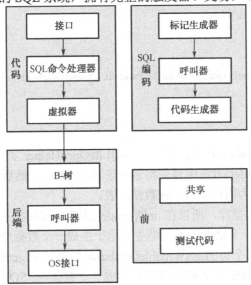

图 7-2 SQLite 内部结构图

Android 在运行时（run-time）集成了 SQLite 数据库，所以每个 Android 应用程序都可以使用 SQLite 数据库。

7.3.2 SQLite 数据库存储分析

Android 数据库编辑中，和传统编程一样，也通常需要使用 SQL 语言来操作数据库中的数据，如添加、删除、修改等操作。在 Android 开发应用中，通过 SQL 可以分别实现对这些数据的添加、删除和修改操作。

以下实例，使用 SQLite 存储方式来保存数据，并通过编程方式来操作数据。

【实现步骤】
（1）在 Eclipse 环境下，建立一个名为 E7_3_2 的工程。
（2）编写 src/com.example.e7_3_2 目录下的 MainActivity.java 文件。
① 定义继承于 SQLiteOpenHelper 的类 DatabaseHelper，代码为：

```java
private static class DatabaseHelper extends SQLiteOpenHelper
{
    DatabaseHelper(Context context)
    {
        super(context, DATABASE_NAME, null, DATABASE_VERSION);
    }
    @Override
    public void onCreate(SQLiteDatabase db)
    {
        String sql = "CREATE TABLE " + TABLE_NAME + " (" + TITLE + " text not null, " + BODY + " text not null " + ");";
        Log.i("haiyang:createDB=", sql);
        db.execSQL(sql);
    }
    @Override
    public void onUpgrade(SQLiteDatabase db, int oldVersion, int newVersion)
    {
    }
}
```

在以上代码中，类 DatabaseHelper 继承了 SQLiteOpenHelper 类，并且重写了 onCreate 和 onUpgrade 方法。在 onCreate()方法中首先构造一条 SQL 语句，然后调用 db.execSQL(sql)执行 SQL 语句。这条 SQL 语句为用户生成一张数据库表。

因为此处不需要升级数据库，所以在 onUpgrade()函数中没有执行任何操作。

SQLiteOpenHelper 为一个辅助类，此类主要用于生成一个数据库，并对数据库的版本进行管理。当在程序当中调用这个类的 getWritableDatabase()方法或者 getReadableDatabase()方法时，如果当时没有数据，那么 Android 系统就会自动生成一个数据库。SQLiteOpenHelper 为一个抽象类，通常需要继承它，并且实现里面的 3 个函数，各个函数的具体说明如下。

- onCreate(SQLiteDatabase)：在数据库第一次生成时调用这个方法，一般在这个方法中生成数据库表。
- onUpgrade(SQLiteDatabase,int,int)：当数据库需要升级时，Android 系统会主动调用这个方法。一般在这个方法中删除数据表，并建立新的数据表，当然是否还需要做其他的操作，完全取决于应用的需要。
- onOpen(SQLiteDatabase)：当打开数据库时的回调函数，一般也不会用到。

② 编写按钮处理事件：单击"插入 2 条数据"按钮后即插入两条新的数据到数据库中的 diary 表中，并且在屏幕的 title（标题）区域就会显示成功的提示，如图 7-3 所示。

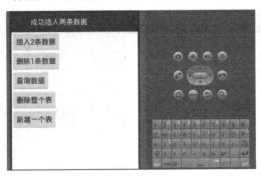

图 7-3　成功插入 2 条数据

单击这个按钮后执行监听器里的 onClick 方法，并最终执行 insertItem 方法以插入两条数据。实现代码为：

```
/*
 * 插入两条数据
 */
private void insertItem()
{
    SQLiteDatabase db = mOpenHelper.getWritableDatabase();
    String sql1 = "insert into " + TABLE_NAME + " (" + TITLE + ", " + BODY
        + ") values('mmm', 'Android 很好');";
    String sql2 = "insert into " + TABLE_NAME + " (" + TITLE + ", " + BODY
        + ") values('nnn', 'Android 很好');";
    try {
        Log.i("haiyang:sql1=", sql1);
        Log.i("haiyang:sql2=", sql2);
        db.execSQL(sql1);
        db.execSQL(sql2);
        setTitle("成功插入两条数据");
    } catch (SQLException e)
    {
        setTitle("插入失败");
    }
}
```

关于以上代码说明如下。
- sql1 和 sql2：构造标准的插入 SQL 语句。
- Log.i()：会将参数内容打印到日志当中，并且打印级别是 Info 级别。
- db.execSQL(sql1)：执行 SQL 语句。

注意：Android 支持 5 种打印输出级别，分别为 Verbose、Debug、Info、Warning、Error，在程序当中一般用到的是 Info 级别。

③ 单击"查询数据"按钮会在屏幕的 title 区别中显示当前数据表中的数据条数，因为刚才插入了 2 条数据，所以此时显示为 2 条记录，如图 7-4 所示。

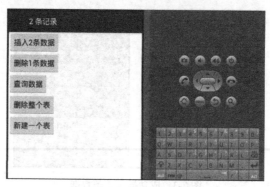

图 7-4　查询数据效果

单击"查询数据"按钮会执行 showItems 方法，实现代码为：

```
/*
 * 在屏幕的 title 区域显示当前数据表当中的数据的条数
 */
private void showItems()
{
    SQLiteDatabase db = mOpenHelper.getReadableDatabase();
    String col[] = { TITLE, BODY };
    Cursor cur = db.query(TABLE_NAME, col, null, null, null, null, null);
    Integer num = cur.getCount();
    setTitle(Integer.toString(num) + " 条记录");
}
```

在以上代码中，语句"Cursor cur = db.query(TABLE_NAME, col, null, null, null, null, null);"比较难以理解，此语句用于将查询到的数据放到一个 Cursor 当中。这个 Cursor 里边封装了这个数据表 TABLE_NAME 当中的所有条列。query()方法相当有用，包含了 7 个参数，各个参数的具体说明如下。

- 第 1 个参数为数据库里表的名字，比如在这个例子中，表的名字即为 TABLE_NAME，也即是"diary"。
- 第 2 个字段是想要返回数据包含的列的信息。在这个例子当中想要得到的列有 title、body。把这两个列的名字放到字符串数组中。
- 第 3 个参数为 selection，相当于 SQL 语句的 where 部分，如果想返回所有的数据，那么就直接置为 null。

- 第 4 个参数为 selectionArgs。在 slection 部分，有可能用到"？"，那么在 selectionArgs 定义的字符串会代替 selection 中的"？"。
- 第 5 个参数为 groupBy。定义查询出来的数据是否分组，如果为 null 则说明不用分组。
- 第 6 个参数为 having，相当于 SQL 语句当中的 having 部分。
- 第 7 个参数为 orderBy，来描述期望的返回值是否需要排序，如果设置为 null 则说明不需要排序。

④ 单击"删除 1 条数据"按钮会删除库中的一条数据，如果成功删除，在屏幕的 title 区域会显示对应的文字提示，如图 7-5 所示。

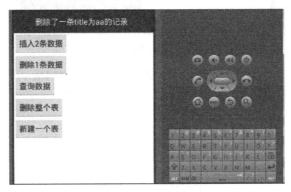

图 7-5　删除一条记录

现在再单击查询数据库按钮，看数据库中的记录是不是少了一条。当单击"删除一条数据"按钮后会执行 deleteItem 方法，其实现代码为：

```
/*
* 删除其中的一条数据
*/
private void deleteItem()
{
    try {
        SQLiteDatabase db = mOpenHelper.getWritableDatabase();
        db.delete(TABLE_NAME, " title = 'mmm'", null);
        setTitle("删除了一条 title 为 aa 的记录");
    } catch (SQLException e)
    {
    }
}
```

在以上代码中，通过"db.delete(TABLE_NAME, " title = 'mmm'", null);"语句删除了一条 title='aa'的数据。当然如果有很多条数据 title 都为'aa'，那么一并删除。Delete 方法各个参数的具体说明如下：

- 第 1 个参数为数据库表名，在这里为 TABLE_NAME，即 diary。
- 第 2 个参数，相当于 SQL 语句当中的 where 部分，也就是描述了删除的条件。

如果在第二个参数当中有"？"符号，那么第三个参数中的字符串会依次替换在第 2 个参数当中出现的"？"符号。

⑤单击"删除整个表"按钮后可以删除数据表 diary,如图 7-6 所示。

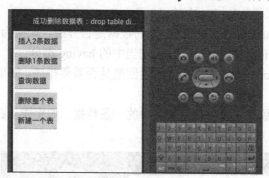

图 7-6　删除表

删除数据表功能是通过方法 dropTable 实现的,在此方法中构造了一个标准的删除数据表的 SQL 语句,然后执行这条语句 db.execSQL(sql)。其实现代码为:

```
/*
 * 删除数据表
 */
private void dropTable()
{
    SQLiteDatabase db = mOpenHelper.getWritableDatabase();
    String sql = "drop table " + TABLE_NAME;
    try {
            db.execSQL(sql);
            setTitle("成功删除数据表: " + sql);
    } catch (SQLException e)
    {
            setTitle("删除错误");
    }
}
```

在以上代码中,构造了一个标准删除数据表的 SQL 语句,然后执行这条语句 db.execSQL(sql)。

⑥ 单击其他按钮,程序会出现异常,在此单击"新建一个表"按钮,如图 7-7 所示。

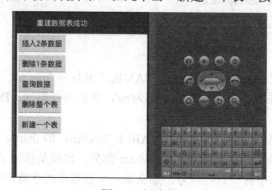

图 7-7　新建表

创建新表功能是通过方法 CreateTable 实现的，其实现代码为：

```
/*
 * 重新建立数据表
 */
private void CreateTable()
{
    SQLiteDatabase db = mOpenHelper.getWritableDatabase();
    String sql = "CREATE TABLE " + TABLE_NAME + " (" + TITLE +" text not null, " + BODY + " text not null " + ");";
    Log.i("aaa:createDB=", sql);
    try {
        db.execSQL("DROP TABLE IF EXISTS diary");
        db.execSQL(sql);
        setTitle("重建数据表成功");
    } catch (SQLException e)
    {
        setTitle("重建错误");
    }
}
```

如果此时单击"查询数据"按钮，则显示有 0 条记录，如图 7-8 所示。

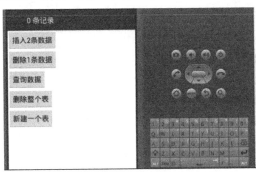

图 7-8　显示 0 条记录

在以上代码中，如果已经存在 diary 表，则需要先删除原有表，因为在同一个数据表中不能出现两张同名的表。

7.3.3　SQLite 数据库存储应用实例

该实例用于允许用户将自己不熟悉的单词添加到系统数据库中，当用户需要查询某个单词或解释时，只要在程序中输入相应的关键词，程序中相应的条目就会显示出来。

【实现步骤】

（1）在 Eclipse 环境下，建立一个名为 E7_3_3 的工程。
（2）打开 res/layout 目录下的 main.xml 文件，代码修改为：

```xml
<?xml version="1.0" encoding="utf-8"?>
<LinearLayout xmlns:android="http://schemas.android.com/apk/res/android"
    android:orientation="vertical"
    android:layout_width="fill_parent"
    android:layout_height="fill_parent">
    <EditText
        android:id="@+id/word"
        android:layout_width="fill_parent"
        android:layout_height="wrap_content" />
    <EditText
        android:id="@+id/detail"
        android:layout_width="fill_parent"
        android:layout_height="wrap_content"
        android:lines="3"/>
    <Button
        android:id="@+id/insert"
        android:layout_width="wrap_content"
        android:layout_height="wrap_content"
        android:text="@string/insert"/>
    <EditText
        android:id="@+id/key"
        android:layout_width="fill_parent"
        android:layout_height="wrap_content" />
    <Button
        android:id="@+id/search"
        android:layout_width="wrap_content"
        android:layout_height="wrap_content"
        android:text="@string/search"/>
    <ListView
        android:id="@+id/show"
        android:layout_width="fill_parent"
        android:layout_height="fill_parent" />
</LinearLayout>
```

（3）在 res/layout 目录下分别新建两个名为 line.xml 及 popup.xml 的文件。
line.xml 文件的代码为：

```xml
<?xml version="1.0" encoding="utf-8"?>
<LinearLayout xmlns:android="http://schemas.android.com/apk/res/android"
    android:orientation="vertical"
    android:layout_width="fill_parent"
    android:layout_height="fill_parent">
```

```xml
<EditText
    android:id="@+id/word"
    android:layout_width="wrap_content"
    android:layout_height="wrap_content"
    android:width="120px"
    android:editable="false"/>
<TextView
    android:layout_width="fill_parent"
    android:layout_height="wrap_content"
    android:text="@string/detail"/>
<EditText
    android:id="@+id/detail"
    android:layout_width="fill_parent"
    android:layout_height="wrap_content"
    android:editable="false"
    android:lines="3"/>
</LinearLayout>
```

popup.xml 文件的代码为：

```xml
<?xml version="1.0" encoding="utf-8"?>
<LinearLayout xmlns:android="http://schemas.android.com/apk/res/android"
    android:orientation="vertical"
    android:layout_width="fill_parent"
    android:layout_height="fill_parent"
    android:gravity="center">
<ImageView
    android:layout_width="fill_parent"
    android:layout_height="wrap_content"
    android:src="@drawable/line" />
<ListView
    android:id="@+id/show"
    android:layout_width="fill_parent"
    android:layout_height="fill_parent" />
</LinearLayout>
```

（4）打开 src/com.example.e7_3_3 目录下的 MainActivity.java 文件，代码修改为：

```java
public class MainActivity extends Activity
{
    E7_4_3DatabaseHelper dbHelper;
    Button insert = null;
    Button search = null;
    @Override
```

```java
public void onCreate(Bundle savedInstanceState)
{
    super.onCreate(savedInstanceState);
    setContentView(R.layout.main);
    // 创建 MyDatabaseHelper 对象，指定数据库版本为1，此处使用相对路径即可
    // 数据库文件自动会保存在程序的数据文件夹的 databases 目录下
    dbHelper = new E7_4_3DatabaseHelper(this, "myDict.db3" , 1);
    insert = (Button)findViewById(R.id.insert);
    search = (Button)findViewById(R.id.search);
    insert.setOnClickListener(new OnClickListener()
    {
        @Override
        public void onClick(View source)
        {
            //获取用户输入
            String word = ((EditText)findViewById(R.id.word))
                .getText().toString();
            String detail = ((EditText)findViewById(R.id.detail))
                .getText().toString();
            //插入生词记录
            insertData(dbHelper.getReadableDatabase() , word , detail);
            //显示提示信息
            Toast.makeText(MainActivity.this, "添加生词成功！ ", 8000).show();
        }
    });
    search.setOnClickListener(new OnClickListener()
    {
        @Override
        public void onClick(View source)
        {
            // 获取用户输入
            String key = ((EditText) findViewById(R.id.key)).getText()
                .toString();
            / 执行查询
            Cursor cursor = dbHelper.getReadableDatabase().rawQuery(
                "select * from dict where word like ? or detail like ?",
                new String[]{"%" + key + "%" , "%" + key + "%"});

            //创建一个 Bundle 对象
            Bundle data = new Bundle();
```

```java
                data.putSerializable("data", converCursorToList(cursor));
                //创建一个 Intent
                Intent intent = new Intent(MainActivity.this, result.class);
                intent.putExtras(data);
                //启动 Activity
                startActivity(intent);
            }
        });
    }
    protected ArrayList<Map<String , String>>
        converCursorToList(Cursor cursor)
    {
        ArrayList<Map<String , String>> result =
            new ArrayList<Map<String , String>>();
        //遍历 Cursor 结果集
        while(cursor.moveToNext())
        {
            //将结果集中的数据存入 ArrayList 中
            Map<String , String> map = new
                HashMap<String , String>();
            //取出查询记录中第 2 列、第 3 列的值
            map.put("word" , cursor.getString(1));
            map.put("detail" , cursor.getString(2));
            result.add(map);
        }
        return result;
    }
    private void insertData(SQLiteDatabase db
        , String word , String detail)
    {
        //执行插入语句
        db.execSQL("insert into dict values(null , ? , ?)" new String[]{word , detail});
    }
    @Override
    public void onDestroy()
    {
        super.onDestroy();
        //退出程序时关闭 MyDatabaseHelper 里的 SQLiteDatabase
        if (dbHelper != null)
```

```
            {
                dbHelper.close();
            }
        }
    }
```

（5）在 src/com.example.e7_3_3 目录下新建两个名为 result.java 及 E7_3_4DatabaseHelper. Java 的文件，代码为：

```
public class E7_3_4DatabaseHelper extends SQLiteOpenHelper
{
    final String CREATE_TABLE_SQL =  "create table dict(_id integer primary key autoincrement , word , detail)";
    public E7_3_4DatabaseHelper(Context context, String name, int version)
    {
        super(context, name, null, version);
    }
    @Override
    public void onCreate(SQLiteDatabase db)
    {
        // 第一次使用数据库时自动建表
        db.execSQL(CREATE_TABLE_SQL);
    }
    @Override
    public void onUpgrade(SQLiteDatabase db, int oldVersion, int newVersion)
    {
        System.out.println("================onUpdate Called================"
            + oldVersion + "--->" + newVersion);
    }
}
```

运行程序，效果如图 7-9 所示。

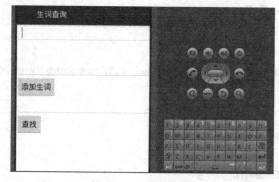

图 7-9　生词查询界面

7.4　ContentProvider 存储

Android 系统和其他的操作系统还不太一样，数据在 Android 当中是私有的，当然这些数据包括文件数据和数据库数据以及一些其他类型的数据。那这个时候有读者就会提出问题，难道两个程序之间就没有办法对数据进行交换？解决这个问题主要靠 ContentProvider。一个 Content Provider 类实现了一组标准的方法接口，从而能够让其他的应用保存或读取此 ContentProvider 的各种数据类型。也就是说，一个程序可以通过实现一个 ContentProvider 的抽象接口将自己的数据暴露出去。外界根本看不到，也不用看到这个应用暴露的数据在应用当中是如何存储的，用数据库存储还是用文件存储，还是通过网上获得，这一切都不重要，重要的是外界可以通过这一套标准及统一的接口和程序里的数据打交道，既可以读取程序的数据，也可以删除程序的数据，当然，中间也会涉及一些权限的问题。

7.4.1　ContentProvider 存储分析

一个程序可以通过实现一个 ContentProvider 的抽象接口将自己的数据完全暴露出去，而且 ContentProvider 是以类似数据库中表的方式将数据暴露的，也就是说 ContentProvider 就像一个"数据库"。那么外界获取其提供的数据，也就应该与从数据库中获取数据的操作基本一样，只不过是采用 URI 来表示外界需要访问的"数据库"。

ContentProvider 提供了一种多应用间数据共享的方式，例如，联系人信息可以被多个应用程序访问。

ContentProvider 是个实现了一组用于提供其他应用程序存取数据的标准方法的类。应用程序可以在 ContentProvider 中执行如下操作：查询数据、修改数据、添加数据、删除数据。Android 提供了一些已经在系统中实现的标准 ContentProvider，如联系人信息，图片库等，你可以用这些 ContentProvider 来访问设备上存储的联系人信息、图片等。

1．查询记录

在 ContentProvider 中使用的查询字符串有别于标准的 SQL 查询。很多诸如 select、add、delete、modify 等操作我们都使用一种特殊的 URI 来实现,这种 URI 由 3 部分组成，"content://"，代表数据的路径，和一个可选的标识数据的 ID。

以下是一些示例 URI：

- content://media/internal/images：这个 URI 将返回设备上存储的所有图片。
- content://contacts/people/：这个 URI 将返回设备上的所有联系人信息。
- content://contacts/people/45：这个 URI 返回单个结果（联系人信息中 ID 为 45 的联系人记录）。

尽管这种查询字符串格式很常见，但是它看起来还是有点令人迷惑。为此，Android 提供一系列的帮助类（在 android.provider 包下），里面包含了很多以类变量形式给出的查询字符串，这种方式更容易让我们理解一点，参见下例：

```
MediaStore.Images.Media.INTERNAL_CONTENT_URI ontacts.People.CONTENT_URI
```

因此，如上面 content://contacts/people/45 这个 URI 就可以写成如下形式：

```
Uri person = ContentUris.withAppendedId(People.CONTENT_URI, 45);
```

然后执行数据查询：

```
Cursor cur = managedQuery(person, null, null, null);
```

这个查询返回一个包含所有数据字段的游标，可以通过迭代这个游标来获取所有的数据。例如，要怎样实现一个依次读取联系人信息表中的指定数据列 name 和 number 呢？代码如下：

```
public class ContentProviderDemo extends Activity
{
    @Override
    public void onCreate(Bundle savedInstanceState)
    {
        super.onCreate(savedInstanceState);
        setContentView(R.layout.main);
        displayRecords();
    }
    private void displayRecords()
    {
        //该数组中包含了所有要返回的字段
        String columns[] = new String[] { People.NAME, Peoplc.NUMBER };
        Uri mContacts = People.CONTENT_URI;
        Cursor cur = managedQuery(
            mContacts,
            columns,            // 要返回的数据字段
            null,               // WHERE 子句
            null,               // WHERE 子句的参数
            null                // Order-by 子句
        );
        if (cur.moveToFirst())
        {
            String name = null;
            String phoneNo = null;
            do{
                // 获取字段的值
            name = cur.getString(cur.getColumnIndex(People.NAME));
                phoneNo = cur.getString(cur.getColumnIndex(People.NUMBER));
                Toast.makeText(this, name + " " + phoneNo, Toast.LENGTH_LONG).show();
            } while (cur.moveToNext());
        }
    }
}
```

2. 修改记录

在 Android 中可以使用 ContentResolver.update()方法来修改数据，实现代码为：

```
private void updateRecord(int recNo, String name)
{
    Uri uri = ContentUris.withAppendedId(People.CONTENT_URI, recNo);
    ContentValues values = new ContentValues();
    values.put(People.NAME, name);
    getContentResolver().update(uri, values, null, null);
}
```

此时可以调用上面的方法来更新指定记录：updateRecord(10, "XYZ"); //更改第 10 条记录的 name 字段值为"XYZ"。

3. 添加记录

要增加记录，可以调用 ContentResolver.insert()方法，该方法接受一个要增加的记录的目标 URI，以及一个包含了新记录值的 Map 对象，调用后的返回值是新记录的 URI，包含记录号。

上面的例子中我们都是基于联系人信息簿这个标准的 ContentProvider，现在继续来创建一个 insertRecord() 方法以对联系人信息簿中进行数据的添加，实现代码为：

```
private void insertRecords(String name, String phoneNo)
{
    ContentValues values = new ContentValues();
    values.put(People.NAME, name);
    Uri uri = getContentResolver().insert(People.CONTENT_URI, values);
    Log.d("ANDROID", uri.toString());
    Uri numberUri = Uri.withAppendedPath(uri, People.Phones.CONTENT_DIRECTORY);
    values.clear();
    values.put(Contacts.Phones.TYPE, People.Phones.TYPE_MOBILE);
    values.put(People.NUMBER, phoneNo);
    getContentResolver().insert(numberUri, values);
}
```

这样就可以调用 insertRecords(name, phoneNo)的方式来向联系人信息簿中添加联系人姓名和电话号码。

4. 删除记录

Content Provider 中的 getContextResolver.delete()方法可以用来删除记录。下面的记录用来删除设备上所有的联系人信息：

```
private void deleteRecords()
{
    Uri uri = People.CONTENT_URI;
    getContentResolver().delete(uri, null, null);
}
```

也可以指定 WHERE 条件语句来删除特定的记录，实现代码为：

```
getContentResolver().delete(uri, "NAME=" + "'XYZ XYZ'", null);
```

这将会删除 name 为"XYZ XYZ"的记录。

7.4.2 ContentProvider 存储创建步骤

总结可知，要创建自己的 ContentProvider 的话，需要遵循以下几步：

（1）创建一个继承了 ContentProvider 父类的类。

（2）定义一个名为 CONTENT_URI，并且是 public static final 的 Uri 类型的类变量，必须为其指定一个唯一的字符串值，最好的命名方式是以类的全名称命名，如：public static final Uri CONTENT_URI = Uri.parse("content://com.google.android.MyContentProvider") 。

（3）创建数据存储系统。大多数 Content Provider 使用 Android 文件系统或 SQLite 数据库来保持数据，但是也可以以任何想要的方式来存储。

（4）定义要返回给客户端的数据列名。如果正在使用 Android 数据库，则数据列的使用方式就和以往所熟悉的其他数据库一样。但是，必须为其定义一个叫 _id 的列，它用来表示每条记录的唯一性。

（5）如果要存储字节型数据，如位图文件等，那保存该数据的数据列其实是一个表示实际保存文件的 URI 字符串，客户端通过它来读取对应的文件数据，处理这种数据类型的 Content Provider 需要实现一个名为 _data 的字段，_data 字段列出了该文件在 Android 文件系统上的精确路径。这个字段不仅供客户端使用，而且也可以供 ContentResolver 使用。客户端可以调用 ContentResolver.openOutputStream()方法来处理该 URI 指向的文件资源，如果是 ContentResolver 本身的话，由于其持有的权限比客户端要高，所以它能直接访问该数据文件。

（6）声明 public static String 类型的变量，用于指定要从游标处返回的数据列。

（7）查询返回一个 Cursor 类型的对象。所有执行写操作的方法如 insert()、update()以及 delete()都将被监听。可以通过使用 ContentResover().notifyChange()方法来通知监听器关于数据更新的信息。

（8）在 AndroidMenifest.xml 中使用标签来设置 Content Provider。

（9）如果要处理的数据类型是一种比较新的类型，就必须先定义一个新的 MIME 类型，以供 ContentProvider.geType(url)来返回。

7.4.3 ContentProvider 应用实例

使用 ContentProvider 实现共享单词本数据。

【实现步骤】

（1）在 Eclipse 环境下，建立一个名为 E7_4_3 的工程。

（2）打开 res/layout 目录下的 main.xml 文件，代码修改为：

```
<?xml version="1.0" encoding="utf-8"?>
<LinearLayout xmlns:android="http://schemas.android.com/apk/res/android"
    android:orientation="vertical"
    android:layout_width="fill_parent"
    android:layout_height="fill_parent">
<EditText
    android:id="@+id/word"
    android:layout_width="fill_parent"
    android:layout_height="wrap_content" />
```

```xml
<EditText
    android:id="@+id/detail"
    android:layout_width="fill_parent"
    android:layout_height="wrap_content"
    android:lines="3"/>
<Button
    android:id="@+id/insert"
    android:layout_width="wrap_content"
    android:layout_height="wrap_content"
    android:text="@string/insert"/>
<EditText
    android:id="@+id/key"
    android:layout_width="fill_parent"
    android:layout_height="wrap_content" />
<Button
    android:id="@+id/search"
    android:layout_width="wrap_content"
    android:layout_height="wrap_content"
    android:text="@string/search"/>
<ListView
    android:id="@+id/show"
    android:layout_width="fill_parent"
    android:layout_height="fill_parent" />
</LinearLayout>
```

（3）在 res/layout 目录下建立的 line.xml 及 popup.xml 文件的代码与 7.3.3 节相同。

（4）在此为该 ContentProvider 定义一个工具类，该类中只是包含一个 public static 的常量（在 src/com.example.e7_4_3 目录下建立一个名为 word.java 的文件）。其代码为：

```java
public final class word
{
    // 定义该 ContentProvider 的 Authority
    public static final String AUTHORITY
        = "org.crazyit.providers.dictprovider";
    //定义一个静态内部类
    public static final class Word implements BaseColumns
    {
        // 定义 Content 所允许操作的 3 个数据列
        public final static String _ID = "_id";
        public final static String WORD = "word";
        public final static String DETAIL = "detail";
        // 定义该 Content 提供服务的两个 Uri
        public final static Uri DICT_CONTENT_URI =
```

```
                    Uri.parse("content://" +   AUTHORITY + "/words");
            public final static Uri WORD_CONTENT_URI =
                    Uri.parse("content://" +   AUTHORITY + "/word");
        }
    }
```

（5）开发一个 ContentProvider 的子类，并重写其中的增、删、改、查等方法（打开 src/com.example.e7_4_3 目录下的 MainActivity.java 文件），代码修改为：

```
    public class MainActivity extends ContentProvider
    {
        private static UriMatcher matcher
            = new UriMatcher(UriMatcher.NO_MATCH);
        private static final int WORDS = 1;
        private static final int WORD = 2;
        private E7_4_3DatabaseHelper dbOpenHelper;
        static
        {
            // 为 UriMatcher 注册两个 Uri
            matcher.addURI(word.AUTHORITY, "words", WORDS);
            matcher.addURI(word.AUTHORITY, "word/#", WORD);
        }
        // 第一次调用该 DictProvider 时，系统先创建 DictProvider 对象，并回调该方法
        @Override
        public boolean onCreate()
        {
            dbOpenHelper = new E7_4_3DatabaseHelper(this.getContext(), "myDict.db3", 1);
            return true;
        }
        // 插入数据方法
        @Override
        public Uri insert(Uri uri, ContentValues values)
        {
            // 获得数据库实例
            SQLiteDatabase db = dbOpenHelper.getReadableDatabase();
            // 插入数据，返回行 ID
            long rowId = db.insert("dict", word.Word._ID, values);
            // 如果插入成功返回 uri
            if (rowId > 0)
            {
                // 在已有的 Uri 的后面追加 ID 数据
                Uri wordUri = ContentUris.withAppendedId(uri, rowId);
```

```java
            // 通知数据已经改变
            getContext().getContentResolver().notifyChange(wordUri, null);
            return wordUri;
        }
        return null;
    }
    // 删除数据的方法
    @Override
    public int delete(Uri uri, String selection, String[] selectionArgs)
    {
        SQLiteDatabase db = dbOpenHelper.getReadableDatabase();
        // 记录所删除的记录数
        int num = 0;
        // 对 uri 进行匹配
        switch (matcher.match(uri))
        {
            case WORDS:
                num = db.delete("dict", selection, selectionArgs);
                break;
            case WORD:
                // 解析出所需要删除的记录 ID
                long id = ContentUris.parseId(uri);
                String where = word.Word._ID + "=" + id;
                // 如果原来的 where 子句存在，拼接 where 子句
                if (selection != null && !selection.equals(""))
                {
                    where = where + " and " + selection;
                }
                num = db.delete("dict", where, selectionArgs);
                break;
            default:
                throw new IllegalArgumentException("未知 Uri:" + uri);
        }
        // 通知数据已经改变
        getContext().getContentResolver().notifyChange(uri, null);
        return num;
    }
    // 修改数据的方法
    @Override
```

```java
public int update(Uri uri, ContentValues values, String selection,
    String[] selectionArgs)
{
    SQLiteDatabase db = dbOpenHelper.getWritableDatabase();
    // 记录所修改的记录数
    int num = 0;
    switch (matcher.match(uri))
    {
        case WORDS:
            num = db.update("dict", values, selection, selectionArgs);
            break;
        case WORD:
            // 解析出想修改的记录 ID
            long id = ContentUris.parseId(uri);
            String where = word.Word._ID + "=" + id;
            // 如果原来的 where 子句存在，拼接 where 子句
            if (selection != null && !selection.equals(""))
            {
                where = where + " and " + selection;
            }
            num = db.update("dict", values, where, selectionArgs);
            break;
        default:
            throw new IllegalArgumentException("未知 Uri:" + uri);
    }
    // 通知数据已经改变
    getContext().getContentResolver().notifyChange(uri, null);
    return num;
}
// 查询数据的方法
@Override
public Cursor query(Uri uri, String[] projection, String selection,
    String[] selectionArgs, String sortOrder)
{
    SQLiteDatabase db = dbOpenHelper.getReadableDatabase();
    switch (matcher.match(uri))
    {
        case WORDS:
            // 执行查询
```

```
            return db.query("dict", projection, selection, selectionArgs,
                null, null, sortOrder);
        case WORD:
            // 解析出想查询的记录 ID
            long id = ContentUris.parseId(uri);
            String where = word.Word._ID + "=" + id;
            // 如果原来的 where 子句存在，拼接 where 子句
            if (selection != null && !"".equals(selection))
            {
                where = where + " and " + selection;
            }
            return db.query("dict", projection, where, selectionArgs, null,
                null, sortOrder);
        default:
            throw new IllegalArgumentException("未知 Uri:" + uri);
        }
    }
    // 返回指定 uri 参数对应的数据的 MIME 类型
    @Override
    public String getType(Uri uri)
    {
        switch (matcher.match(uri))
        {
            // 如果操作的数据是多项记录
            case WORDS:
                return "vnd.android.cursor.dir/org.crazyit.dict";
            // 如果操作的数据是单项记录
            case WORD:
                return "vnd.android.cursor.item/org.crazyit.dict";
            default:
                throw new IllegalArgumentException("未知 Uri:" + uri);
        }
    }
}
```

（6）在 AndroidManifest.xml 文件中注册该 ContentProvider，这就需要在 Android 的 Manifest.xml 文件中增加如下配置：

```xml
<!-- 注册一个 ContentProvider -->
<provider android:name=".DictProvider"
    android:authorities="com.example.e7_4_3.MainActivity"/>
```

(7) 提供添加单词、查询单词功能（在 src/com.example.e7_4_3 目录下建立一个名为 E7_4_3 Database Helper.java 的文件），代码为：

```java
public class DictResolver extends Activity
{
    ContentResolver contentResolver;
    Button insert = null;
    Button search = null;
    @Override
    public void onCreate(Bundle savedInstanceState)
    {
        super.onCreate(savedInstanceState);
        setContentView(R.layout.main);
        // 获取系统的 ContentResolver 对象
        contentResolver = getContentResolver();
        insert = (Button)findViewById(R.id.insert);
        search = (Button)findViewById(R.id.search);
        // 为 insert 按钮的单击事件绑定事件监听器
        insert.setOnClickListener(new OnClickListener()
        {
            @Override
            public void onClick(View source)
            {
                //获取用户输入
                String words = ((EditText)findViewById(R.id.word))
                    .getText().toString();
                String detail = ((EditText)findViewById(R.id.detail))
                    .getText().toString();
                //插入生词记录
                ContentValues values = new ContentValues();
                values.put(word.Word.WORD , words);
                values.put(word.Word.DETAIL , detail);
                contentResolver.insert(word.Word.DICT_CONTENT_URI , values);
                //显示提示信息
                Toast.makeText(DictResolver.this, "添加生词成功！" , 8000)
                    .show();
            }
        });
        // 为 search 按钮的单击事件绑定事件监听器
        search.setOnClickListener(new OnClickListener()
        {
```

```java
            @Override
            public void onClick(View source)
            {
                // 获取用户输入
                String key = ((EditText) findViewById(R.id.key)).getText()
                    .toString();
                // 执行查询
                Cursor cursor = contentResolver.query(
                    word.Word.DICT_CONTENT_URI, null
                    , "word like ? or detail like ?"
                    , new String[]{"%" + key + "%" , "%" + key + "%"}
                    , null);
                //创建一个 Bundle 对象
                Bundle data = new Bundle();
                data.putSerializable("data", converCursorToList(cursor));
                //创建一个 Intent
                Intent intent = new Intent(DictResolver.this, result.class);
                intent.putExtras(data);
                //启动 Activity
                startActivity(intent);
            }
        });
    }
    private ArrayList<Map<String, String>> converCursorToList(
        Cursor cursor)
    {
        ArrayList<Map<String, String>> result
            = new ArrayList<Map<String, String>>();
        // 遍历 Cursor 结果集
        while (cursor.moveToNext())
        {
            // 将结果集中的数据存入 ArrayList 中
            Map<String, String> map = new HashMap<String, String>();
            // 取出查询记录中第 2 列、第 3 列的值
            map.put(word.Word.WORD, cursor.getString(1));
            map.put(word.Word.DETAIL, cursor.getString(2));
            result.add(map);
        }
        return result;
    }
}
```

7.5 网络存储

前面介绍的几种存储都是将数据存储在本地设备上，除此之外，还有一种存储（获取）数据的方式，通过网络来实现数据的存储和获取。可以调用 WebService 返回的数据或解析 HTTP 协议实现网络数据交互。

具体需要熟悉 java.net.*和 Android.net.*这两个包的内容，在此不做详细介绍。

本实例的功能是通过邮政编码查询该地区的天气预报，以 POST 发送的方式发送请求到 webservicex.net 站点，访问 WebService.webservicex.net 站点上提供查询天气预报的服务。具体请参考其 WSDL 文档，网址如下：

　　　　http://www.webservicex.net/WeatherForecast.asmx?WSDL

输入：美国某个城市的邮政编码。

输出：该邮政编码对应城市的天气预报。

实现该示例的代码为：

```java
public class MyAndroidWeatherActivity extends Activity
{
    //定义需要获取的内容来源地址
    private static final String SERVER_URL = "http://www.webservicex.net/ WeatherForecast.asmx/GetWeatherByPlaceName";
    /* 第一次调用活动 */
    @Override
    public void onCreate(Bundle savedInstanceState)
    {
        super.onCreate(savedInstanceState);
        setContentView(R.layout.main);
        //根据内容来源地址创建一个 Http 请求
        HttpPost request = new HttpPost(SERVER_URL);
        // 添加一个变量
        List<NameValuePair> params = new ArrayList<NameValuePair>();
        // 设置一个地区名称
        params.add(new BasicNameValuePair("PlaceName", "NewYork")); //添加必须的参数
        try {
            //设置参数的编码
            request.setEntity(new UrlEncodedFormEntity(params, HTTP.UTF_8));
            //发送请求并获取反馈
            HttpResponse httpResponse = new DefaultHttpClient().execute(request);

            // 解析返回的内容
            if(httpResponse.getStatusLine().getStatusCode() != 404)
            {
                String result = EntityUtils.toString(httpResponse.getEntity());
```

```
                    System.out.println(result);
                }
            } catch (Exception e)
            {
                e.printStackTrace();
            }
        }
    }
}
```

注意： 在配置文件中设置访问网络权限如下：

`<uses-permission android:name="android.permission.INTERNET" />`

第 8 章 Android 传递消息与联网

作为一个移动手机设备,当然需要具备交互通信的功能。手机中的交互通信方式有多种,例如,常见的通话、短信、邮件及 Wi-Fi 等。

8.1 电话管理器

TelephonyManager 为一个管理手机通话状态、电话网络信息的服务类,该类提供了大量的 getXXX()方法来获取电话网络的相关信息。

8.1.1 网络与 SIM 卡获取信息

通过 TelephonyManager 提供的系列方法即可获取手机网络、SIM 卡的相关信息,该程序使用了一个 ListView 来显示网络和 SIM 卡的相关信息。

【实现步骤】

(1) 在 Eclipse 环境下建立一个名为 E8_1_1 的工程。
(2) 打开 res/layout 目录下的 main.xml 文件,代码修改为:

```xml
<?xml version="1.0" encoding="utf-8"?>
<LinearLayout xmlns:android="http://schemas.android.com/apk/res/android"
    android:layout_width="fill_parent"
    android:layout_height="fill_parent"
    android:orientation="vertical" >
    <ListView
        android:id="@+id/listview"
        android:layout_width="fill_parent"
        android:layout_height="wrap_content"
        android:text="@string/hello_world" />
</LinearLayout>
```

(3) 在 res/layout 目录下新建一个名为 line.xml 的文件,代码为:

```xml
<?xml version="1.0" encoding="utf-8"?>
<LinearLayout xmlns:android="http://schemas.android.com/apk/res/android"
    android:orientation="horizontal"
    android:layout_width="fill_parent"
    android:layout_height="wrap_content">
    <TextView
```

```xml
    android:id="@+id/name"
    android:layout_width="wrap_content"
    android:layout_height="wrap_content"
    android:width="120px"
    android:textSize="16dip"/>
<TextView
    android:id="@+id/value"
    android:layout_width="fill_parent"
    android:layout_height="wrap_content"
    android:paddingLeft="8px"
    android:textSize="16dip"/>
</LinearLayout>
```

（4）在 res/values 目录下新建一个名为 array.xml 的文件，代码为：

```xml
<?xml version="1.0" encoding="utf-8"?>
<resources>
    <!-- 声明一个名为 statusNames 的字符串数组 -->
    <string-array name="statusNames">
        <item>设备编号</item>
        <item>软件版本</item>
        <item>网络运营商代号</item>
        <item>网络运营商名称</item>
        <item>手机制式</item>
        <item>设备当前位置</item>
        <item>SIM 卡的国别</item>
        <item>SIM 卡序列号</item>
        <item>SIM 卡状态</item>
    </string-array>
    <!-- 声明一个名为 simState 的字符串数组 -->
    <string-array name="simState">
        <item>状态未知</item>
        <item>无 SIM 卡</item>
        <item>被 PIN 加锁</item>
        <item>被 PUK 加锁</item>
        <item>被 NetWork PIN 加锁</item>
        <item>已准备好</item>
    </string-array>
    <!-- 声明一个名为 phoneType 的字符串数组 -->
    <string-array name="phoneType">
        <item>未知</item>
        <item>GSM</item>
        <item>CDMA</item>
```

　　　　</string-array>
　</resources>

(5) 打开 src/com.example.e8_1_1 目录下的 MainActivity.java 文件，代码修改为：

```java
public class MainActivity extends Activity
{
    /** 第一次调用活动 */
    ListView showView;
    //状态名数组
    String[] statusNames;
    //手机状态的集合
    ArrayList<String> statusValues = new ArrayList<String>();
    @Override
    public void onCreate(Bundle savedInstanceState)
    {
        super.onCreate(savedInstanceState);
        setContentView(R.layout.main);
        //获取系统的 TelePhonyManager
        TelephonyManager phoneManger = (TelephonyManager)
            getSystemService(Context.TELEPHONY_SERVICE);
        //自己定义:各种状态的数组
        statusNames = getResources().getStringArray(R.array.statusNames);
        //获取 SIM 卡的状态数组
        String[] simState = getResources().getStringArray(R.array.simState);
        //获取代表电话网络类型的数组
        String[] phoneType = getResources().getStringArray(R.array.phoneType);
        //获取设备编号
        statusValues.add(phoneManger.getDeviceId());
        //系统平台版本
        statusValues.add(phoneManger.getDeviceSoftwareVersion() != null ?
            phoneManger.getDeviceSoftwareVersion()
                : "未知");
        //网络运营商代号
        statusValues.add(phoneManger.getNetworkOperator());
        // 获取手机网络类型
        statusValues.add(phoneType[phoneManger.getPhoneType()]);
        //网络运营商名称
        statusValues.add(phoneManger.getNetworkOperatorName());
        //获取设备所在位置
        statusValues.add(phoneManger.getCellLocation() != null ?
            phoneManger.getCellLocation().toString()
                : "未知位置");
```

```
//获取 SIM 卡的国别
statusValues.add(phoneManger.getSimCountryIso());
//获取 SIM 卡的序列号
statusValues.add(phoneManger.getSimSerialNumber());
//获取 SIM 卡的状态
statusValues.add(simState[phoneManger.getSimState()]);
showView = (ListView) findViewById(R.id.listview);
ArrayList<Map<String, String>> status = new ArrayList<Map<String, String>>();
//遍历 statusValues 集合，将 statusNames、statusValues
//的数据封装到 List<Map<String , String>>集合中
for (int i = 0; i < statusValues.size(); i++)
{
    HashMap<String, String> map = new HashMap<String, String>();
    map.put("name", statusNames[i]);
    map.put("value", statusValues.get(i));
    status.add(map);
}
// 使用 SimpleAdapter 封装 List 数据
SimpleAdapter adapter = new SimpleAdapter(this, status, R.layout.line,
        new String[] { "name", "value" }, new int[] { R.id.name,
            R.id.value });
// 为 ListView 设置 Adapter
showView.setAdapter(adapter);
    }
}
```

（6）在 AndroidManidfest.xml 文件中添加权限，不加会报错，添加代码为：

```
<!-- 添加访问手机位置的权限 -->
<uses-permission android:name="android.permission.ACCESS_COARSE_LOCATION"/>
<!-- 添加访问手机状态的权限 -->
<uses-permission android:name="android.permission.READ_PHONE_STATE"/>
```

运行程序，效果如图 8-1 所示。

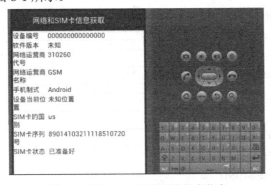

图 8-1　获取 SIM 卡和网络状态信息

8.1.2 拨打电话

在拨打电话时，首先要在 AndroidManifest 中添加 uses-permission，这样就实现了对拨打电话的声明，然后通过自定义的 Intent 对象，通过"ACTION_CALL"键和 Uri.parse()方法将用户输入的电话号码输入，最后通过 startActivity 方法即可完成程序拨打电话的功能。

【实现步骤】

（1）在 Eclipse 环境下，建立一个名为 E8_1_2 的工程。

（2）打开 res/layout 目录下的 main.xml 文件，代码修改为：

```xml
<?xml version="1.0" encoding="utf-8"?>
<AbsoluteLayout
    xmlns:android="http://schemas.android.com/apk/res/android"
    android:id="@+id/myWidget1"
    android:layout_width="fill_parent"
    android:layout_height="fill_parent" >
    <EditText
        android:id="@+id/myEditText1"
        android:layout_width="wrap_content"
        android:layout_height="wrap_content"
        android:text=""
        android:textSize="18sp"
        android:layout_x="12px"
        android:layout_y="15px" >
    </EditText>
    <Button
        android:id="@+id/myButton1"
        android:layout_width="wrap_content"
        android:layout_height="wrap_content"
        android:text="@string/str_button1"
        android:layout_x="10px"
        android:layout_y="65px" >
    </Button>
</AbsoluteLayout>
```

（3）打开 src/com.example.e8_1_2 目录下的 MainActivity.java 文件，代码修改为：

```java
public class MainActivity extends Activity
{
    /*声明 Button 与 EditText 对象名称*/
    private Button mButton1;
    private EditText mEditText1;
```

```java
/** 第一次调用活动 */
@Override
public void onCreate(Bundle savedInstanceState)
{
    super.onCreate(savedInstanceState);
    setContentView(R.layout.main);
    /*通过 findViewById 构造器来构造 EditText 与 Button 对象*/
    mEditText1 = (EditText) findViewById(R.id.myEditText1);
    mButton1 = (Button) findViewById(R.id.myButton1);
    /*设置 Button 对象的 OnClickListener 来聆听 OnClick 事件*/
    mButton1.setOnClickListener(new Button.OnClickListener()
    {
        @Override
        public void onClick(View v)
        {
            try
            {
                /*取得 EditText 中用户输入的字符串*/
                String strInput = mEditText1.getText().toString();
                if (isPhoneNumberValid(strInput)==true)
                {
                    /*建构一个新的 Intent 用于运行 action.CALL 的常数，可以通过 Uri 载入字符串*/
                    Intent myIntentDial = new
                    Intent
                    (
                        "android.intent.action.CALL",
                        Uri.parse("tel:"+strInput)
                    );
                    /*在 startActivity()方法中带入自定义的 Intent 对象以运行拨打电话的工作 */
                    startActivity(myIntentDial);
                    mEditText1.setText("");
                }
                else
                {
                    mEditText1.setText("");
                    Toast.makeText(
                    MainActivity.this, "输入的电话号码格式不符，请重新输入",
                    Toast.LENGTH_LONG).show();
                }
            }
```

```
                    catch(Exception e)
                    {
                        e.printStackTrace();
                    }
                }
            });
        }
        /*检查字符串是否为电话号码的方法,并返回 true or false 的判断值*/
        public static boolean isPhoneNumberValid(String phoneNumber)
        {
            boolean isValid = false;
            /* 可接受的电话号码格式有:
             * ^\\(? : 可以使用 "(" 作为开头
             * (\\d{3}): 紧接着三个数字
             * \\)? : 可以使用")"接续
             * [- ]? : 在上述格式后可以使用具选择性的  "-"
             * (\\d{4}) : 再紧接着三个数字
             * [- ]? : 可以使用具选择性的  "-"  接续
             * (\\d{4})$: 以四个数字结束
             * 可以比较下列数字格式:
             * (123)456-78900, 123-4560-7890, 12345678900, (123)-4560-7890
             */
            String expression = "^\\(?(\\d{3})\\)?[- ]?(\\d{3})[- ]?(\\d{5})$";
            String expression2 ="^\\(?(\\d{3})\\)?[- ]?(\\d{4})[- ]?(\\d{4})$";
            CharSequence inputStr = phoneNumber;
            /*创建 Pattern*/
            Pattern pattern = Pattern.compile(expression);
            /*将 Pattern  以参数传入 Matcher 当做 Regular expression*/
            Matcher matcher = pattern.matcher(inputStr);
            /*创建 Pattern2*/
            Pattern pattern2 =Pattern.compile(expression2);
            /*将 Pattern2  以参数传入 Matcher2 当做 Regular expression*/
            Matcher matcher2= pattern2.matcher(inputStr);
            if(matcher.matches()||matcher2.matches())
            {
                isValid = true;
            }
            return isValid;
        }
    }
```

（4）设置拨打电话权限，打开 AndroidManifest.xml 文件，代码修改为：

```xml
<?xml version="1.0" encoding="utf-8"?>
<manifest xmlns:android="http://schemas.android.com/apk/res/android"
    package="com.example.e8_1_2"
    android:versionCode="1"
    android:versionName="1.0.0">
    <application android:icon="@drawable/icon" android:label="@string/app_name">
        <activity android:name="com.example.e8_1_2.MainActivity"
            android:label="@string/app_name">
            <intent-filter>
                <action android:name="android.intent.action.MAIN" />
                <category android:name="android.intent.category.LAUNCHER" />
            </intent-filter>
        </activity>
    </application>
    <!-- 拨打电话权限 -->
    <uses-permission android:name="android.permission.CALL_PHONE"></uses-permission>
</manifest>
```

运行程序，效果如图 8-2 所示。

图 8-2　执行效果

如果输入不规范的号码会输出对应的提示，如图 8-3 所示。

当输入规范的号码并单击"拨打"按钮时，会实现拨号通话处理并显示界面，如图 8-4 所示。

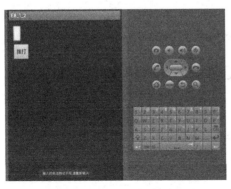

图 8-3　显示提示信息

图 8-4　通话界面

8.1.3 监听手机来电

在 Android 中提供了 PhoneStateListener 作为通话状态的监听器，该监听器可用于对 TelephonyManager 进行监听。当手机来电铃响时，程序将会把号码记录到文件中。

【实现步骤】

（1）在 Eclipse 环境下，建立一个名为 E8_1_3 的工程。

（2）打开 src/com.example.e8_1_3 目录下的 MainActivity.java 文件，代码修改为：

```java
public class MainActivity extends Activity
{
    TelephonyManager tManager;
    @Override
    public void onCreate(Bundle savedInstanceState)
    {
        super.onCreate(savedInstanceState);
        setContentView(R.layout.main);
        // 取得 TelephonyManager 对象
        tManager = (TelephonyManager) getSystemService
            (Context.TELEPHONY_SERVICE);
        // 创建一个通话状态监听器
        PhoneStateListener listener = new PhoneStateListener()
        {
            @Override
            public void onCallStateChanged(int state
                , String incomingNumber)
            {
                switch (state)
                {
                    // 无任何状态
                    case TelephonyManager.CALL_STATE_IDLE:
                        break;
                    case TelephonyManager.CALL_STATE_OFFHOOK:
                        break;
                    // 来电铃响时
                    case TelephonyManager.CALL_STATE_RINGING:
                        OutputStream os = null;
                        try
                        {
                            os = openFileOutput("phoneList", MODE_APPEND);
                        }
```

```
                    catch (FileNotFoundException e)
                    {
                        e.printStackTrace();
                    }
                    PrintStream ps = new PrintStream(os);
                    // 将来电号码记录到文件中
                    ps.println(new Date() + "     来电："+ incomingNumber);
                    ps.close();
                    break;
                default:
                    break;
                }
                super.onCallStateChanged(state, incomingNumber);
            }
        };
        //监听电话通话状态的改变
        tManager.listen(listener, PhoneStateListener.LISTEN_CALL_STATE);
    }
}
```

运行程序，在保证该程序运行的状态下，再启动另一个模拟器呼叫该电话。接下来就可以在 DDMS 的 File Explorer 面板的根目录下看到一个 phoneList 文件，将该文件导出到计算机上。

8.2 信息处理

短信是任何一款手机不可或缺的基本应用，是使用频率最高的程序之一。在此将用程序代码来介绍信息处理。

8.2.1 发送短信

这个例子中不是简单地使用 Intent 激活 Android 自带的短信程序，而是使用 SmsManager 类完成发送短信的功能。

【实现步骤】
（1）在 Eclipse 环境下，建立一个名为 E8_2 的工程。
（2）打开 res/layout 目录下的 main.xml 文件，代码修改为：

```
<?xml version="1.0" encoding="utf-8"?>
<LinearLayout xmlns:android="http://schemas.android.com/apk/res/android"
    android:orientation="vertical"
    android:layout_width="fill_parent"
    android:layout_height="fill_parent">
```

```xml
<TextView
    android:layout_width="fill_parent"
    android:layout_height="wrap_content"
    android:text="输入发送短信号码"/>
<EditText
    android:id="@+id/txtPhoneNo"
    android:layout_width="fill_parent"
    android:layout_height="wrap_content"/>
<TextView
    android:layout_width="fill_parent"
    android:layout_height="wrap_content"
    android:text="短信"/>
<EditText
    android:id="@+id/txtMessage"
    android:layout_width="fill_parent"
    android:layout_height="150px"
    android:gravity="top" />
<Button
    android:id="@+id/btnSendSMS"
    android:layout_width="fill_parent"
    android:layout_height="wrap_content"
    android:text="发送短信"/>
</LinearLayout>
```

(3) 设置发送短信的权限声明，打开 AndroidManifest.xml 文件，添加如下代码：

```xml
</application>
    <!--设置短信权限-->
    <uses-permission android:name="android.permission.SEND_SMS" />
</manifest>
```

(4) 打开 src/com.example.e8_2 目录下的 MainActivity.java 文件，代码修改为：

```java
public class MainActivity extends Activity
{
    /** 第一次调用活动. */
    @Override
    public void onCreate(Bundle savedInstanceState)
    {
        super.onCreate(savedInstanceState);
        setContentView(R.layout.main);
        final Button btnSendSMS = (Button) findViewById(R.id.btnSendSMS);
        final EditText txtPhoneNo = (EditText) findViewById(R.id.txtPhoneNo);
        final EditText txtMessage = (EditText) findViewById(R.id.txtMessage);
        btnSendSMS.setOnClickListener(new View.OnClickListener()
```

```java
            {
                public void onClick(View v)
                {
                    String phoneNo = txtPhoneNo.getText().toString();
                    String message = txtMessage.getText().toString();
                    if (phoneNo.length()>0 && message.length()>0)
                    {
                            Log.v("ROGER", "will begin sendSMS");
                            sendSMS(phoneNo, message);
                    }
                    else
                        Toast.makeText(MainActivity.this, "请重新输入电话号码和短信内容",
                            Toast.LENGTH_LONG).show();
                }
        });
    }
    private void sendSMS(String phoneNumber, String message)
    {
        PendingIntent pi = PendingIntent.getActivity(this, 0,
            new Intent(this, MainActivity.class), 0);
        Log.v("ROGER", "will init SMS Manager");
        SmsManager sms = SmsManager.getDefault();
        Log.v("ROGER", "will send SMS");
        sms.sendTextMessage(phoneNumber, null, message, pi, null);
    }
}
```

运行程序，效果如图 8-5 所示。

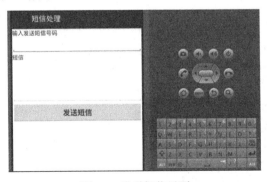

图 8-5 发送短信界面

SmsManager 是 android.telephony.gsm.SmsManager 中定义的用户管理短信应用的类。它的用法有点特殊，开发人员不用直接实例化 SmsManager 类，而只需要调用静态方法 getDefault()获得 SmsManager 对象，方法 sendTextMessage()用于发送短信到指定号码。

8.2.2 群发短信

短信群发是十分实用的功能，逢年过节，很多人都喜欢通过群发短信向自己的朋友表示祝福。短信群发可以将一条短信同时向多个人发送。短信群发的实现机制十分简单，只要让程序遍历每个收件人号码并依次给每个收件人发送短信即可。

本示例用于实现短信的群发。

【实现步骤】

（1）在 Eclipse 环境下，建立一个名为 E8_2_2 的工程。

（2）打开 res/layout 目录下的 main.xml 文件，代码修改为：

```xml
<?xml version="1.0" encoding="utf-8"?>
<LinearLayout xmlns:android="http://schemas.android.com/apk/res/android"
    android:orientation="vertical"
    android:layout_width="fill_parent"
    android:layout_height="fill_parent">
    <TextView
        android:layout_width="fill_parent"
        android:layout_height="wrap_content"
        android:text="@string/numbers"/>
    <EditText
        android:id="@+id/numbers"
        android:layout_width="fill_parent"
        android:layout_height="wrap_content"
        android:editable="false"
        android:cursorVisible="false"
        android:lines="2"/>
    <TextView
        android:layout_width="fill_parent"
        android:layout_height="wrap_content"
        android:text="@string/content"/>
    <EditText
        android:id="@+id/content"
        android:layout_width="fill_parent"
        android:layout_height="wrap_content"
        android:lines="2"/>
    <LinearLayout
        android:orientation="horizontal"
        android:layout_width="fill_parent"
        android:layout_height="fill_parent"
        android:gravity="center_horizontal">
```

```xml
<Button
    android:id="@+id/select"
    android:layout_width="wrap_content"
    android:layout_height="wrap_content"
    android:text="@string/select"/>
<Button
    android:id="@+id/send"
    android:layout_width="wrap_content"
    android:layout_height="wrap_content"
    android:text="@string/send"/>
</LinearLayout>
</LinearLayout>
```

（3）在 res/layout 目录下新建一个名为 list.xml 的文件，代码为：

```xml
<?xml version="1.0" encoding="UTF-8"?>
<LinearLayout xmlns:android="http://schemas.android.com/apk/res/android"
    android:orientation="vertical"
    android:layout_width="fill_parent"
    android:layout_height="fill_parent">
<ListView
    android:id="@+id/list"
    android:layout_width="fill_parent"
    android:layout_height="wrap_content"/>
</LinearLayout>
```

（4）设置短信群发权限，打开 AndroidManifest.xml 文件，添加代码如下：

```xml
</application>
    <!-- 授予读联系人 ContentProvider 的权限 -->
    <uses-permission android:name="android.permission.READ_CONTACTS"/>
    <!-- 授予发送短信的权限 -->
    <uses-permission android:name="android.permission.SEND_SMS"/>
</manifest>
```

（5）打开 src/com.example.e8_2_2 目录下的 MainActivity.java 文件，代码修改为：

```java
public class MainActivity extends Activity
{
    EditText numbers, content;
    Button select, send;
    SmsManager sManager;
    // 记录需要群发的号码列表
    ArrayList<String> sendList = new ArrayList<String>();
    @Override
    public void onCreate(Bundle savedInstanceState)
    {
```

```java
super.onCreate(savedInstanceState);
setContentView(R.layout.main);
sManager = SmsManager.getDefault();
// 获取界面上的文本框、按钮组件
numbers = (EditText) findViewById(R.id.numbers);
content = (EditText) findViewById(R.id.content);
select = (Button) findViewById(R.id.select);
send = (Button) findViewById(R.id.send);
// 为 send 按钮的单击事件绑定监听器
send.setOnClickListener(new OnClickListener()
{
    @Override
    public void onClick(View v)
    {
        for (String number : sendList)
        {
            // 创建一个 PendingIntent 对象
            PendingIntent pi = PendingIntent.getActivity(
                MainActivity.this, 0, new Intent(), 0);
            // 发送短信
            sManager.sendTextMessage(number, null
                , content.getText()
                .toString(), pi, null);
        }
        // 提示短信群发完成
        Toast.makeText(MainActivity.this, "短信群发完成", 8000)
            .show();
    }
});
// 为 select 按钮的单击事件绑定监听器
select.setOnClickListener(new OnClickListener()
{
    @Override
    public void onClick(View v)
    {
        // 查询联系人的电话号码
        final Cursor cursor = getContentResolver().query(
            ContactsContract.CommonDataKinds.Phone.CONTENT_URI
            , null, null, null, null);
        BaseAdapter adapter = new BaseAdapter()
```

```java
    {
        @Override
        public int getCount()
        {
            return cursor.getCount();
        }
        @Override
        public Object getItem(int position)
        {
            return position;
        }
        @Override
        public long getItemId(int position)
        {
            return position;
        }
        @Override
        public View getView(int position, View convertView,
            ViewGroup parent)
        {
            cursor.moveToPosition(position);
            CheckBox rb = new CheckBox(MainActivity.this);
            // 获取联系人的电话号码，并去掉中间的连线
            String number = cursor
                .getString(
                cursor.getColumnIndex(ContactsContract.CommonDataKinds.
                Phone.NUMBER))
                .replace("-", "");
            rb.setText(number);
            // 如果该号码已经被加入发送人名单，默认勾选该号码
            if (isChecked(number))
            {
                rb.setChecked(true);
            }
            return rb;
        }
    };
    // 加载 list.xml 布局文件对应的 View
    View selectView = getLayoutInflater().inflate(R.layout.list,null);
    // 获取 selectView 中的名为 list 的 ListView 组件
```

```java
                    final ListView listView = (ListView) selectView
                        .findViewById(R.id.list);
                    listView.setAdapter(adapter);
                    new AlertDialog.Builder(MainActivity.this)
                        .setView(selectView)
                        .setPositiveButton("确定",
                            new DialogInterface.OnClickListener()
                            {
                                @Override
                                public void onClick(DialogInterface dialog,
                                    int which)
                                {
                                    // 清空 sendList 集合
                                    sendList.clear();
                                    // 遍历 listView 组件的每个列表项
                                    for (int i = 0; i < listView.getCount(); i++)
                                    {
                                        CheckBox checkBox = (CheckBox) listView
                                            .getChildAt(i);
                                        // 如果该列表项被勾选
                                        if (checkBox.isChecked())
                                        {
                                            // 添加该列表项的电话号码
                                            sendList.add(checkBox.getText()
                                                .toString());
                                        }
                                    }
                                    numbers.setText(sendList.toString());
                                }
                            }).show();
                }
            });
        }
        // 判断某个电话号码是否已在群发范围内
        public boolean isChecked(String phone)
        {
            for (String s1 : sendList)
            {
                if (s1.equals(phone))
                {
                    return true;
```

```
            }
        }
        return false;
    }
}
```

运行程序，效果如图 8-6 所示。

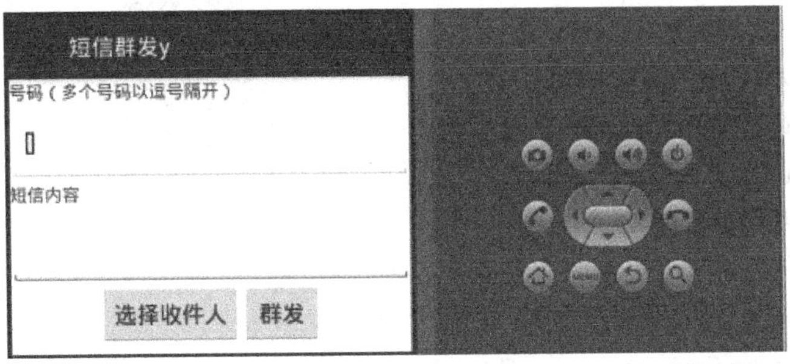

图 8-6 短信群发界面

程序中提供了一个带列表的对话框供用户选择群发短信的收件人号码，程序则使用了一个 ArrayList<String>集合来保存所有的收件人号码。为了实现群发功能，程序使用循环遍历 ArrayList<String>中的每个号码，并使用 SmsManager 依次向每个号码发送短信即可。

8.3 发送邮件

手机除了拨打电话和发送短信外，发送邮件也是其另外一个极为重要的功能。在具体实现上，邮件的收发过程是通过 Android 内置的 Gmail 程序实现的，而并不是使用 SMTP 的 Protocol 实现的。为了确保邮件能够发出，必须在收件人字段上输入标准的邮件地址格式，如果格式不规范，则发送按钮处于不可用状态。

下面通过示例来介绍在 Android 中怎样发送邮件。

【实现步骤】

（1）在 Eclipse 环境下，建立一个名为 E8_3_3 的工程。
（2）打开 res/layout 目录下的 main.xml 文件，代码修改为：

```xml
<?xml version="1.0" encoding="utf-8" ?>
<LinearLayout xmlns:android="http://schemas.android.com/apk/res/android"
    android:orientation="vertical"
    android:layout_width="fill_parent"
    android:layout_height="fill_parent">
<TextView android:layout_width="fill_parent"
    android:layout_height="wrap_content"
    android:text="@string/hello" />
<TextView android:layout_width="fill_parent"
    android:layout_height="30px"
    android:text="收件人" />
```

```xml
<EditText android:id="@+id/text01"
    android:layout_width="fill_parent"
    android:layout_height="wrap_content"
    android:text="503112801@qq.com" />
<TextView android:layout_width="fill_parent"
    android:layout_height="30px"
    android:text="邮件内容" />
<EditText android:id="@+id/text02"
    android:layout_width="fill_parent"
    android:layout_height="wrap_content"
    android:text="bbbbbbbbbbb" />
<Button android:id="@+id/send"
    android:layout_width="fill_parent"
    android:layout_height="wrap_content"
    android:layout_centerInParent="true"
    android:text="发送"/>
</LinearLayout>
```

（3）打开 src/com.example.e8_3_3 目录下的 MainActivity.java 文件，代码修改为：

```java
public class MainActivity extends Activity
{
    /**第一次调用活动*/
    private EditText text01;
    private EditText text02;
    private Button send;
    private String [] strEmailReciver;
    private String strEmailBody;
    @Override
    public void onCreate(Bundle savedInstanceState)
    {
        super.onCreate(savedInstanceState);
        setContentView(R.layout.main);
        send = (Button)findViewById(R.id.send);
        text01 = (EditText)findViewById(R.id.text01);
        text02 = (EditText)findViewById(R.id.text02);
        text01.setOnKeyListener(new EditText.OnKeyListener()
        {
            @Override
            public boolean onKey(View v, int keyCode, KeyEvent event)
            {
                // TODO：自动存根法
                if(true)
```

```
                {
                    send.setEnabled(true);
                }else{
                    send.setEnabled(false);
                }
                return false;
            }
        });
        send.setOnClickListener(new Button.OnClickListener()
        {
            @Override
            public void onClick(View v)
            {
                Intent intent = new Intent(android.content.Intent.ACTION_SEND);
                intent.setType("plain/text");
                strEmailReciver = new String[]{text01.getText().toString()};
                strEmailBody = text02.getText().toString();
                intent.putExtra(android.content.Intent.EXTRA_EMAIL, strEmailReciver);
                intent.putExtra(android.content.Intent.EXTRA_TEXT, strEmailBody);
                startActivity(Intent.createChooser(intent,getResources().getString(R.string.str_message)));
            }
        });
    }
}
```

运行程序，效果如图 8-7 所示。

图 8-7　邮件发送界面

8.4　实现震动

使用手机的震动函数针对 Notification 来让手机执行特定样式的震动。Android 可以控制震动的样式，我们可以使用震动来传达信息以获取用户的注意。为了设置震动样式，给 Notification 的 vibrate

属性设定一个时间数组。构建一个数组,每个间隔的数字相应地代表震动或暂停的时间长度。

下面通过示例来介绍在 Android 中如何实现手机震动。

【实现步骤】

(1) 在 Eclipse 环境下,建立一个名为 E8_4 的工程。

(2) 打开 res/layout 目录下的 main.xml 文件,代码修改为:

```xml
<?xml version="1.0" encoding="utf-8"?>
<AbsoluteLayout xmlns:android="http://schemas.android.com/apk/res/android"
    android:orientation="vertical"
    android:layout_width="fill_parent"
    android:layout_height="fill_parent">
    <!-- 建立第一个 TextView -->
    <TextView
        android:id="@+id/myTextView1"
        android:layout_width="fill_parent"
        android:layout_height="wrap_content" />
    <!-- 建立第二个 TextView -->
    <TextView
        android:id="@+id/myTextView2"
        android:layout_width="127px"
        android:layout_height="35px"
        android:layout_x="90px"
        android:layout_y="33px"
        android:text="@string/str_text1"/>
    <!-- 建立第三个 TextView -->
    <TextView
        android:id="@+id/myTextView3"
        android:layout_width="127px"
        android:layout_height="35px"
        android:layout_x="90px"
        android:layout_y="115px"
        android:text="@string/str_text2"/>
    <!-- 建立第四个 TextView -->
    <TextView
        android:id="@+id/myTextView4"
        android:layout_width="127px"
        android:layout_height="35px"
        android:layout_x="90px"
        android:layout_y="216px"
        android:text="@string/str_text3"/>
    <!-- 建立第一个 ToggleButton -->
```

```xml
<ToggleButton
    android:id="@+id/myTogglebutton1"
    android:layout_width="wrap_content"
    android:layout_height="wrap_content"
    android:layout_x="29px"
    android:layout_y="31px" />
<!-- 建立第二个 ToggleButton -->
<ToggleButton
    android:id="@+id/myTogglebutton2"
    android:layout_width="wrap_content"
    android:layout_height="wrap_content"
    android:layout_x="29px"
    android:layout_y="114px"/>
<!-- 建立第三个 ToggleButton -->
<ToggleButton
    android:id="@+id/myTogglebutton3"
    android:layout_width="wrap_content"
    android:layout_height="wrap_content"
    android:layout_x="29px"
    android:layout_y="214px"/>
</AbsoluteLayout>
```

(3) 打开 res/values 目录下的 strings.xml 文件，代码修改为：

```xml
<resources>
    <string name="app_name">E8_4</string>
    <string name="menu_settings">Settings</string>
    <string name="title_activity_main">震动效果</string>
    <string name="str_text1">短时震动</string>
    <string name="str_text2">长时震动</string>
    <string name="str_text3">节奏震动</string>
    <string name="str_ok">震动中....</string>
    <string name="str_end">震动结束</string>
</resources>
```

(4) 打开 src/com.example.e8_4 目录下的 MainActivity.java 文件，代码修改为：

```java
public class MainActivity extends Activity
{
    private Vibrator mVibrator01;
    /** 第一次调用活动 */
    @Override
    public void onCreate(Bundle savedInstanceState)
    {
        super.onCreate(savedInstanceState);
```

```java
setContentView(R.layout.main);
/*设置 ToggleButton 的对象*/
mVibrator01 = ( Vibrator )getApplication().getSystemService
(Service.VIBRATOR_SERVICE);
final ToggleButton mtogglebutton1 =
(ToggleButton) findViewById(R.id.myTogglebutton1);
final ToggleButton mtogglebutton2 =
(ToggleButton) findViewById(R.id.myTogglebutton2);
final ToggleButton mtogglebutton3 =
(ToggleButton) findViewById(R.id.myTogglebutton3);
/* 短震动 */
mtogglebutton1.setOnClickListener(new OnClickListener()
{
  public void onClick(View v)
  {
    if (mtogglebutton1.isChecked())
    {
      /* 设置震动的周期 */
      mVibrator01.vibrate( new long[]{100,10,100,1000},-1);
      /*用 Toast 显示震动启动*/
      Toast.makeText
      (
        MainActivity.this,
        getString(R.string.str_ok),
        Toast.LENGTH_SHORT
      ).show();
    }
    else
    {
      /* 取消震动 */
      mVibrator01.cancel();
      /*用 Toast 显示震动已被取消*/
      Toast.makeText
      (
        MainActivity.this,
        getString(R.string.str_end),
        Toast.LENGTH_SHORT
      ).show();
    }
  }
});
```

```
/* 长震动 */
mtogglebutton2.setOnClickListener(new OnClickListener()
{
  public void onClick(View v)
   {
     if (mtogglebutton2.isChecked())
      {
        /*设置震动的周期*/
        mVibrator01.vibrate(new long[]{100,100,100,1000},0);

        /*用 Toast 显示震动启动*/
        Toast.makeText
        (
           MainActivity.this,
           getString(R.string.str_ok),
           Toast.LENGTH_SHORT
        ).show();
      }
      else
      {
        /* 取消震动 */
        mVibrator01.cancel();

        /* 用 Toast 显示震动取消 */
        Toast.makeText
        (
           MainActivity.this,
           getString(R.string.str_end),
           Toast.LENGTH_SHORT
        ).show();
      }
   }
});
/* 节奏震动 */
mtogglebutton3.setOnClickListener(new OnClickListener()
{
  public void onClick(View v)
   {
     if (mtogglebutton3.isChecked())
      {
        /* 设置震动的周期 */
```

```
                mVibrator01.vibrate( new long[]{1000,50,1000,50,1000},0);

                /*用 Toast 显示震动启动*/
                Toast.makeText
                (
                    MainActivity.this, getString(R.string.str_ok),
                    Toast.LENGTH_SHORT
                ).show();
            }
            else
            {
                /* 取消震动 */
                mVibrator01.cancel();
                /* 用 Toast 显示震动取消 */
                Toast.makeText
                (
                    MainActivity.this,
                    getString(R.string.str_end),
                    Toast.LENGTH_SHORT
                ).show();
            }
        }
    });
  }
}
```

（5）设置权限，打开 AndroidManifest.xml 文件，添加如下代码：

```
    </application>
        <users-permission android:name="android.permission.VIBRATE"/>
    </manifest>
```

运行程序，效果如图 8-8 所示。

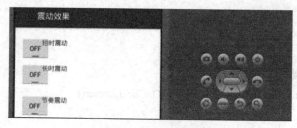

图 8-8 震动效果设置

8.5 闹钟

Android API 中提供了 AlarmManager 类，可以设置在指定的时间运行某些动作，其主要用来开发手机闹钟。

8.5.1 AlarmManager 类概述

AlarmManager 不仅可用于开发闹钟应用，还可作为一个全局定时器使用，Android 的程序中也通过 Context 的 getSystemService()方法来获取 AlarmManager 对象，一旦程序获取了 AlarmManager 对象，就可调用它的如下方法来设置定时启动指定组件。

（1）void set(int type,long triggerAtTime,PendingIntent operation)：设置在 triggerAtTime 时间启动由 operation 参数指定的组件。其中第一个参数指定定时服务的类型，该参数可接受如下值。

- ELAPSED_REALTIME：指定从现在开始时间过了一定时间后启动 operation 所对应的组件。
- ELAPSED_REALTIME_WAKEUP：指定从现在开始时间过了一定时间后启动 operation 所对应的组件，即使系统关机也会执行 operation 所对应的组件。
- RTC：指定当系统调用 System.currentTimeMillis()方法返回值与 triggerAtTime 相等时启动 operation 所对应的组件。
- RTC_WAKEUP：指定当系统调用 System.currentTimeMillis()方法返回值与 triggerAtTime 相等时启动 operation 所对应的组件。即使系统关机也会执行 operation 所对应的组件。

（2）void setInexactRepeating(int type,long triggerAtTime,long interval,PendingIntent operation)：设置一个非精确的周期性任务。

（3）void setRepeating(int type,long triggerAtTime,long interval,PendingIntent operation)：设置一个周期性执行的定时服务。

（4）void cance(PendingIntent operation)：取消 AlarmManager 的定时服务。

8.5.2 设定闹钟实例

本实例的程序中以 getSystemService(ALARM_SERVICE)取得 AlarmManager 并利用 set()和 setRepeating()两个方法来实现两种不同类型的闹钟。

（1）只响一次的闹钟。当单击"设置闹钟"按钮时，会触发 Button 的 onClick 事件，跳出 TimePickerDialog 来设置时间，设置完成后，就以 PendingIntent.getBroadcast()生成 PendingIntent，再利用 AlarmManager 的 set()将设置的时间与 PendingIntent 传入 AlarmManager，最后以 Toast 提示用户已完成设置，并改变屏幕上的设置时间。

（2）重复响起的闹钟。当单击"设置闹钟"按钮时，会触发 Button 的 onClick 事件，跳出自定义的 Layout 的 AlertDialog 来设置开始时间与重复间隔。设置完成后，以 PendingIntent.getBroadcast()生成 PendingIntent，再利用 AlarmManager 的 setReapting()来设置开始时间、重复间隔等，最后以 Toast 提示用户已完成设置,并改变屏幕上的设置时间。

(3) 最后程序中用 cancel()方法来删除 AlarmManager 中设置的闹钟。

【实现步骤】

(1) 在 Eclipse 环境下,建立一个名为 E8_5_1 的工程。

(2) 打开 res/layout 目录下的 main.xml 文件,代码修改为:

```xml
<?xml version="1.0" encoding="utf-8"?>
<AbsoluteLayout xmlns:android="http://schemas.android.com/apk/res/android"
    android:id="@+id/layout1"
    android:layout_width="fill_parent"
    android:layout_height="fill_parent">
    <DigitalClock android:id="@+id/dClock"
        android:layout_width="wrap_content"
        android:layout_height="wrap_content"
        android:layout_x="30dp"
        android:layout_y="32dp"
        android:textSize="40sp" />
    <TextView android:id="@+id/text1"
        android:layout_width="wrap_content"
        android:layout_height="wrap_content"
        android:layout_x="10dp"
        android:layout_y="90dp"
        android:text="@string/str_title1"
        android:textSize="20sp" />
    <Button android:id="@+id/mButton1"
        android:layout_width="wrap_content"
        android:layout_height="40dp"
        android:layout_x="160dp"
        android:layout_y="86dp"
        android:text="@string/str_button1"
        android:textSize="18sp" />
    <TextView android:id="@+id/text2"
        android:layout_width="wrap_content"
        android:layout_height="wrap_content"
        android:layout_x="10dp"
        android:layout_y="227dp"
        android:text="@string/str_title2"
        android:textSize="20sp" />
    <Button android:id="@+id/mButton4"
        android:layout_width="wrap_content"
        android:layout_height="40dp"
        android:layout_x="160dp"
```

```
            android:layout_y="269dp"
            android:text="@string/str_button2"
            android:textSize="18sp" />
        <Button android:id="@+id/mButton2"
            android:layout_width="wrap_content"
            android:layout_height="40dp"
            android:layout_x="160dp"
            android:layout_y="140dp"
            android:text="@string/str_button2"
            android:textSize="18sp" />
        <Button android:id="@+id/mButton3"
            android:layout_width="wrap_content"
            android:layout_height="40dp"
            android:layout_x="160dp"
            android:layout_y="220dp"
            android:text="@string/str_button1"
            android:textSize="18sp" />
        <TextView android:id="@+id/setTime2"
            android:layout_width="135dp"
            android:layout_height="wrap_content"
            android:layout_x="10dp"
            android:layout_y="278dp"
            android:text="@string/str_default"
            android:textSize="16sp" />
        <TextView android:id="@+id/setTime1"
            android:layout_width="wrap_content"
            android:layout_height="wrap_content"
            android:layout_x="10dp"
            android:layout_y="150dp"
            android:text="@string/str_default"
            android:textSize="16sp" />
</AbsoluteLayout>
```

（3）在 res/layout 目录下新建一个名为 timeset.xml 的文件，代码修改为：

```
<?xml version="1.0" encoding="utf-8"?>
<AbsoluteLayout
    android:id="@+id/layout2"
    android:layout_width="fill_parent"
    android:layout_height="fill_parent"
    xmlns:android="http://schemas.android.com/apk/res/android">
    <TextView
        android:id="@+id/text1"
```

```xml
    android:layout_width="wrap_content"
    android:layout_height="wrap_content"
    android:text="@string/str_text1"
    android:textColor="@drawable/white"
    android:textSize="16sp"
    android:layout_x="10px"
    android:layout_y="32px">
</TextView>
<TextView
    android:id="@+id/text2"
    android:layout_width="wrap_content"
    android:layout_height="wrap_content"
    android:text="@string/str_text2"
    android:textColor="@drawable/white"
    android:textSize="16sp"
    android:layout_x="10px"
    android:layout_y="172px">
</TextView>
<TimePicker
    android:id="@+id/tPicker"
    android:layout_width="wrap_content"
    android:layout_height="wrap_content"
    android:layout_x="100px"
    android:layout_y="12px">
</TimePicker>
<EditText
    android:id="@+id/mEdit"
    android:layout_width="52px"
    android:layout_height="40px"
    android:text="15"
    android:textSize="16sp"
    android:layout_x="120px"
    android:layout_y="162px">
</EditText>
<TextView
    android:id="@+id/text3"
    android:layout_width="wrap_content"
    android:layout_height="wrap_content"
    android:text="@string/str_text3"
    android:textColor="@drawable/white"
    android:textSize="16sp"
```

```
            android:layout_x="180px"
            android:layout_y="172px">
    </TextView>
</AbsoluteLayout>
```

（4）打开 src/com.example.e8_5_1 目录下的 MainActivity.java 文件，代码修改为：

```java
public class MainActivity extends Activity
{
    /* 声明对象变量 */
    TextView setTime1;
    TextView setTime2;
    Button mButton1;
    Button mButton2;
    Button mButton3;
    Button mButton4;
    Calendar c=Calendar.getInstance();
    @Override
    public void onCreate(Bundle savedInstanceState)
    {
        super.onCreate(savedInstanceState);
        /* 载入 main.xml Layout */
        setContentView(R.layout.main);
        /* 以下为只响一次的闹钟的设置 */
        setTime1=(TextView) findViewById(R.id.setTime1);
        /* 只响一次闹钟的设置 Button */
        mButton1=(Button)findViewById(R.id.mButton1);
        mButton1.setOnClickListener(new View.OnClickListener()
        {
            public void onClick(View v)
            {
                /* 取得单击按钮时的时间作为 TimePickerDialog 的默认值 */
                c.setTimeInMillis(System.currentTimeMillis());
                int mHour=c.get(Calendar.HOUR_OF_DAY);
                int mMinute=c.get(Calendar.MINUTE);

                /* 跳出 TimePickerDialog 来设置时间 */
                new TimePickerDialog(MainActivity.this,
                    new TimePickerDialog.OnTimeSetListener()
                    {
                        public void onTimeSet(TimePicker view,int hourOfDay,int minute)
                        {
                            /* 取得设置后的时间，秒跟毫秒设为 0 */
```

```java
                    c.setTimeInMillis(System.currentTimeMillis());
                    c.set(Calendar.HOUR_OF_DAY,hourOfDay);
                    c.set(Calendar.MINUTE,minute);
                    c.set(Calendar.SECOND,0);
                    c.set(Calendar.MILLISECOND,0);
                    /* 指定闹钟设置时间到时要运行 CallAlarm.class */
                    Intent intent = new Intent(MainActivity.this, AlarmService.class);
                    /* 创建 PendingIntent */
                    PendingIntent sender=PendingIntent.getBroadcast(MainActivity.this,0, intent, 0);
                    /* AlarmManager.RTC_WAKEUP 设置服务在系统休眠时同样会运行
                     * 以 set()设置的 PendingIntent 只会运行一次
                     **/
                    AlarmManager am;
                    am = (AlarmManager)getSystemService(ALARM_SERVICE);
                    am.set(AlarmManager.RTC_WAKEUP,
                            c.getTimeInMillis(),
                            sender
                            );
                    /* 更新显示的设置闹钟时间 */
                    String tmpS=format(hourOfDay)+": "+format(minute);
                    setTime1.setText(tmpS);
                    /* 以 Toast 提示设置已完成 */
                    Toast.makeText(MainActivity.this,"设置闹钟时间为"+tmpS,
                        Toast.LENGTH_SHORT)
                        .show();
                }
            },mHour,mMinute,true).show();
    }
});
/* 只响一次闹钟的删除 Button */
mButton2=(Button) findViewById(R.id.mButton2);
mButton2.setOnClickListener(new View.OnClickListener()
{
    public void onClick(View v)
    {
        Intent intent = new Intent(MainActivity.this, AlarmService.class);
        PendingIntent sender=PendingIntent.getBroadcast(
            MainActivity.this,0, intent, 0);
        /* 由 AlarmManager 中删除 */
        AlarmManager am;
```

```java
            am =(AlarmManager)getSystemService(ALARM_SERVICE);
            am.cancel(sender);
            /* 以 Toast 提示已删除设置, 并更新显示的闹钟时间 */
            Toast.makeText(MainActivity.this,"闹钟时间解除",
                            Toast.LENGTH_SHORT).show();
            setTime1.setText("目前无设置");
        }
    });
    /* 以下为重复响起闹钟的设置 */
    setTime2=(TextView) findViewById(R.id.setTime2);
    /* create 重复响起闹钟的设置画面 */
    /* 引用 timeset.xml 为 Layout */
    LayoutInflater factory = LayoutInflater.from(this);
    final View setView = factory.inflate(R.layout.timeset,null);
    final TimePicker tPicker=(TimePicker)setView
                                    .findViewById(R.id.tPicker);
    tPicker.setIs24HourView(true);
    /* create 重复响起闹钟的设置 Dialog */
    final AlertDialog di=new AlertDialog.Builder(MainActivity.this)
            .setIcon(R.drawable.clock)
            .setTitle("设置")
            .setView(setView)
            .setPositiveButton("确定",
                new DialogInterface.OnClickListener()
                {
                    public void onClick(DialogInterface dialog, int which)
                    {
                        /* 取得设置的间隔秒数 */
                        EditText ed=(EditText)setView.findViewById(R.id.mEdit);
                        int times=Integer.parseInt(ed.getText().toString())*1000;
                        /* 取得设置的开始时间, 秒及毫秒设为 0 */
                        c.setTimeInMillis(System.currentTimeMillis());
                        c.set(Calendar.HOUR_OF_DAY,tPicker.getCurrentHour());
                        c.set(Calendar.MINUTE,tPicker.getCurrentMinute());
                        c.set(Calendar.SECOND,0);
                        c.set(Calendar.MILLISECOND,0);
                        /* 指定闹钟设置时间到时要运行 CallAlarm.class */
                        Intent intent = new Intent(MainActivity.this,
                                AlarmService.class);
                        PendingIntent sender = PendingIntent.getBroadcast(
                                MainActivity.this,1, intent, 0);
```

```
                    /* setRepeating()可让闹钟重复运行 */
                    AlarmManager am;
                    am = (AlarmManager)getSystemService(ALARM_SERVICE);
                    am.setRepeating(AlarmManager.RTC_WAKEUP,
                            c.getTimeInMillis(),times,sender);
                    /* 更新显示的设置闹钟时间 */
                    String tmpS=format(tPicker.getCurrentHour())+":"+
                            format(tPicker.getCurrentMinute());
                    setTime2.setText("设置闹钟时间为"+tmpS+
                            "开始，重复间隔为"+times/1000+"秒");
                    /* 以 Toast 提示设置已完成 */
                    Toast.makeText(MainActivity.this,"设置闹钟时间为"+tmpS+
                            "开始，重复间隔为"+times/1000+"秒",
                            Toast.LENGTH_SHORT).show();
                }
            })
            .setNegativeButton("取消",
              new DialogInterface.OnClickListener()
              {
                  public void onClick(DialogInterface dialog, int which)
                  {
                  }
              }).create();
        /* 重复响起闹钟的设置 Button */
        mButton3=(Button) findViewById(R.id.mButton3);
        mButton3.setOnClickListener(new View.OnClickListener()
        {
            public void onClick(View v)
            {
                /* 取得单击按钮时的时间作为 tPicker 的默认值 */
                c.setTimeInMillis(System.currentTimeMillis());
                tPicker.setCurrentHour(c.get(Calendar.HOUR_OF_DAY));
                tPicker.setCurrentMinute(c.get(Calendar.MINUTE));
                /* 跳出设置画面 di */
                di.show();
            }
        });
        /* 重复响起闹钟的删除 Button */
        mButton4=(Button) findViewById(R.id.mButton4);
        mButton4.setOnClickListener(new View.OnClickListener()
        {
```

```java
    public void onClick(View v)
    {
        Intent intent = new Intent(MainActivity.this, AlarmService.class);
        PendingIntent sender = PendingIntent.getBroadcast(
                MainActivity.this,1, intent, 0);
        /* 由 AlarmManager 中删除 */
        AlarmManager am;
        am = (AlarmManager)getSystemService(ALARM_SERVICE);
        am.cancel(sender);
        /* 以 Toast 提示已删除设置，并更新显示的闹钟时间 */
        Toast.makeText(MainActivity.this,"闹钟时间解除",
                        Toast.LENGTH_SHORT).show();
        setTime2.setText("目前无设置");
    }
});
}
/* 日期时间显示两位数的方法 */
private String format(int x)
{
    String s=""+x;
    if(s.length()==1) s="0"+s;
    return s;
}
}
```

（5）在 src/com.example.e8_5_1 目录下新建一个名为 AlarmReceiver.java 的文件（该程序中引入相关 class 类，并调用闹钟 Alert 的 Receiver，创建 Intent）用于调用 AlarmAlet.class。其代码为：

```java
public class AlermReceiver extends Activity
{
    @Override
    protected void onCreate(Bundle savedInstanceState)
    {
        super.onCreate(savedInstanceState);
        /* 跳出的闹铃警示 */
        new AlertDialog.Builder(AlermReceiver.this)
            .setIcon(R.drawable.clock)
            .setTitle("闹钟响!!")
            .setMessage("起床吧!起床吧!!")
            .setPositiveButton("关掉它")
             new DialogInterface.OnClickListener()
            {
                public void onClick(DialogInterface dialog, int whichButton)
```

```
            {
                /* 关闭 Activity */
                AlermReceiver.this.finish();
            }
        });
        .show();
    }
}
```

(6) 在 src/com.example.e8_5_1 目录下新建一个名为 AlarmService.java 的文件（实现实际跳出闹铃 Dialog 的 Activity，通过 AlertDialog.Builder(AlarmService.this)实现弹出闹钟警示框。通过 onClick(DialogInterface dialog,int whichButton)关闭 Activity）。其代码为：

```
public class AlarmService extends BroadcastReceiver
{
    @Override
    public void onReceive(Context context, Intent intent)
    {
        /* create Intent，调用 AlarmAlert.class */
        Intent i = new Intent(context, AlermReceiver.class);

        Bundle bundleRet = new Bundle();
        bundleRet.putString("STR_CALLER", "");
        i.putExtras(bundleRet);
        i.addFlags(Intent.FLAG_ACTIVITY_NEW_TASK);
        context.startActivity(i);
    }
}
```

(7) 设置权限，打开 AndroidManifest.xml 中添加对 CallAlarm 的 receiver 设置。

```
<ractivity android:name=".AlarmService" android:process=":remote"/>
<ractivity android:name=".AlarmReceiver" android:label="@string/app_name"/>
```

运行程序，效果如图 8-9 所示。

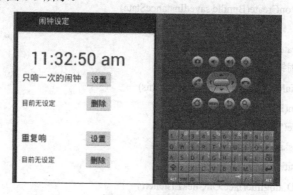

图 8-9　闹钟界面

8.5.3 更换墙纸实例

本实例将通过 AlarmManager 来周期性地调用某个 Service，从而让系统实现定时更换墙纸的功能。

更换墙纸的 API 为 WallpaperManager，它提供了 clear()方法来清除墙纸，还提供了如下方法来设置系统的墙纸。

- setBitmap(Bitmap bitmap)：将墙纸设置为 bitmap 所代表的位图。
- setResource(int resid)：将墙纸设置为 resid 资源所代表的图片。
- setStream(InputStream data)：将墙纸设置为 data 数据所代表的图片。

下面实例的界面中有两个按钮：一个按钮用于启动定时更换墙纸；另一个按钮用于关闭定时更换墙纸。

【实现步骤】

（1）在 Eclipse 环境下，建立一个名为 E8_5_2 的工程。

（2）打开 res/layout 目录下的 main.xml 文件，代码修改为：

```xml
<?xml version="1.0" encoding="utf-8"?>
<LinearLayout xmlns:android="http://schemas.android.com/apk/res/android"
    android:orientation="horizontal"
    android:layout_width="fill_parent"
    android:layout_height="fill_parent"
    android:gravity="center">
    <Button
        android:id="@+id/start"
        android:layout_width="wrap_content"
        android:layout_height="wrap_content"
        android:text="@string/start"/>
    <Button
        android:id="@+id/stop"
        android:layout_width="wrap_content"
        android:layout_height="wrap_content"
        android:text="@string/stop"/>
</LinearLayout>
```

（3）打开 src/com.example.e8_5_2 目录下的 MainActivity.java 文件，代码修改为：

```java
public class MainActivity extends Activity
{
    // 定义 AlarmManager 对象
    AlarmManager aManager;
    Button start, stop;
    @Override
    public void onCreate(Bundle savedInstanceState)
    {
```

```java
super.onCreate(savedInstanceState);
setContentView(R.layout.main);
start = (Button)findViewById(R.id.start);
stop = (Button)findViewById(R.id.stop);
aManager = (AlarmManager) getSystemService(Service.ALARM_SERVICE);
// 指定启动 ChangeService 组件
Intent intent = new Intent(MainActivity.this, setTime.class);
// 创建 PendingIntent 对象
final PendingIntent pi = PendingIntent.getService(
        MainActivity.this, 0, intent, 0);
start.setOnClickListener(new OnClickListener()
{
    @Override
    public void onClick(View arg0)
    {
        // 设置每隔 5s 执行 pi 代表的组件一次
        aManager.setRepeating(AlarmManager.RTC_WAKEUP
                , 0 , 5000, pi);
        start.setEnabled(false);
        stop.setEnabled(true);
        Toast.makeText(MainActivity.this, "墙纸定时更换启动成功啦"
                , 5000).show();
    }
});
stop.setOnClickListener(new OnClickListener()
{
    @Override
    public void onClick(View arg0)
    {
        start.setEnabled(true);
        stop.setEnabled(false);
        // 取消对 pi 的调度
        aManager.cancel(pi);
    }
});
}
}
```

（4）在 src/com.example.e8_5_2 目录下新建一个名为 setTime.java 的文件，代码为：

```java
public class setTime extends Service
{
    // 定义定时更换的墙纸资源
```

```java
int[] wallpapers = new int[]
{
    R.drawable.j1,
    R.drawable.j2,
    R.drawable.j3,
    R.drawable.j4
};
// 定义系统的墙纸管理服务
WallpaperManager wManager;
// 定义当前所显示的墙纸
int current = 0;
@Override
public void onStart(Intent intent, int startId)
{
    // 如果到了最后一张，系统重新开始
    if(current >= 4)
        current = 0;
    try
    {
        // 改变墙纸
        wManager.setResource(wallpapers[current++]);
    }
    catch (Exception e)
    {
        e.printStackTrace();
    }
    super.onStart(intent, startId);
}
@Override
public void onCreate()
{
    super.onCreate();
    // 初始化 WallpaperManager
    wManager = WallpaperManager.getInstance(this);
}
@Override
public IBinder onBind(Intent intent)
{
    return null;
}
}
```

（5）为程序授予权限，打开 AndroidManifest.xml 文件，添加如下代码：

```
</application>
    <!-- 授予用户修改墙纸的权限 -->
    <uses-permission android:name="android.permission.SET_WALLPAPER"/>
</manifest>
```

运行程序，并单击程序界面中的"启动定时更换"按钮，接着退出程序，返回程序桌面，将会看到系统桌面每 5s 更换一次墙纸。

8.6 自动显示电量

在使用过程中，最担心的是手机没电而影响业务，所以及时显示电池容量功能是非常必要的。

可以使用 Android API 中的 BroadcastReseiver 类和 Button 的 Listener 类，当 Reseiver 被注册后会在后台等待被其他程序调用。当指定要捕捉的 Action 发生时，Reseiver 就会被调用，并运行 onReseiver 来实现里面的程序。

在实例中，将利用 BroadcastReseiver 的特性来获取手机电池的容量。通过注册 BroadcastReseiver 时设置的 IntentFiler 来获取系统发出的 Intent.ACTION_BATTERY_CHANGED，然后以此来获取电池的容量。

【实现步骤】

（1）在 Eclipse 环境下，建立一个名为 E8_6 的工程。

（2）打开 res/layout 目录下的 main.xml 文件，代码修改为：

```xml
<?xml version="1.0" encoding="utf-8"?>
<AbsoluteLayout
    xmlns:android="http://schemas.android.com/apk/res/android"
    android:id="@+id/layout1"
    android:layout_width="fill_parent"
    android:layout_height="fill_parent">
    <TextView
        android:id="@+id/myTextView1"
        android:layout_width="fill_parent"
        android:layout_height="wrap_content"
        android:textSize="20sp"
        android:text="@string/str_title"
        android:layout_x="60px"
        android:layout_y="40px" >
    </TextView>
    <Button
        android:id="@+id/myButton1"
        android:layout_width="wrap_content"
```

```
          android:layout_height="wrap_content"
          android:text="@string/str_button1"
          android:textSize="14sp"
          android:layout_x="80px"
          android:layout_y="90px">
       </Button>
   </AbsoluteLayout>
```

（3）在 res/layout 目录下新建一个名为 exadialog.xml 的文件，代码为：

```
<?xml version="1.0" encoding="utf-8"?>
<LinearLayout
    xmlns:android="http://schemas.android.com/apk/res/android"
    android:orientation="vertical"
    android:layout_width="fill_parent"
    android:layout_height="fill_parent">
    <TextView
       android:id="@+id/myTextView2"
       android:layout_width="fill_parent"
       android:layout_height="wrap_content"
       android:textSize="16sp"
       android:gravity="center"
       android:padding="10px">
    </TextView>
    <Button
       android:id="@+id/myButton2"
       android:layout_width="wrap_content"
       android:layout_height="wrap_content"
       android:text="@string/str_button2">
    </Button>
</LinearLayout>
```

（4）打开 src/com.example.e8_6 目录下的 MainActivity.java 文件，代码修改为：

```
public class MainActivity extends Activity
{
   /* 变量声明 */
   private int intLevel;
   private int intScale;
   private Button mButton01;
   /* 创建 BroadcastReceiver */
   private BroadcastReceiver mBatInfoReceiver=new BroadcastReceiver()
   {
      public void onReceive(Context context, Intent intent)
      {
```

```java
        String action = intent.getAction();
        /* 如果捕捉到的 Action 是 ACTION_BATTERY_CHANGED,
         * 就运行 onBatteryInfoReceiver() */
        if (Intent.ACTION_BATTERY_CHANGED.equals(action))
        {
          intLevel = intent.getIntExtra("level", 0);
          intScale = intent.getIntExtra("scale", 100);
          onBatteryInfoReceiver(intLevel,intScale);
        }
      }
    };
    /** 第一次调用活动 */
    @Override
    public void onCreate(Bundle savedInstanceState)
    {
      super.onCreate(savedInstanceState);
      /* 载入 main.xml Layout */
      setContentView(R.layout.main);

      /* 初始化 Button, 并设置单击后的动作 */
      mButton01 = (Button)findViewById(R.id.myButton1);
      mButton01.setOnClickListener(new Button.OnClickListener()
      {
        @Override
        public void onClick(View v)
        {
          /* 注册一个系统 BroadcastReceiver, 作为访问电池电量之用 */
          registerReceiver
          (
            mBatInfoReceiver,
            new IntentFilter(Intent.ACTION_BATTERY_CHANGED)
          );
        }
      });
    }
    /* 捕捉到 ACTION_BATTERY_CHANGED 时要运行的 method */
    public void onBatteryInfoReceiver(int intLevel, int intScale)
    {
      /* create 跳出的对话窗口 */
      final Dialog d = new Dialog(MainActivity.this);
      d.setTitle(R.string.str_dialog_title);
```

```
d.setContentView(R.layout.exadialog);

/* 创建一个背景模糊的 Window，且将对话窗口放在前景中 */
Window window = d.getWindow();
window.setFlags
(
    WindowManager.LayoutParams.FLAG_BLUR_BEHIND,
    WindowManager.LayoutParams.FLAG_BLUR_BEHIND
);
/* 将取得的电池计量显示于 Dialog 中 */
TextView mTextView02=(TextView)d.findViewById(R.id.myTextView2);
mTextView02.setText
(
    getResources().getText(R.string.str_dialog_body)+
    String.valueOf(intLevel * 100 / intScale) + "%"
);
/* 设置返回主画面的按钮 */
Button mButton02 = (Button)d.findViewById(R.id.myButton2);
mButton02.setOnClickListener(new Button.OnClickListener()
{
    @Override
    public void onClick(View v)
    {
        /* 反注册 Receiver，并关闭对话窗口 */
        unregisterReceiver(mBatInfoReceiver);
        d.dismiss();
    }
});
d.show();
}
}
```

在程序中的 onCreate 方法中载入主布局文件 main.xml，然后初始化 Button 并设置单击后的动作，再注册一个系统 BrodacastReceiver，用于访问电池电量。当捕捉到 ACTION_BATTERY_CHANGED 要运行的 method 时即要定义 OnBatteryInfoReceiver，首先创建一个背景模糊的 Window，且将对话框窗口放在前景中，然后将取得的电池电量显示于 Dialog 中，最后设置返回主画面的按钮。

运行程序，并单击界面中的"获取"按钮，得到如图 8-10 所示效果。

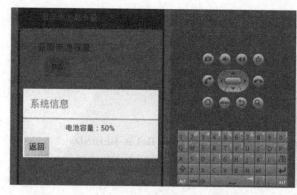

图 8-10 获取电池剩余电量效果

8.7 Wi-Fi 使用

Wi-Fi 是一种可以将 PC、手持设备（如 PDA、手机）等终端以无线方式互相连接的技术。Wi-Fi 是一个无线网络通信技术的品牌，由 Wi-Fi 联盟（Wi-Fi Alliance）所持有。目的是改善基于 IEEE 802.11 标准的无线网络产品之间的互通性。现在一般会把 Wi-Fi 及 IEEE 802.11 混为一谈，甚至把 Wi-Fi 等同于无线网际网络。

Wi-Fi 是一种短程无线传输技术，能够在数百英尺范围内支持互联网接入的无线电信号。随着技术的发展，以及 IEEE 802.11a 及 IEEE 802.11g 等标准的出现，现在 IEEE 802.11 这个标准已被统称为 Wi-Fi。从应用层面来说，要使用 Wi-Fi，用户首先要有 Wi-Fi 兼容的用户端装置。

本示例演示了使用 Wi-Fi 的基本流程。

【实现步骤】

（1）在 Eclipse 环境下，建立一个名为 E8_7 的工程。

（2）打开 res/layout 目录下的 main.xml 文件，代码修改为：

```xml
<?xml version="1.0" encoding="utf-8"?>
<LinearLayout
    xmlns:android="http://schemas.android.com/apk/res/android"
    android:orientation="vertical"
    android:layout_width="fill_parent"
    android:layout_height="fill_parent">
    <TextView
        android:id="@+id/myTextView1"
        android:layout_width="fill_parent"
        android:layout_height="wrap_content"
        android:text="@string/hello_world"/>
    <CheckBox
        android:id="@+id/myCheckBox1"
        android:layout_width="wrap_content"
        android:layout_height="wrap_content"
```

```
            android:text="@string/str_checked"/>
        </LinearLayout>
```

（3）打开 res/values 目录下的 strings.xml 文件，添加如下代码：

```xml
<string name="str_checked">打开 WiFi</string>
<string name="str_uncheck">关闭 WiFi</string>
<string name="str_start_wifi_failed">打开失败</string>
<string name="str_start_wifi_done">打开成功</string>
<string name="str_stop_wifi_failed">关闭失败</string>
<string name="str_stop_wifi_done">关闭成功</string>
<string name="str_wifi_enabling">正在启动...</string>
<string name="str_wifi_disabling">正在关闭...</string>
<string name="str_wifi_disabled">已关闭</string>
<string name="str_wifi_unknow">未知...</string>
```

（4）打开 src/com.example.e8_7 目录下的 MainActivity 文件，代码修改为：

```java
public class MainActivity extends Activity
{
    private TextView mTextView1;
    private CheckBox mCheckBox1;
    /* 创建 WiFiManager 对象 */
    private WifiManager mWiFiManager1;
    //定义了 mTextView1 和 mCheckBox1，分别用于显示提示文本和获取复选框的选择状态
    /**第一次调用活动 */
    @Override
    public void onCreate(Bundle savedInstanceState)
    {
        super.onCreate(savedInstanceState);
        setContentView(R.layout.main);
        mTextView1 = (TextView) findViewById(R.id.myTextView1);
        mCheckBox1 = (CheckBox) findViewById(R.id.myCheckBox1);
        /*以 getSystemServices 取得 WIFI_SERVICE，然后通过 if 语句来判断动
         * 作程序后的 Wi-Fi 状态是否打开或已打开，这样即可显示对应的提示信息
         */
        mWiFiManager1 = (WifiManager)
                        this.getSystemService(Context.WIFI_SERVICE);
        /* 判断运行程序后的 Wi-Fi 状态是否打开或已打开 */
        if(mWiFiManager1.isWifiEnabled())
        {
            /* 判断 Wi-Fi 状态是否"已打开" */
            if(mWiFiManager1.getWifiState()==WifiManager.WIFI_STATE_ENABLED)
            {
                /* 若 Wi-Fi 已打开，将选项对钩取消 */
```

```
            mCheckBox1.setChecked(true);
            /* 更改选项文字为关闭 Wi-Fi*/
            mCheckBox1.setText(R.string.str_uncheck);
          }
          else
          {
            /* 若 Wi-Fi 未打开，将选项对钩取消 */
            mCheckBox1.setChecked(false);
            /* 更改选项文字为打开 Wi-Fi*/
            mCheckBox1.setText(R.string.str_checked);
          }
        }
        else
        {
          mCheckBox1.setChecked(false);
          mCheckBox1.setText(R.string.str_checked);
        }
        /* 捕捉 CheckBox 的单击事件 */
        mCheckBox1.setOnClickListener(
        new CheckBox.OnClickListener()
        {
          @Override
          public void onClick(View v)
          {
            // TODO Auto-generated method stub

            /* 当选项为取消选取状态 */
            if(mCheckBox1.isChecked()==false)
            {
              /* 尝试关闭 Wi-Fi 服务 */
              try
              {
                /* 判断 Wi-Fi 状态是否为已打开 */
                if(mWiFiManager1.isWifiEnabled())
                {
                  /* 关闭 Wi-Fi */
                  if(mWiFiManager1.setWifiEnabled(false))
                  {
                    mTextView1.setText(R.string.str_stop_wifi_done);
                  }
                  else
```

```
          {
            mTextView1.setText(R.string.str_stop_wifi_failed);
          }
        }
        else
        {
          /* Wi-Fi 状态不是已打开状态时 */
          switch(mWiFiManager1.getWifiState())
          {
            /* Wi-Fi 正在打开过程中，导致无法关闭... */
            case WifiManager.WIFI_STATE_ENABLING:
              mTextView1.setText
              (
                getResources().getText
                (R.string.str_stop_wifi_failed)+":"+
                getResources().getText
                (R.string.str_wifi_enabling)
              );
              break;
            /* Wi-Fi 正在关闭过程中，导致无法关闭... */
            case WifiManager.WIFI_STATE_DISABLING:
              mTextView1.setText
              (
                getResources().getText
                (R.string.str_stop_wifi_failed)+":"+
                getResources().getText
                (R.string.str_wifi_disabling)
              );
              break;
            /* Wi-Fi 已经关闭 */
            case WifiManager.WIFI_STATE_DISABLED:
              mTextView1.setText
              (
                getResources().getText
                (R.string.str_stop_wifi_failed)+":"+
                getResources().getText
                (R.string.str_wifi_disabled)
              );
              break;
            /* 无法取得或辨识 Wi-Fi 状态 */
```

```
                    case WifiManager.WIFI_STATE_UNKNOWN:
                    default:
                      mTextView1.setText
                      (
                        getResources().getText
                        (R.string.str_stop_wifi_failed)+":"+
                        getResources().getText
                        (R.string.str_wifi_unknow)
                      );
                      break;
                  }
                  mCheckBox1.setText(R.string.str_checked);
                }
              }
              catch (Exception e)
              {
                Log.i("HIPPO", e.toString());
                e.printStackTrace();
              }
            }
            else if(mCheckBox1.isChecked()==true)
            {
              /* 尝试打开 Wi-Fi 服务 */
              try
              {
                /* 确认 Wi-Fi 服务是关闭的且不在打开作业中 */
                if(!mWiFiManager1.isWifiEnabled() &&
                   mWiFiManager1.getWifiState()!=
                   WifiManager.WIFI_STATE_ENABLING )
                {
                  if(mWiFiManager1.setWifiEnabled(true))
                  {
                    switch(mWiFiManager1.getWifiState())
                    {
                      /* Wi-Fi 正在打开过程中，导致无法打开... */
                      case WifiManager.WIFI_STATE_ENABLING:
                        mTextView1.setText
                        (
                          getResources().getText
                          (R.string.str_wifi_enabling)
```

```java
            );
            break;
        /* Wi-Fi 已经为打开，无法再次打开... */
        case WifiManager.WIFI_STATE_ENABLED:
            mTextView1.setText
            (
                getResources().getText
                (R.string.str_start_wifi_done)
            );
            break;
        /* 其他未知的错误 */
        default:
            mTextView1.setText
            (
                getResources().getText
                (R.string.str_start_wifi_failed)+":"+
                getResources().getText
                (R.string.str_wifi_unknow)
            );
            break;
        }
    }
    else
    {
        mTextView1.setText(R.string.str_start_wifi_failed);
    }
}
else
{
    switch(mWiFiManager1.getWifiState())
    {
        /* Wi-Fi 正在打开过程中，导致无法打开... */
        case WifiManager.WIFI_STATE_ENABLING:
            mTextView1.setText
            (
                getResources().getText
                (R.string.str_start_wifi_failed)+":"+
                getResources().getText
                (R.string.str_wifi_enabling)
            );
            break;
```

```java
                    /* Wi-Fi 正在关闭过程中，导致无法打开... */
                    case WifiManager.WIFI_STATE_DISABLING:
                        mTextView1.setText
                        (
                            getResources().getText
                            (R.string.str_start_wifi_failed)+":"+
                            getResources().getText
                            (R.string.str_wifi_disabling)
                        );
                        break;
                    /* Wi-Fi 已经关闭 */
                    case WifiManager.WIFI_STATE_DISABLED:
                        mTextView1.setText
                        (
                            getResources().getText
                            (R.string.str_start_wifi_failed)+":"+
                            getResources().getText
                            (R.string.str_wifi_disabled)
                        );
                        break;
                    /* 无法取得或识别 Wi-Fi 状态 */
                    case WifiManager.WIFI_STATE_UNKNOWN:
                    default:
                        mTextView1.setText
                        (
                            getResources().getText
                            (R.string.str_start_wifi_failed)+":"+
                            getResources().getText
                            (R.string.str_wifi_unknow)
                        );
                        break;
                }
            }
            mCheckBox1.setText(R.string.str_uncheck);
        }
        catch (Exception e)
        {
            Log.i("HIPPO", e.toString());
            e.printStackTrace();
        }
    }
}
```

```java
        }
    });
}
//定义 mMakeTextToast()方法来根据当前操作显示对应的提示性信息
public void mMakeTextToast(String str, boolean isLong)
{
    if(isLong==true)
    {
        Toast.makeText(MainActivity.this, str, Toast.LENGTH_LONG).show();
    }
    else
    {
        Toast.makeText(MainActivity.this, str, Toast.LENGTH_SHORT).show();
    }
}
@Override
protected void onResume()
{
    /* 在 onResume 重写事件为获取打开程序 Wi-Fi 状态 */
    try
    {
        switch(mWiFiManager1.getWifiState())
        {
            /* Wi-Fi 已经在打开状态... */
            case WifiManager.WIFI_STATE_ENABLED:
                mTextView1.setText
                (
                    getResources().getText(R.string.str_wifi_enabling)
                );
                break;
            /* Wi-Fi 正在打开过程中... */
            case WifiManager.WIFI_STATE_ENABLING:
                mTextView1.setText
                (
                    getResources().getText(R.string.str_wifi_enabling)
                );
                break;
            /* Wi-Fi 正在关闭过程中... */
            case WifiManager.WIFI_STATE_DISABLING:
                mTextView1.setText
                (
```

```java
                getResources().getText(R.string.str_wifi_disabling)
            );
            break;
            /* Wi-Fi 已经关闭 */
            case WifiManager.WIFI_STATE_DISABLED:
                mTextView1.setText
                (
                    getResources().getText(R.string.str_wifi_disabled)
                );
                break;
            /* 无法取得或识别 Wi-Fi 状态 */
            case WifiManager.WIFI_STATE_UNKNOWN:
            default:
                mTextView1.setText
                (
                    getResources().getText(R.string.str_wifi_unknow)
                );
                break;
            }
        }
        catch(Exception e)
        {
            mTextView1.setText(e.toString());
            e.getStackTrace();
        }
        super.onResume();
    }
    @Override
    protected void onPause()
    {
        super.onPause();
    }
}
```

(5) 授予权限。打开 AndroidManifest.xml 文件，代码如下：

```xml
</application>
<!-- 声明 Wi-Fi 以及网络等相关权限 -->
<uses-permission
    android:name="android.permission.CHANGE_NETWORK_STATE"></uses-permission>
<uses-permission android:name="android.permission.CHANGE_WIFI_STATE"></uses-permission>
<uses-permission
```

```
            android:name="android.permission.ACCESS_NETWORK_STATE"></uses-permission>
        <uses-permission android:name="android.permission.ACCESS_WIFI_STATE"></uses-permission>
        <uses-permission android:name="android.permission.INTERNET"></uses-permission>
        <uses-permission android:name="android.permission.WAKE_LOCK"></uses-permission>
    </manifest>
```

运行程序，效果如图 8-11 所示。当选择复选框后会执行对应的操作，并显示对应的提示信息。

图 8-11　Wi-Fi 页面

8.8　联网

本节主要介绍使用 HTTP 协议与外界进行连接。使用 HTTP 协议可以执行各种任务，如从 Web 服务器上下载 Web 页面、下载二进制数据等。

如下实例使用 HTTP 协议连接到 Web 上来下载各种数据。

【实现步骤】

（1）在 Eclipse 环境下，建立一个名为 E8_8 的工程。

（2）授予权限，打开 AndroidManifest.xml 文件，添加如下代码：

```
    </application>
        <uses-sdk android:minSdkVersion="8" />
    <uses-permission android:name="android.permission.INTERNET"></uses-permission>
    </manifest>
```

（3）在 MainActivity.java 文件中添加如下代码：

```
        package com.example.e8_8;
        import java.io.IOException;
        import java.io.InputStream;
        import java.io.InputStreamReader;
        import java.net.HttpURLConnection;
        import java.net.URL;
        import java.net.URLConnection;
        import android.graphics.Bitmap;
        import android.graphics.BitmapFactory;
        import android.util.Log;
        import android.view.View;
        import android.widget.Button;
```

```java
import android.widget.ImageView;
import android.widget.Toast;
import javax.xml.parsers.DocumentBuilder;
import javax.xml.parsers.DocumentBuilderFactory;
import javax.xml.parsers.ParserConfigurationException;
import org.w3c.dom.Document;
import org.w3c.dom.Element;
import org.w3c.dom.Node;
import org.w3c.dom.NodeList;
public class MainActivity extends Activity
{
    /* 第一次调用活动*/
    @Override
    public void onCreate(Bundle savedInstanceState)
    {
        super.onCreate(savedInstanceState);
        setContentView(R.layout.main);
    }
}
```

（4）在 MainActivity.java 文件中定义 OpenHttpConnection()方法：

```java
public class MainActivity extends Activity
{
    private InputStream OpenHttpConnection(String urlString)
    throws IOException
    {
        InputStream in = null;
        int response = -1;

        URL url = new URL(urlString);
        URLConnection conn = url.openConnection();

        if (!(conn instanceof HttpURLConnection))
            throw new IOException("Not an HTTP connection");
        try{
            HttpURLConnection httpConn = (HttpURLConnection) conn;
            httpConn.setAllowUserInteraction(false);
            httpConn.setInstanceFollowRedirects(true);
            httpConn.setRequestMethod("GET");
            httpConn.connect();
            response = httpConn.getResponseCode();
            if (response == HttpURLConnection.HTTP_OK)
```

```
                {
                    in = httpConn.getInputStream();
                }
            }
            catch (Exception ex)
            {
                throw new IOException("Error connecting");
            }
            return in;
    }
    /* 第一次调用活动*/
    @Override
    public void onCreate(Bundle savedInstanceState)
    {
        super.onCreate(savedInstanceState);
        setContentView(R.layout.main);
    }
}
```

8.8.1 下载二进制数据

HTTP 协议需要执行的常见任务之一是从 Web 上下载二进制数据，例如，可能想从一台服务器上下载一幅图像以在应用程序中显示。

【实现步骤】

（1）在前面建立的 E8_8 工程中，打开 res/layout 目录下的 main.xml 文件，代码修改为：

```
<?xml version="1.0" encoding="utf-8"?>
<LinearLayout xmlns:android="http://schemas.android.com/apk/res/android"
    android:orientation="vertical"
    android:layout_width="fill_parent"
    android:layout_height="fill_parent">
<ImageView
    android:id="@+id/img"
    android:layout_width="wrap_content"
    android:layout_height="wrap_content"
    android:layout_gravity="center" />
</LinearLayout>
```

（2）打开 src/com.example.e8_8 目录下的 MainActivity.java 文件，代码修改为：

```
public class MainActivity extends Activity
{
    ImageView img;
```

```java
private InputStream OpenHttpConnection(String urlString) throws IOException
{
    //...
}
private Bitmap DownloadImage(String URL)
{
    Bitmap bitmap = null;
    InputStream in = null;
    try {
        in = OpenHttpConnection(URL);
        bitmap = BitmapFactory.decodeStream(in);
        in.close();
    } catch (IOException e1) {
        Toast.makeText(this, e1.getLocalizedMessage(),
            Toast.LENGTH_LONG).show();
        e1.printStackTrace();
    }
    return bitmap;
}
/* 第一次调用活动*/
@Override
public void onCreate(Bundle savedInstanceState)
{
    super.onCreate(savedInstanceState);
    setContentView(R.layout.main);
    //下载一幅图像
    Bitmap bitmap = DownloadImage("http://www.streetcar.org/mim/cable/images/bj.jpg");
    img=(ImageView)findViewById(R.id.img);
    img.setImageBitmap(bitmap);
}
```

在以上代码中，DownloadImage()方法接受要下载的图像的 URL，然后使用前面定义过的 OpenHttpConnection()方法打开到服务器的连接。利用连接返回的 InputStream 对象，使用 BitmapFactory 类的 decodeStream()方法来下载和解码数据，使之成为一个 Bitmap 对象。DownloadImage()返回一个 Bitmap 对象。

8.8.2 下载文本文件

除了下载二进制数据外，还可以下载纯文本文件，例如，可能会写一个 RSS Reader 应用程序，因此需要下载 RSS XML 源来进行处理。下面演示怎样在应用程序中下载纯文本文件。

【实现步骤】

在前面建立的 E8_8 工程中，打开 src/com.example.e8_8 目录下的 MainActivity.java 文件，代码修改为：

```java
public class MainActivity extends Activity
{
    ImageView img;
    private InputStream OpenHttpConnection(String urlString)throws IOException
    {
        //...
    }
    private Bitmap DownloadImage(String URL)
    {
        //...
    }
    private String DownloadText(String URL)
    {
        int BUFFER_SIZE = 2000;
        InputStream in = null;
        try {
            in = OpenHttpConnection(URL);
        } catch (IOException e1)
        {
            // TOD: 自动存根法
            e1.printStackTrace();
            return "";
        }

        InputStreamReader isr = new InputStreamReader(in);
        int charRead;
        String str = "";
        char[] inputBuffer = new char[BUFFER_SIZE];
        try {
            while ((charRead = isr.read(inputBuffer))>0)
            {
                //字符转换
                String readString =
                    String.copyValueOf(inputBuffer, 0, charRead);
                str += readString;
                inputBuffer = new char[BUFFER_SIZE];
            }
            in.close();
```

```
            } catch (IOException e)
            {
                e.printStackTrace();
                return "";
            }
        return str;
    }
    /* 第一次调用活动*/
    @Override
    public void onCreate(Bundle savedInstanceState)
    {
        super.onCreate(savedInstanceState);
        setContentView(R.layout.main);
        //下载一幅图像
        Bitmap bitmap = DownloadImage("http://www.streetcar.org/mim/cable/images/bj.jpg");
        img=(ImageView)findViewById(R.id.img);
        img.setImageBitmap(bitmap);
        //下载一个 RSS 源
        String str=DownloadText("http://www.appleinsider.com/appleinsider.rss");
        Toast.makeText(getBassContext(),str,Toast.LENGTE_SHORT).show();
    }
}
```

以上程序中，DownloadText()方法接受要下载的文本文件的 URL，然后返回下载的文本文件的字符串。它基本上是打开一个到服务器的 HTTP 连接，然后使用一个 InputStreamReader 对象从流中读取每个字符，将它们保存在一个 String 对象中。

8.8.3 在线播放音乐

为了节约手机的存储空间，可以从网络中下载一个 MP3 然后再播放。在本实例中将插入 4 个按钮，分别用于播放、暂停、重新播放和停止处理。执行后，通过 Runnable 发起云行线程，在线程中远程下载指定的 MP3 文件，是通过网络传输方式下载的。下载完毕后，临时保存到 SD 卡中，这样可以通过 4 个按钮对其进行控制。当程序关闭后，删除 SD 卡中的临时性文件。

【实现步骤】
（1）在 Eclipse 环境下，建立一个名为 E8_8_3 的工程。
（2）打开 res/layout 目录下的 main.xml 文件，代码修改为：

```
<?xml version="1.0" encoding="utf-8"?>
<LinearLayout
    xmlns:android="http://schemas.android.com/apk/res/android"
    android:orientation="vertical"
    android:layout_width="fill_parent"
```

```xml
    android:layout_height="fill_parent">
<TextView
    android:id="@+id/myTextView1"
    android:layout_width="fill_parent"
    android:layout_height="wrap_content"
    android:text="@string/hello_world"/>
<LinearLayout
    android:orientation="horizontal"
    android:layout_height="wrap_content"
    android:layout_width="fill_parent"
    android:padding="10dip" >
<ImageButton android:id="@+id/play"
    android:layout_height="wrap_content"
    android:layout_width="wrap_content"
    android:src="@drawable/play"/>
<ImageButton android:id="@+id/pause"
    android:layout_height="wrap_content"
    android:layout_width="wrap_content"
    android:src="@drawable/pause" />
<ImageButton android:id="@+id/reset"
    android:layout_height="wrap_content"
    android:layout_width="wrap_content"
    android:src="@drawable/reset"/>
<ImageButton android:id="@+id/stop"
    android:layout_height="wrap_content"
    android:layout_width="wrap_content"
    android:src="@drawable/stop"/>
</LinearLayout>
</LinearLayout>
```

（3）打开 src/com.example.e8_8_3 目录下的 MainActivity.java 文件，代码修改为：

```java
public class MainActivity extends Activity
{
    //调用 private 声明系统中需要的对象
    private TextView mTextView01;
    private MediaPlayer mMediaPlayer01;
    private ImageButton mPlay, mReset, mPause, mStop;
    private boolean bIsReleased = false;
    private boolean bIsPaused = false;
    //currentFilePath 用于记录当前正在播放 MP3 的地址 URL，定义 currentTempFilePath 表示当前
    //播放 MP3 的路径
    private static final String TAG = "Hippo_URL_MP3_Player";
```

```java
private String currentFilePath = "";
private String currentTempFilePath = "";
private String strVideoURL = "";
/* 第一次调用活动 */
@Override
//引用主布局文件 main.xml，通过 strVideoURL 设置要播放 MP3 文件的网址，并设置透明度
public void onCreate(Bundle savedInstanceState)
{
    super.onCreate(savedInstanceState);
    setContentView(R.layout.main);
    strVideoURL =
"http://box.zhangmen.baidu.com/m?rf=top-top100&l_id=2&gate=10&ct=134217728&tn=baidumt&c_n=mp3order&l_n=%E6%96%B0%E6%AD%8CTOP100&s_o=5";
    mTextView01 = (TextView)findViewById(R.id.myTextView1);
    getWindow().setFormat(PixelFormat.TRANSPARENT);
    mPlay = (ImageButton)findViewById(R.id.play);
    mReset = (ImageButton)findViewById(R.id.reset);
    mPause = (ImageButton)findViewById(R.id.pause);
    mStop = (ImageButton)findViewById(R.id.stop);
    //播放按钮所触发的处理事件
    mPlay.setOnClickListener(new ImageButton.OnClickListener()
    {
        public void onClick(View view)
        {
            playVideo(strVideoURL);
            mTextView01.setText
            (
                getResources().getText(R.string.str_play).toString()+
                "\n"+ strVideoURL
            );
        }
    });
    //重新播放按钮所触发的处理事件
    mReset.setOnClickListener(new ImageButton.OnClickListener()
    {
        public void onClick(View view)
        {
            if(bIsReleased == false)
            {
                if (mMediaPlayer01 != null)
                {
```

```java
              mMediaPlayer01.seekTo(0);
              mTextView01.setText(R.string.str_play);
          }
        }
      }
  });
  //暂停所触发的处理事件
  mPause.setOnClickListener(new ImageButton.OnClickListener()
  {
      public void onClick(View view)
      {
          if (mMediaPlayer01 != null)
          {
              if(bIsReleased == false)
              {
                  if(bIsPaused==false)
                  {
                      mMediaPlayer01.pause();
                      bIsPaused = true;
                      mTextView01.setText(R.string.str_pause);
                  }
                  else if(bIsPaused==true)
                  {
                      mMediaPlayer01.start();
                      bIsPaused = false;
                      mTextView01.setText(R.string.str_play);
                  }
              }
          }
      }
  });
  //终止按钮所触发的事件
  mStop.setOnClickListener(new ImageButton.OnClickListener()
  {
      public void onClick(View view)
      {
          try
          {
              if (mMediaPlayer01 != null)
              {
                  if(bIsReleased==false)
```

```java
            {
                mMediaPlayer01.seekTo(0);
                mMediaPlayer01.pause();
                mTextView01.setText(R.string.str_stop);
            }
          }
        }
        catch(Exception e)
        {
          mTextView01.setText(e.toString());
          Log.e(TAG, e.toString());
          e.printStackTrace();
        }
      }
    });
}
//定义方法 playVideo(final String strPath)，用于播放指定的 MP3，其播放的是存储卡中暂时保存
//的 MP3 文件
private void playVideo(final String strPath)
{
  try
  {
    if (strPath.equals(currentFilePath)&& mMediaPlayer01 != null)
    {
      mMediaPlayer01.start();
      return;
    }
    currentFilePath = strPath;
    mMediaPlayer01 = new MediaPlayer();
    mMediaPlayer01.setAudioStreamType(2);
    //定义 setOnErrorListener 处理事件来处理错误
    mMediaPlayer01.setOnErrorListener(new MediaPlayer.OnErrorListener()
    {
      @Override
      public boolean onError(MediaPlayer mp, int what, int extra)
      {
        Log.i(TAG, "Error on Listener, what: " + what + "extra: " + extra);
        return false;
      }
    });
```

```java
//定义 setOnBufferingUpdateListener 来捕捉使用 MediaPlayer 缓冲区的更新
mMediaPlayer01.setOnBufferingUpdateListener(new MediaPlayer.OnBufferingUpdateListener()
{
  @Override
  public void onBufferingUpdate(MediaPlayer mp, int percent)
  {
    Log.i(TAG, "Update buffer: " + Integer.toString(percent)+ "%");
  }
});
//定义 setOnCompletionListener 用于实现播放完毕所触发的事件
mMediaPlayer01.setOnCompletionListener(new MediaPlayer.OnCompletionListener()
{
  @Override
  public void onCompletion(MediaPlayer mp)
  {
    Log.i(TAG,"mMediaPlayer01 Listener Completed");
  }
});
//定义 setOnPreparedListener 用于开始阶段的监听
mMediaPlayer01.setOnPreparedListener(new MediaPlayer.OnPreparedListener()
{
  @Override
  public void onPrepared(MediaPlayer mp)
  {
    Log.i(TAG,"Prepared Listener");
  }
});
//用 Runnable 来确保文件存储完毕后才开始 start()
Runnable r = new Runnable()
{
  public void run()
  {
    try
    {
      setDataSource(strPath);
      mMediaPlayer01.prepare();
      Log.i(TAG, "Duration: " + mMediaPlayer01.getDuration());
      mMediaPlayer01.start();
      bIsReleased = false;
    }
```

```java
                    catch (Exception e)
                    {
                        Log.e(TAG, e.getMessage(), e);
                    }
                }
            };
            new Thread(r).start();
        }
        catch(Exception e)
        {
            if (mMediaPlayer01 != null)
            {
                mMediaPlayer01.stop();
                mMediaPlayer01.release();
            }
            e.printStackTrace();
        }
    }
    //定义函数 setDataSource 用于存储 URL 的 MP3 文件到存储卡
    private void setDataSource(String strPath) throws Exception
    {
        if (!URLUtil.isNetworkUrl(strPath))
        {
            mMediaPlayer01.setDataSource(strPath);
        }
        else
        {
            if(bIsReleased == false)
            {
                URL myURL = new URL(strPath);
                URLConnection conn = myURL.openConnection();
                conn.connect();
                InputStream is = conn.getInputStream();
                if (is == null)
                {
                    throw new RuntimeException("stream is null");
                }
                File myTempFile = File.createTempFile("yinyue", "."+getFileExtension(strPath));
                currentTempFilePath = myTempFile.getAbsolutePath();
                FileOutputStream fos = new FileOutputStream(myTempFile);
```

```
            byte buf[] = new byte[128];
            do
            {
                int numread = is.read(buf);
                if (numread <= 0)
                {
                    break;
                }
                fos.write(buf, 0, numread);
            }while (true);

            mMediaPlayer01.setDataSource(currentTempFilePath);
            try
            {
                is.close();
            }
            catch (Exception ex)
            {
                Log.e(TAG, "error: " + ex.getMessage(), ex);
            }
        }
    }
}
//用于获取音乐文件扩展名自定义函数
private String getFileExtension(String strFileName)
{
    File myFile = new File(strFileName);
    String strFileExtension=myFile.getName();
    strFileExtension=(strFileExtension.substring(strFileExtension.lastIndexOf(".")+1)).toLowerCase();
    if(strFileExtension=="")
    {
        strFileExtension = "dat";
    }
    return strFileExtension;
}
//离开程序时需要调用自定义函数删除临时音乐文件
private void delFile(String strFileName)
{
    File myFile = new File(strFileName);
    if(myFile.exists())
```

```
            {
                myFile.delete();
            }
        }
        @Override
        protected void onPause()
        {
            try
            {
                delFile(currentTempFilePath);
            }
            catch(Exception e)
            {
                e.printStackTrace();
            }
            super.onPause();
        }
    }
```

运行程序，效果如图 8-12 所示。

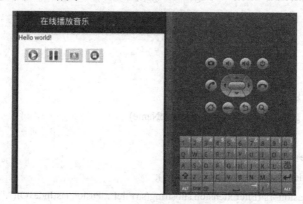

图 8-12　在线播放音乐控件

第 9 章 Android 多媒体

Android 应用面向的是普通个人用户，这些用户往往会更加关注用户体验，因此为 Android 应用增加动画、视频、音乐等多媒体功能十分必要。就目前的手机发展趋势而言，手机已经不再是单一的通信工具，其已经发展为集成照相机、音乐播放器、视频播放器、个人小型终端于一体的智能设备，因此为手机提供音频录制、播放、视频录制、播放的功能十分必要。

9.1 音频/视频的播放

在 Android 系统提供了几种常见的音频/视频播放器，下面给以介绍。

9.1.1 MediaPlay 类

在 Android 系统中，MediaPlayer 类包含了 Audio 和 Video 的播放功能，用于播放音频、视频以及网络上的流媒体。MediaPlayer 类中提供了相关方法。

1．create 方法

create 方法用于创建一个要播放的多媒体，可以通过指定源文件中的多媒体文件来实现，也可以通过指定 URI 连接来播放互联网上的多媒体文件。Creat 方法是一种创建 MediaPlayer 的静态方法。其语法格式为：

```
public static MediaPlayer create(Context context,int resid)
public static MediaPlayer create(Context context,Uri uri)
```

其中，参数 context 为上下文，可以通过 getApplicationContext()方法获取；参数 resid 为多媒体文件的资源 ID；参数 uri 为多媒体文件的 URI 地址。

2．start 方法

start 方法用于开始播放一个多媒体对象，这是播放多媒体文件的主要方法。如果在播放前执行了 puase 方法暂停播放，则再次执行 start 方法时将从暂停点继续播放。其语法格式为：

```
public void start()
```

3．stop 方法

stop 方法用于停止播放多媒体文件。使用该方法后，多媒体对象将停止播放。如果再次调用 start 方法，则将从头开始播放。其语法格式为：

```
public void stop()
```

4. pause 方法

pause 方法用于暂停播放多媒体文件。使用该方法后，多媒体对象将暂停播放，如果再次调用 start 方法，则将从暂停的地方继续播放。其语法格式为：

```
public void pause()
```

5. reset 方法

reset 方法用于复位多媒体实例，执行该方法后，将使 MediaPlayer 对象变为未初始化的状态。复位多媒体之后，必须重新设置数据源并执行 prepare()方法，方可继续播放。该方法还可以使播放器从 Error 状态中恢复过来。其语法格式为：

```
public void reset()
```

6. setDataSource 方法

setDataSource 方法用于为一个 MediaPlayer 对象设置播放的数据源，使用该方法，可以在程序中动态更新播放的音乐文件，其经常用于音乐播放软件的设计中。其语法格式为：

```
public void setDataSource(FileDescriptor fd)
public void setDataSource(String path)
public void setDataSource(Context context,Uri uri)
```

其中，参数 fd 为数据源文件的描述；参数 path 为数据源文件的路径，其为字符串形式；参数 context 为上下文，可以通过 getApplicationContext()方法获取；参数 uri 为多媒体文件的 URI 地址。

7. prepare 方法

prepare 方法用于执行播放前的准备工作，这是一种同步的调用方法。如果需要使用异步方式播放，则可以使用 prepareAsync 方法，如果 MediaPlayer 实例是由 create 方法创建的，那么第一次启动播放前不需要再调用 prepare 方法，因为 create 方法里已经调用过。其语法格式为：

```
public void prepare()
```

通过以下代码来演示前面介绍的几种方法。

```
public class MainActivity extends Activity
{
    /* 第一次调用活动*/
    @Override
    public void onCreate(Bundle savedInstanceState)
    {
        super.onCreate(savedInstanceState);
        setContentView(R.layout.main);
        Button btn1=(Button)findViewById(R.id.button1);
        Button btn2=(Button)findViewById(R.id.button2);
        Button btn3=(Button)findViewById(R.id.button3);
        final MediaPlayer mp=new MediaPlayer();
        final String path="/sdcard/yudie.mp3";
        btn1.setOanClickListener(new View.OnClickListener()
```

```java
{
    //按钮监听器
    @Override
    public void onClick(View v)
    {
        try {
            mp.reset();
            mp.setDataSource(path);
            mp.prepare();
        } catch (IllegalArgumentException e)
        {
            e.printStackTrace();
        } catch (IllegalStateException e)
        {
            e.printStackTrace();
        } catch (IOException e)
        {
            e.printStackTrace();
        }
        mp.start();
        Toast.makeText(getApplicationContext(), "开始播放", Toast.LENGTH_LONG).show();
    }
});
btn2.setOnClickListener(new View.OnClickListener()
{
    @Override
    public void onClick(View v)
    {
        mp.stop();
        Toast.makeText(getApplicationContext(), "停止播放", Toast.LENGTH_LONG).show();
    }
});
btn3.setOnClickListener(new View.OnClickListener()
{
    @Override
    public void onClick(View v)
    {
        mp.pause();
        Toast.makeText(getApplicationContext(), "暂停播放", Toast.LENGTH_LONG).show();
    }
});
```

```
            }
        }
```

程序中，先创建了 MediaPlayer 实例，然后在按钮监听器中调用 reset 方法来复位多媒体，调用 setDataSource 设置播放的数据源为 SD 卡上的文件，调用 prepare 方法来准备播放，最后使用 start 方法开始播放。

运行程序，效果如图 9-1 所示。

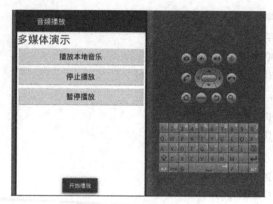

图 9-1　播放 SD 上的音乐

注意：在播放音乐前，先将音乐文件放置到 SD 上。

8．seekTo 方法

seekTo 方法用于指定播放的位置。在默认情况下，都是从 0 时刻开始播放的。使用该方法，可指定从哪一时刻开始播放，以毫秒为单位。其语法格式为：

```
public void seekTo(int msec)
```

其中，参数 msec 为指定的开始播放时间，以毫秒为单位。

9．setVolume 方法

setVolume 方法用于设置播放多媒体文件时的音量，可以分别设置左声道和右声道的音量大小。需要注意的是，使用该方法设置的仅是当前播放对象的音量，而不是系统播放音量。其语法格式为：

```
public void setVolume(float leftVolume,float rightVolume)
```

其中，参数 leftVolume 为左声道的音量，取值范围为 0～1；参数 rightVolume 为右声道的音量，取值范围为 0～1。

通过以下代码来演示怎样设置音乐播放的声道。

```
public class MainActivity extends Activity
{
    private float leftVolume,rightVolume;
    /*第一次调用活动 */
    @Override
    public void onCreate(Bundle savedInstanceState)
    {
```

```java
super.onCreate(savedInstanceState);
setContentView(R.layout.main);
Button btn1=(Button)findViewById(R.id.button1);
Button btn2=(Button)findViewById(R.id.button2);
Button btn3=(Button)findViewById(R.id.button3);
Button btn4=(Button)findViewById(R.id.button4);
Button btn5=(Button)findViewById(R.id.button5);
final MediaPlayer mp=new MediaPlayer();
final String path="/sdcard/yudie.mp3";
leftVolume=0;
rightVolume=0;
btn1.setOnClickListener(new View.OnClickListener()
{
    //按钮监听器
    @Override
    public void onClick(View v)
    {
        try {
            mp.reset();          //复位
            mp.setDataSource(path);
            mp.prepare();
        } catch (IllegalArgumentException e)
        {
            e.printStackTrace();
        } catch (IllegalStateException e)
        {
            e.printStackTrace();
        } catch (IOException e)
        {
            e.printStackTrace();
        }
        mp.start();
        Toast.makeText(getApplicationContext(), "开始播放", Toast.LENGTH_LONG).show();
    }
});
btn2.setOnClickListener(new View.OnClickListener()
{
    @Override
    public void onClick(View v)
    {
        mp.stop();
```

```java
            Toast.makeText(getApplicationContext(), "停止播放", Toast.LENGTH_LONG).show();
        }
    });
    btn3.setOnClickListener(new View.OnClickListener()
    {
        @Override
        public void onClick(View v)
        {
            mp.pause();
            Toast.makeText(getApplicationContext(), "暂停播放", Toast.LENGTH_LONG).
                show();
        }
    });
    btn4.setOnClickListener(new View.OnClickListener()
    {
        @Override
        public void onClick(View v)
        {

            leftVolume=leftVolume+0.1f;
            rightVolume=rightVolume+0.1f;
            mp.setVolume(leftVolume, rightVolume);
            Toast.makeText(getApplicationContext(), "当前音量为"+leftVolume,
                Toast.LENGTH_LONG).show();
        }
    });
    btn5.setOnClickListener(new View.OnClickListener()
    {
        @Override
        public void onClick(View v)
        {
            leftVolume=leftVolume-0.1f;
            rightVolume=rightVolume-0.1f;
            mp.setVolume(leftVolume, rightVolume);
            Toast.makeText(getApplicationContext(), "当前音量为"+leftVolume,
                Toast.LENGTH_LONG).show();
        }
    });
    }
}
```

以上代码中，先创建了 MediaPlayer 实例，然后分别在不同的按钮监听器中实现了播放、暂停和停止的功能。在第 4 个和第 5 个按钮监听器中，调用 setVolume 方法设置播放的音量，分别实现声音的右声道与左声道。

运行程序，效果如图 9-2 所示。

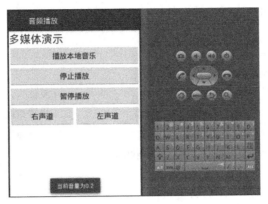

图 9-2　设置音量效果

10．setLooping 方法

setLooping 方法用于设置是否循环播放多媒体文件。当其取值为 true 时表示循环播放，当其取值为 false 时表示仅播放一次。其语法格式为：

  public void setLooping(Boolean looping)

其中，参数 looping 表示是否循环播放，当其取值为 true 时表示循环播放，取值为 false 时表示仅播放一次。

通过以下代码来演示怎样设置循环播放音乐。

```java
public class MainActivity extends Activity
{
    Boolean looping;
    /* 第一次调用活动 */
    @Override
    public void onCreate(Bundle savedInstanceState)
    {
        super.onCreate(savedInstanceState);
        setContentView(R.layout.main);
        Button btn1=(Button)findViewById(R.id.button1);
        Button btn2=(Button)findViewById(R.id.button2);
        Button btn3=(Button)findViewById(R.id.button3);
        final Button btn4=(Button)findViewById(R.id.button4);
        looping=false;
        btn4.setText("单曲播放");
        final MediaPlayer mp=new MediaPlayer();          //MediaPlayer 实例
        final String path="/sdcard/yudie.mp3";           //SD 卡文件路径
        btn1.setOnClickListener(new View.OnClickListener()
```

```java
        {
            //按钮监听器
            @Override
            public void onClick(View v)
            {
                //自动存根法
                try {
                    mp.reset();                          //复位
                    mp.setDataSource(path);              //设置数据源
                    mp.prepare();                        //准备播放
                } catch (IllegalArgumentException e)
                {
                    //错误处理
                    e.printStackTrace();
                } catch (IllegalStateException e)
                {
                    e.printStackTrace();
                } catch (IOException e)
                {
                    e.printStackTrace();
                }
                mp.start();
                Toast.makeText(getApplicationContext(), "开始播放", Toast.LENGTH_LONG).show();
            }
        });
        btn2.setOnClickListener(new View.OnClickListener()
        {
            @Override
            public void onClick(View v)
            {
                mp.stop();
                Toast.makeText(getApplicationContext(), "停止播放", Toast.LENGTH_LONG).show();
            }
        });
        btn3.setOnClickListener(new View.OnClickListener()
        {
            @Override
            public void onClick(View v)
            {
                mp.pause();
                Toast.makeText(getApplicationContext(), "暂停播放", Toast.LENGTH_LONG).show();
```

```
            }
        });
        btn4.setOnClickListener(new View.OnClickListener()
        {
            @Override
            public void onClick(View v)
            {
                if(looping)
                {
                    looping=false;
                    btn4.setText("单曲播放");
                }
                else
                {
                    looping=true;
                    btn4.setText("循环播放");
                }
                mp.setLooping(looping);
            }
        });
    }
}
```

运行程序，效果如图 9-3 所示。

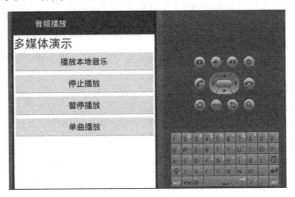

图 9-3　设置歌曲播放状态

11．setOnCompletionListener 方法

setOnCompletionListener 方法用于设置播放完成监听器。当 MediaPlayer 对象播放完毕时，将触发该监听器。通过重载 onCompletion 方法即可以完成一系列用户功能。该方法常常用于播放列表功能的场合，当一个音乐文件播放完毕时，自动播放下一个音乐文件。其语法格式为：

public void setOnCompletionListener(MediaPlayer.OnCompletionListener listener)

其中，参数 listener 为监听器对象。

12. setOnSeekCompleteListener 方法

setOnSeekCompleteListener 方法用于设置寻址完成监听器。当 MediaPlayer 对象通过 seekTo 方法寻址完毕后，将触发该监听器。通过重载 onSeekComplete 方法即可完成一系列用户功能。该方法常常用于像 A-B 复读这样功能的场合。其语法格式为：

    public void setOnSeekCompleteListener(MediaPlayer.OnSeekCompleteListener listener)

其中，参数 listener 为监听器对象。

通过以下代码来实现怎样设置寻址完成监听器。

```java
public class MainActivity extends Activity
{
    Boolean looping;
    /* 第一次调用活动 */
    @Override
    public void onCreate(Bundle savedInstanceState)
    {
        super.onCreate(savedInstanceState);
        setContentView(R.layout.main);
        Button btn1=(Button)findViewById(R.id.button1);
        Button btn2=(Button)findViewById(R.id.button2);
        Button btn3=(Button)findViewById(R.id.button3);
        Button btn4=(Button)findViewById(R.id.button4);
        final MediaPlayer mp=new MediaPlayer();            //MediaPlayer 实例
        final String path="/sdcard/yudie.mp3";             //SD 卡文件路径
        btn1.setOnClickListener(new View.OnClickListener()
        {
            //按钮监听器
            @Override
            public void onClick(View v)
            {
                // TODO：自动存根法
                try {
                    mp.reset();                            //复位
                    mp.setDataSource(path);                //设置数据源
                    mp.prepare();                          //准备播放
                } catch (IllegalArgumentException e)
                {
                    //错误处理
                    e.printStackTrace();
                } catch (IllegalStateException e)
                {
                    e.printStackTrace();
```

```java
            } catch (IOException e)
            {
                e.printStackTrace();
            }
            mp.start();
            Toast.makeText(getApplicationContext(), "开始播放", Toast.LENGTH_LONG).show();
        }
    });
    btn2.setOnClickListener(new View.OnClickListener()
    {
        @Override
        public void onClick(View v)
        {
            mp.stop();
            Toast.makeText(getApplicationContext(), "停止播放", Toast.LENGTH_LONG).show();
        }
    });
    btn3.setOnClickListener(new View.OnClickListener()
    {
        @Override
        public void onClick(View v)
        {
            mp.pause();
            Toast.makeText(getApplicationContext(), "暂停播放", Toast.LENGTH_LONG).show();
        }
    });
    btn4.setOnClickListener(new View.OnClickListener()
    {
        @Override
        public void onClick(View v)
        {
            mp.seekTo(10000);
        }
    });
    mp.setOnSeekCompleteListener(new MediaPlayer.OnSeekCompleteListener()
    {
        @Override
        public void onSeekComplete(MediaPlayer mp)
        {
            Toast.makeText(getApplicationContext(), "已寻址到指定播放位置",
                Toast.LENGTH_LONG).show();
```

```
                    }
                });
            }
        }
```

运行程序，效果如图 9-4 所示。

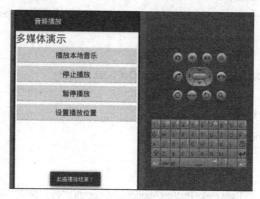

图 9-4 寻址到指定播放位置效果

13. setOnPreparedListener 方法

setOnPreparedListener 方法用于设置准备完毕监听器。当 MediaPlayer 对象通过 prepare 方法准备完毕时，将触发该监听器。通过重载 onPrepared 方法即可完成一系列用户功能。其语法格式为：

```
public void setOnPreparedListener(MediaPlayer.OnPreparedListener listener)
```

其中，参数 listener 为监听器对象。

14. setOnErrorListener 方法

setOnErrorListener 方法用于设置播放错误监听器。当 MediaPlayer 对象播放多媒体文件发生错误时，将触发该监听器。通过重载 onError 方法即可完成一系列用户功能。其语法格式为：

```
public void setOnErrorListener(MediaPlayer.OnErrorListener listener)
```

其中，参数 listener 为监听器对象。

15. setOnBufferingUpdateListener 方法

setOnBufferingUpdateListener 方法用于设置网络流媒体的缓冲监听器。当 MediaPlayer 对象播放网络上的多媒体文件并发生缓冲时，将触发该监听器。通过重载 onBufferingUpdate 方法即可完成一系列用户功能，这样当缓冲播放时可以给用户提示信息等。其语法格式为：

```
public void setOnBufferingUpdateListener(MediaPlayer.OnBufferingUpdateListener listener)
```

其中，参数 listener 为监听器对象。

16. release 方法

release 方法用于释放 MediaPlayer 多媒体对象。当一个 MediaPlayer 对象播放完毕不再使用时，需要调用该方法来释放其占用的系统资源。其语法格式为：

```
public void release()
```

9.1.2 SoundPool 类

如果应用程序经常需要播放密集、短促的音效,这时还使用 MediaPlayer 类就有些不合适了。MediaPlayer 类存在如下缺点:
- 资源占用较多、延迟时间较长。
- 不支持多个音频同时播放。

除了前面介绍的 MediaPlayer 播放音频之外,Android 还提供了 SoundPool 来播放音效。SoundPool 使用音效池的概念来管理多个短促的音效,例如,它可以开始就加载 20 个音效,以后在程序中按音效的 ID 进行播放。

SoundPool 主要用于播放一些较短的声音片段,与 MediaPlayer 相比,SoundPool 的优势在于 CPU 资源占用量低和反应延迟小。另外,SoundPool 还支持自行设置声音的品质、音量、播放比率等参数。

SoundPool 提供了一个构造器,该构造器可指定它总共支持多个声音(也就是池的大小)、声音的品质等。构造器如下:

SoundPool(int maxStreams,int streamType,int srcQuality)

其中,第一个参数指定支持多个声音;第二个参数指定声音类型;第三个参数指定声音品质。

一旦得到了 SoundPool 对象后,接下来就可调用 SoundPool 的多个重载的 load 方法来加载声音了,SoundPool 提供了如下 4 个 load 方法。

- int load(Context context,int resId,int priority):从 resId 所对应的资源加载声音。
- int load(FileDescriptor fd,long offset,long length,int priority):加载 fd 所对应的文件的 offset 开始、长度为 length 的声音。
- int load(AssetFileDescriptor afd,int priority):从 afd 所对应的文件中加载声音。
- int load(String path,int priority):从 path 对应的文件去加载声音。

上面 4 个方法中有一个 priority 参数,该参数目前还没有任何作用,Android 建议将该参数设为 1,保持和未来的兼容性。

上面 4 个方法加载声音后,都会返回该声音的 ID,以后程序就可以通过该声音的 ID 来播放指定声音,SoundPool 提供的播放指定声音的方法如下:

int play(int soundID,float leftVolume,float rightVolume,int priority,int loop,float rate)

其中,第一个参数指定播放哪个声音;参数 leftVolume、rightVolume 分别指定左、右声道;参数 priority 指定播放声音的优先级,数值越大,优先级越高;参数 loop 指定是否循环,当取值为 0 时即不循环,取值为-1 时即循环;参数 rate 为指定播放的比率,数值可从 0.5 到 2,1 为正常比率。

为了更好地管理 SoundPool 所加载的每个声音的 ID,程序一般会使用一个 HashMap(Integer,Integer)对象来管理声音。

归纳起来,使用 SoundPool 播放声音的步骤如下:
① 调用 SoundPool 的构造器创建 SoundPool 的对象。
② 调用 SoundPool 对象的 load()方法从指定资源、文件中加载声音。最好使用 HashMap(Integer,Integer)来管理所加载的声音。
③ 调用 SoundPool 的 play 方法播放声音。

通过以下代码来演示如何使用 SoundPool 来播放音效。

```java
public class MainActivity extends Activity
{
    private Button bPlay;
    private Button bPause;
    private SoundPool sp;
    private HashMap<Integer,Integer> spMap;
    /** 第一次调用活动 */
    public void onCreate(Bundle savedInstanceState)
    {
        super.onCreate(savedInstanceState);
        setContentView(R.layout.main);
        bPlay = (Button)findViewById(R.id.button1);
        bPause = (Button)findViewById(R.id.button2);
        sp = new SoundPool(2,AudioManager.STREAM_MUSIC,0);
        spMap = new HashMap<Integer,Integer>();
        spMap.put(1, sp.load(this, R.raw.shot, 1));
        spMap.put(2, sp.load(this, R.raw.bomb, 2));
        spMap.put(3, sp.load(this, R.raw.arrow, 3));
        bPlay.setOnClickListener(new OnClickListener()
        {
            public void onClick(View v)
            {
                // TODO：自动存根法
                playSounds(1,1);
                playSounds(2,2);
                playSounds(3,3);
            }
        });
        bPause.setOnClickListener(new OnClickListener()
        {
            @Override
            public void onClick(View v)
            {
                // TODO：自动存根法
                sp.pause(spMap.get(1));
                sp.pause(spMap.get(2));
                sp.pause(spMap.get(3));
            }
        });
    }
    public void playSounds(int sound, int number)
```

```
        {
            AudioManager am = (AudioManager)this.getSystemService(this.AUDIO_SERVICE);
            float audioMaxVolumn = am.getStreamMaxVolume(AudioManager.STREAM_MUSIC);
            float audioCurrentVolumn = am.getStreamVolume(AudioManager.STREAM_MUSIC);
            float volumnRatio = audioCurrentVolumn/audioMaxVolumn;
            sp.play(spMap.get(sound), volumnRatio, volumnRatio, 1, number, 1);
            sp.play(spMap.get(sound), volumnRatio, volumnRatio, 2, number, 2);
            sp.play(spMap.get(sound), volumnRatio, volumnRatio, 3, number, 3);
        }
}
```

在使用 SoundPool 播放声音时有如下几点需要注意：
- SoundPool 虽然可以一次性加载多个声音，但由于受内存容量限制，因此避免使用 SoundPool 来播放歌曲或者游戏背景音乐，只有那些短促、密集的声音才考虑使用 SoundPool 播放。
- 虽然 SoundPool 比 MediaPlayer 的效率高，但也不是绝对不存在延迟问题，尤其在那些性能不太好的手机中，SoundPool 的延迟问题会更严重。

9.1.3 VideoView 类

在 Android 系统中，VideoView 用于 VideoView 控件的显示和控制。VideoView 控件可以用于视频播放的显示控制。VideoView 类提供了视频播放的开始、暂停等功能。通过 VideoView 类可以播放本地的视频文件，也可以用来播放网络上的视频文件。

Android 系统支持 MP4 的 H.264、3GP 和 WMV 视频的解析。但是，由于模拟器性能的限制，在模拟器上能正常播放的视频分辨率和码率都比较低。为了获取比较高的效率，能够在 Android 手机上进行调试会更加方便。

实现 VideoView 类的相关方法主要介绍如下。

1. setMediaController 方法

setMediaController 方法用于设置 VideoView 的播放控制器模式（播放进度条）。这是使用 VideoView 播放视频前必须调用的方法。其语法格式为：

```
public void setMediaController(MediaController controller)
```

其中，参数 controller 为 MediaController 对象。

2. setVideoPath 方法

setVideoPath 方法用于设置 VideoView 的播放视频路径，也就是指定播放文件。这是使用 VideoView 播放视频前必须调用的方法。其语法格式为：

```
public void setVideoPath(String path)
```

其中，path 为视频文件的路径信息。

3. start 方法

start 方法用于开始播放一个视频对象，这是播放视频文件的主要方法。如果在播放前执行了 pause 方法暂停播放，则再次执行 start 方法时将从暂停点继续播放。其语法格式为：

```
public void start()
```

4. pause 方法

pause 方法用于暂停播放视频文件。使用该方法后，视频对象将暂停播放，如果再次调用 start 方法，则将从暂停的地方继续播放。其语法格式为：

```
public void pause()
```

5. setVideoURI 方法

setVideoURI 方法用于设置 VideoView 的播放视频源地址，也就是指定播放文件。这里的源地址可以是本地文件，也可以是网络上的视频文件地址。其语法格式为：

```
public void setVideoURI(Uri uri)
```

其中，参数 uri 为视频文件的地址信息。

通过以下代码来演示怎样设置视频源地址。

```java
public class MainActivity extends Activity
{
    Boolean looping;
    /* 第一次调用活动 */
    @Override
    public void onCreate(Bundle savedInstanceState)
    {
        super.onCreate(savedInstanceState);
        setContentView(R.layout.main);
        Button btn1=(Button)findViewById(R.id.button1);
        Button btn2=(Button)findViewById(R.id.button2);
        final VideoView video = (VideoView) findViewById(R.id.videoView1);
        MediaController mediaController = new MediaController(this);
        mediaController.setAnchorView(video);    //把 MediaController 对象绑定到 VideoView 上
        video.setMediaController(mediaController);
        Uri uri=Uri.parse("/sdcard/kflyb.flv");;
        video.setVideoURI(uri);
        btn1.setOnClickListener(new View.OnClickListener()
        {
            @Override
            public void onClick(View v)
            {
                // TODO：自动存根法
                video.start();
            }
        });
        btn2.setOnClickListener(new View.OnClickListener()
```

```
            {
                @Override
                public void onClick(View v)
                {
                    video.pause();
                }
            });
        }
    }
```

在以上代码中，首先获取按钮和 VideoView 对象，然后为 VideoView 对象设置播放控制器，并通过 setVideoURI 方法设置播放文件的地址。最后，在第一个按钮监听器中通过 start 方法开始播放视频，在第二个按钮监听器中通过 pause 方法暂停播放视频。

运行程序，得到如图 9-5 所示的界面。

图 9-5　视频播放界面

6．setVisibility 方法

setVisibility 方法用于设置 VideoView 的播放视频源地址，也就是指定播放文件。这里的源地址可以是本地文件，也可以是网络上的视频文件地址。该方法与 setVideoPath 方法类似，是使用 VideoView 播放视频前经常调用的方法。其语法格式为：

```
public void setVisibility(int visibility)
```

其中，参数 visibility 为 VideoView 对象是否可见，其可以取如下几种值：

- VISIBLE：为可见状态。
- INVISIBLE：不可见状态。
- GONE：不可见状态，并且控件从布局中消失。

通过以下代码来演示怎样设置视频播放控件不可见。

```
    public class MainActivity extends Activity
    {
        Boolean looping;
        /* 第一次调用活动 */
        @Override
        public void onCreate(Bundle savedInstanceState)
        {
            super.onCreate(savedInstanceState);
            setContentView(R.layout.main);
```

```java
Button btn1=(Button)findViewById(R.id.button1);
Button btn2=(Button)findViewById(R.id.button2);
Button btn3=(Button)findViewById(R.id.button3);
final VideoView video = (VideoView) findViewById(R.id.videoView1);
MediaController mediaController = new MediaController(this);
mediaController.setAnchorView(video);    //把 MediaController 对象绑定到 VideoView 上
video.setMediaController(mediaController);
Uri uri=Uri.parse("/sdcard/kflyb.flv");
video.setVideoURI(uri);
btn1.setOnClickListener(new View.OnClickListener()
{
    @Override
    public void onClick(View v)
    {
        // TODO：自动存根法
        video.start();
    }
});
btn2.setOnClickListener(new View.OnClickListener()
{
    @Override
    public void onClick(View v)
    {
        // TODO：自动存根法
        video.pause();
    }
});
btn3.setOnClickListener(new View.OnClickListener()
{
    //按钮监听器
    @Override
    public void onClick(View v)
    {
        // TODO：自动存根法
        video.setVisibility(VideoView.GONE);   //视频控件占据的位置消失
    }
});
}
}
```

运行程序，界面如图 9-6 所示，当单击界面中的"播放本地视频"按钮后，再单击界面中的"不可见"按钮，效果如图 9-7 所示。

图 9-6　播放视频界面　　　　图 9-7　视频不可见界面

7. setOnCompletionListener 方法

setOnCompletionListener 方法用于设置视频播放完成监听器。当视频对象播放完毕时，将触发该监听器。通过重载 onCompletion 方法即可完成一系列用户功能。该方法常常用于播放列表功能的场合，当一个视频文件播放完毕后，自动播放下一个视频文件。其语法格式为：

    public void setOnCompletionListener(MediaPlayer.OnCompletionListener listener)

其中，参数 listener 为监听器对象。

在以上代码基础上添加如下代码用于设置视频播放完成监听器。

```
……
video.setOnCompletionListener(new MediaPlayer.OnCompletionListener()
{
        @Override
        public void onCompletion(MediaPlayer mp)
        {
            // TODO Auto-generated method stub
            Toast.makeText(getApplicationContext(), "视频播放结束！", Toast.LENGTH_LONG).show();
        }
    });
……
```

8. setOnPreparedListener 方法

setOnPreparedListener 方法用于设置视频准备完毕监听器。当视频加载完毕并且可以播放时，将触发该监听器。通过重载 onPrepared 方法即可完成一系列用户功能。其语法格式为：

    public void setOnPreparedListener(MediaPlayer.OnPreparedListener listener)

其中，参数 listener 为监听器对象。

9. setOnErrorListener 方法

setOnErrorListener 方法用于设置视频播放错误监听器。当视频对象播放视频文件发生错误时，将触发该监听器。通过重载 onError 方法即可完成一系列用户功能。其语法格式为：

    public void setOnErrorListener(MediaPlayer.OnErrorListener listener)

其中,参数 listener 为监听器对象。

10. seekTo 方法

seekTo 方法用于指定视频播放的位置。在默认情况下,都是从 0 时刻开始播放视频的。使用该方法,可以指定从哪一时刻来开始播放,以毫秒为单位。其语法格式为:

```
public void seekTo(int msec)
```

其中,参数 msec 为指定的视频开始播放时间,以毫秒为单位。

通过以下代码来演示怎样指定视频放置位置。

```java
public class MainActivity extends Activity
{
    Boolean looping;
    /* 第一次调用活动 */
    @Override
    public void onCreate(Bundle savedInstanceState)
    {
        super.onCreate(savedInstanceState);
        setContentView(R.layout.main);
        Button btn1=(Button)findViewById(R.id.button1);
        Button btn2=(Button)findViewById(R.id.button2);
        Button btn3=(Button)findViewById(R.id.button3);
        final VideoView video = (VideoView) findViewById(R.id.videoView1);
        MediaController mediaController = new MediaController(this);
        mediaController.setAnchorView(video); //把 MediaController 对象绑定到 VideoView 上
        video.setMediaController(mediaController);
        Uri uri=Uri.parse("/sdcard/test.mp4");
        video.setVideoURI(uri);
        btn1.setOnClickListener(new View.OnClickListener()
        {
            @Override
            public void onClick(View v)
            {
                // TODO：自动存根法
                video.start();
            }
        });
        btn2.setOnClickListener(new View.OnClickListener()
        {
            @Override
            public void onClick(View v)
            {
                // TODO：自动存根法
```

```
                video.pause();
            }
        });
        btn3.setOnClickListener(new View.OnClickListener()
        {
            @Override
            public void onClick(View v)
            {
                // TODO：自动存根法
                int seekpos=20000;
                video.seekTo(seekpos);                    //指定播放位置
                video.start();                            //开始播放
                Toast.makeText(getApplicationContext(), "从"+seekpos+"毫秒开始播放",
                    Toast.LENGTH_LONG).show();
            }
        });
    }
}
```

11．stopPlayback 方法

stopPlayback 方法用于停止回放视频操作。其语法格式为：

```
public void stopPlayback()
```

12．canSeekForward 方法

canSeekForward 方法用于获取 VideoView 对象能否进行视频。如果返回值为 true 则表示可以快进视频，如果返回值为 false 则表示不能快进视频。其语法格式为：

```
public Boolean canSeekForward()
```

通过以下代码来演示怎样获取视频播放的基本信息。

```
public class firstActivity extends Activity
{
    Boolean looping;
    /* 第一次调用活动 */
    @Override
    public void onCreate(Bundle savedInstanceState)
    {
        super.onCreate(savedInstanceState);
        setContentView(R.layout.main);
        Button btn1=(Button)findViewById(R.id.button1);
        Button btn2=(Button)findViewById(R.id.button2);
        Button btn3=(Button)findViewById(R.id.button3);
        final VideoView video = (VideoView) findViewById(R.id.videoView1);
```

```java
MediaController mediaController = new MediaController(this);
mediaController.setAnchorView(video);    //把 MediaController 对象绑定到 VideoView 上
video.setMediaController(mediaController);
Uri uri=Uri.parse("/sdcard/sss.mp4");
video.setVideoURI(uri);
btn1.setOnClickListener(new View.OnClickListener()
{
    @Override
    public void onClick(View v)
    {
        // TODO: 自动存根法
        video.start();
    }
});
btn2.setOnClickListener(new View.OnClickListener()
{
    @Override
    public void onClick(View v)
    {
        // TODO: 自动存根法
        video.pause();
    }
});
btn3.setOnClickListener(new View.OnClickListener()
{
    @Override
    public void onClick(View v)
    {
        // TODO: 自动存根法
        Boolean isPlaying=video.isPlaying();
        Boolean canPause=video.canPause();
        Boolean canSeekBackward=video.canSeekBackward();
        Boolean canSeekForward=video.canSeekForward();
        String str;
        str="该视频播放器信息如下：\n";
        if(isPlaying)
            str=str+"视频正在播放！\n";
        else
            str=str+"视频停止播放！\n";
        if(canPause)
            str=str+"视频可以播放！\n";
```

```
                    else
                        str=str+"视频不可以播放！\n";
                    if(canSeekBackward)
                        str=str+"视频可以倒退播放！\n";
                    else
                        str=str+"视频不可以倒退播放！\n";
                    if(canSeekForward)
                        str=str+"视频可以快进播放！\n";
                    else
                        str=str+"视频不可以快进播放！\n";
                    Toast.makeText(getApplicationContext(), str, Toast.LENGTH_LONG).show();
                }
            });
        }
    }
```

9.1.4 Android 的多媒体播放器综合实例

MediaPlayer 可以播放音频和视频，另外也可以通过 VideoView 来播放视频，虽然 VideoView 比 MediaPlayer 简单易用，但定制性不如 MediaPlayer，要视情况选择。MediaPlayer 播放音频比较简单，但是要播放视频就需要 SurfaceView。SurfaceView 比普通的自定义 View 更有绘图上的优势，它支持完全的 OpenGL ES 库。

通过一个示例来综合实现 Android 的多媒体播放器。

【实现步骤】

（1）在 Elipse 环境下，建立一个名为 E9_1_4 的工程。

（2）打开 res/layout 目录下的 main.xml 文件，代码修改为：

```
<?xml version="1.0" encoding="utf-8"?>
<LinearLayout android:id="@+id/LinearLayout01"
    android:layout_width="fill_parent"
    android:layout_height="fill_parent"
    xmlns:android="http://schemas.android.com/apk/res/android"
    android:orientation="vertical">
    <SeekBar android:id="@+id/SeekBar01"
        android:layout_height="wrap_content"
        android:layout_width="fill_parent"></SeekBar>
    <LinearLayout android:id="@+id/LinearLayout02"
        android:layout_width="wrap_content"
        android:layout_height="wrap_content">
        <Button android:id="@+id/Button01"
            android:layout_width="wrap_content"
```

```xml
            android:layout_height="wrap_content"
            android:text="播放音频"></Button>
        <Button android:id="@+id/Button02"
      android:layout_width="wrap_content"
       android:layout_height="wrap_content"
       android:text="停止播放"></Button>
       </LinearLayout>
    <SeekBar android:id="@+id/SeekBar02"
    android:layout_height="wrap_content"
     android:layout_width="fill_parent"></SeekBar>
      <SurfaceView android:id="@+id/SurfaceView01"
          android:layout_width="fill_parent"
          android:layout_height="250px"></SurfaceView>
    <LinearLayout android:id="@+id/LinearLayout02"
      android:layout_width="wrap_content"
      android:layout_height="wrap_content">
        <Button android:layout_width="wrap_content"
       android:layout_height="wrap_content"
       android:id="@+id/Button03"
       android:text="播放视频"></Button>
        <Button android:layout_width="wrap_content"
       android:layout_height="wrap_content"
       android:text="停止播放"
       android:id="@+id/Button04"></Button>
       </LinearLayout>
    </LinearLayout>
```

（3）打开 src/com.example.e9_1_4 目录下的 MainActivity.java 文件，代码修改为：

```java
public class MainActivity extends Activity
{
    /**第一次调用活动 */
    private SeekBar skb_audio=null;
    private Button btn_start_audio = null;
    private Button btn_stop_audio = null;
    private SeekBar skb_video=null;
    private Button btn_start_video = null;
    private Button btn_stop_video = null;
    private SurfaceView surfaceView;
    private SurfaceHolder surfaceHolder;
    private MediaPlayer m = null;
    private Timer mTimer;
    private TimerTask mTimerTask;
```

```java
//互斥变量,防止定时器与 SeekBar 拖动时进度冲突
private boolean isChanging=false;
@Override
public void onCreate(Bundle savedInstanceState)
{
    super.onCreate(savedInstanceState);
    setContentView(R.layout.main);
    //Media 控件设置
    m=new MediaPlayer();
    //播放结束之后弹出提示
    m.setOnCompletionListener(new MediaPlayer.OnCompletionListener()
    {
        @Override
        public void onCompletion(MediaPlayer arg0)
        {
            Toast.makeText(MainActivity.this, "结束", 1000).show();
            m.release();
        }
    });
    //-定时器记录播放进度
    mTimer = new Timer();
    mTimerTask = new TimerTask()
    {
        @Override
        public void run()
        {
            if(isChanging==true)
                return;
            if(m.getVideoHeight()==0)
                skb_audio.setProgress(m.getCurrentPosition());
            else
                skb_video.setProgress(m.getCurrentPosition());
        }
    };
    mTimer.schedule(mTimerTask, 0, 10);
    btn_start_audio = (Button) this.findViewById(R.id.Button01);
    btn_stop_audio = (Button) this.findViewById(R.id.Button02);
    btn_start_audio.setOnClickListener(new ClickEvent());
    btn_stop_audio.setOnClickListener(new ClickEvent());
    skb_audio=(SeekBar)this.findViewById(R.id.SeekBar01);
    skb_audio.setOnSeekBarChangeListener(new SeekBarChangeEvent());
```

```java
btn_start_video = (Button) this.findViewById(R.id.Button03);
btn_stop_video = (Button) this.findViewById(R.id.Button04);
btn_start_video.setOnClickListener(new ClickEvent());
btn_stop_video.setOnClickListener(new ClickEvent());
skb_video=(SeekBar)this.findViewById(R.id.SeekBar02);
skb_video.setOnSeekBarChangeListener(new SeekBarChangeEvent());
surfaceView = (SurfaceView) findViewById(R.id.SurfaceView01);
surfaceHolder = surfaceView.getHolder();
surfaceHolder.setFixedSize(100, 100);
surfaceHolder.setType(SurfaceHolder.SURFACE_TYPE_PUSH_BUFFERS);
}
/**
 * 按键事件处理
 */
class ClickEvent implements View.OnClickListener
{
    @Override
    public void onClick(View v)
    {
        if(v==btn_start_audio)
        {
            m.reset();                                            //恢复到未初始化的状态
            m=MediaPlayer.create(MainActivity.this, R.raw.yudie);
                                                                  //读取音频
            skb_audio.setMax(m.getDuration());                    //设置 SeekBar 的长度
            try
            {
                m.prepare();                                      //准备
            } catch (IllegalStateException e)
            {
                // TODO:自动存根法
                e.printStackTrace();
            } catch (IOException e)
            {
                // TODO:自动存根法
                e.printStackTrace();
            }
            m.start();                                            //播放
        }
        else if(v==btn_stop_audio || v==btn_stop_video)
        {
```

```java
                        m.stop();
                }
                else if(v==btn_start_video)
                {
                        m.reset();                          //恢复到未初始化的状态
                        m=MediaPlayer.create(MainActivity.this, R.raw.sss);
                                                            //读取视频
                        skb_video.setMax(m.getDuration());
                                                            //设置 SeekBar 的长度
                        m.setAudioStreamType(AudioManager.STREAM_MUSIC);
                        m.setDisplay(surfaceHolder);//设置屏幕
                        try {
                                m.prepare();
                        } catch (IllegalArgumentException e)
                        {
                                //TODO:自动存根法
                                e.printStackTrace();
                        } catch (IllegalStateException e)
                        {
                                //TODO:自动存根法
                                e.printStackTrace();
                        } catch (IOException e)
                        {
                                //TODO:自动存根法
                                e.printStackTrace();
                        }
                        m.start();
                }
        }
        /**
         * SeekBar 进度改变事件
         */
        class SeekBarChangeEvent implements SeekBar.OnSeekBarChangeListener
        {
                @Override
                public void onProgressChanged(SeekBar seekBar, int progress, boolean fromUser)
                {
                        //TODO:自动存根法
                }
                @Override
                public void onStartTrackingTouch(SeekBar seekBar)
                {
```

```
                                        isChanging=true;
                    }
                    @Override
                    public void onStopTrackingTouch(SeekBar seekBar)
                    {
                        m.seekTo(seekBar.getProgress());
                        isChanging=false;
                    }
                }
```
运行程序，效果如图 9-8 所示。

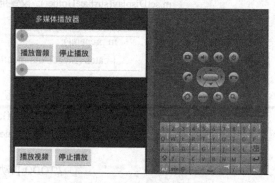

图 9-8 Android 多媒体播放器页面

9.2 录制音频

手机一般都提供了麦克风硬件，而 Android 系统就可以利用该硬件来录制音频了。为了在应用中录制音频，Android 提供了 MediaRecorder 类。使用 MediaRecorder 录制音频的过程很简单。

（1）创建 MediaRecorder 对象。

（2）调用 MediaReorder 对象的 setAudioSource()方法设置声音来源，一般传入 MediaRecorder.AudioSource.MIC 参数指定录制来自麦克风的声音。

（3）调用 MediaRecorder 对象的 setOutputFormatFormat()设置所录制的音频文件的格式。

（4）调用 MediaRecorder 对象的 setAudioEncoder()、setAudioEncodingBitRate(int bitRate)、setAudioSamplingRate(int samplingRate)设置所录制的声音的编码格式、编码位率、采样率等，这样参数将可以控制所录制的声音的品质、文件的大小。一般来说，声音品质越好，声音文件越大。

（5）调用 MediaRecoder 的 setOutputFile(String path)方法设置录制的音频文件的保存位置。

（6）调用 MediaRecoder 的 prepare()方法准备录制。

（7）调用 MediaRecoder 对象的 start()方法开始录制。

（8）录制完后，调用 MediaRecoder 对象的 stop()方法停止录制，并调用 release()方法释放资源。

下面通过一个例子来说明怎样使用 MediaRecoder 来录制声音。

【实现步骤】

（1）在 Eclipse 环境下，建立一个名为 E9_2 的工程。

（2）打开 res/layout 目录下的 main.xml 文件，代码修改为：

```xml
<?xml version="1.0" encoding="utf-8"?>
<LinearLayout
    xmlns:android="http://schemas.android.com/apk/res/android"
    android:orientation="horizontal"
    android:layout_width="fill_parent"
    android:layout_height="fill_parent"
    android:gravity="center_horizontal">
<ImageButton
    android:id="@+id/record"
    android:layout_width="wrap_content"
    android:layout_height="wrap_content"
    android:src="@drawable/record"/>
<ImageButton
    android:id="@+id/stop"
    android:layout_width="wrap_content"
    android:layout_height="wrap_content"
    android:src="@drawable/stop"/>
</LinearLayout>
```

（3）打开 src/com.example.e9_2 目录下的 MainActivity.java 文件，代码修改为：

```java
public class MainActivity extends Activity
    implements OnClickListener
{
    // 程序中的两个按钮
    ImageButton record , stop;
    // 系统的音频文件
    File soundFile ;
    MediaRecorder mRecorder;
    @Override
    public void onCreate(Bundle savedInstanceState)
    {
        super.onCreate(savedInstanceState);
        setContentView(R.layout.main);
        // 获取程序界面中的两个按钮
        record = (ImageButton) findViewById(R.id.record);
        stop = (ImageButton) findViewById(R.id.stop);
        // 为两个按钮的单击事件绑定监听器
        record.setOnClickListener(this);
        stop.setOnClickListener(this);
```

```java
        }
        @Override
        public void onDestroy()
        {
            if (soundFile != null && soundFile.exists())
            {
                // 停止录音
                mRecorder.stop();
                // 释放资源
                mRecorder.release();
                mRecorder = null;
            }
            super.onDestroy();
        }
        @Override
        public void onClick(View source)
        {
            switch (source.getId())
            {
                // 单击录音按钮
                case R.id.record:
                    if (!Environment.getExternalStorageState().equals(
                        android.os.Environment.MEDIA_MOUNTED))
                    {
                        Toast.makeText(MainActivity.this
                            , "SD卡不存在，请插入SD卡！"
                            , 5000)
                            .show();
                        return;
                    }
                    try
                    {
                        // 创建保存录音的音频文件
                        soundFile = new File(Environment
                            .getExternalStorageDirectory()
                            .getCanonicalFile() + "/sound.amr");
                        mRecorder = new MediaRecorder();
                        // 设置录音的声音来源
                        mRecorder.setAudioSource(MediaRecorder.AudioSource.MIC);
                        // 设置录制的声音的输出格式（必须在设置声音编码格式之前设置）
                        mRecorder.setOutputFormat(MediaRecorder
                            .OutputFormat.THREE_GPP);
```

```
                    // 设置声音编码的格式
                    mRecorder.setAudioEncoder(MediaRecorder
                        .AudioEncoder.AMR_NB);
                    mRecorder.setOutputFile(soundFile.getAbsolutePath());
                    mRecorder.prepare();
                    // 开始录音
                    mRecorder.start();
                }
                catch (Exception e)
                {
                    e.printStackTrace();
                }
                break;
            // 单击停止按钮
            case R.id.stop:
                if (soundFile != null && soundFile.exists())
                {
                    // 停止录音
                    mRecorder.stop();
                    // 释放资源
                    mRecorder.release();
                    mRecorder = null;
                }
                break;
        }
    }
}
```

（4）授予权限。打开 AndroidManifest.xml 文件，添加权限代码为：

```
    </application>
    <!-- 授予录音权限 -->
    <uses-permission android:name="permission.RECORD_AUDIO"/>
</manifest>
```

运行程序，效果如图 9-9 所示。

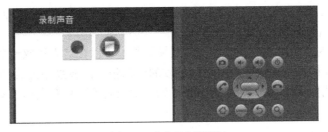

图 9-9　声音录制界面

9.3 照相机

现在的手机一般都会提供相机功能，有些相机的镜头甚至支持 800 万以上像素，有些甚至支持光学变焦，这些手机已经变成了专业数据相机。

9.3.1 照相机常用方法

Camera 类位于 android.hardware 软件包，其是 Android 照相机拍照操作的主要类。其中提供了多种函数，包括打开相机、预览、设置相机参数和关闭相机等。通过这些方法，应用程序可很方便地实现照相机的拍照操作。

要实现照相机效果即要设置授予权限，即在 AndroidManifest.xml 文件中添加如下代码：

```
</application>
<!-- 授予相机权限 -->
<uses-permission android:name="android.permission.CAMERA"/>
<uses-feature android:name="android.hardware.comera"/>
<uses-feature android:name="android.hardware.comera.autofocus"/>
<uses-permission android:name="android.permission.WRITE_EXTERNAL_STORAGE"/>
</manifest>
```

1. open 方法

open 方法用于创建一个新的照相机对象，用于控制手机上的第一个照相机。如果手机上没有照相机组件，该方法将返回 null。其语法格式为：

```
public static Camera open()
public static Camera open(int cameraId)
```

其中，参数 cameraId 为照相机的编号，其取值为 0～getNumberOfCameras()-1。

2. takePicture 方法

takePicture 方法用于使用照相机执行拍照，这是使用照相机功能最主要的方法。其中，需要指定一些重要的回调处理函数，如图像捕获回调函数、raw 图像回调函数、jpeg 图像回调函数等。其语法格式为：

```
public final void takePicture(Camera.ShutterCallback shutter,Camera.PictureCallback raw,Camera.PictureCallback postview,Camera.PictureCallback jpeg)
```

其中，参数 shutter 为当图像被捕获时触发的回调函数，可以在这里播放一个快门声音以提醒用户；参数 raw 为 raw 格式的图像数据可用时触发的回调函数；参数 jpeg 为当 jpeg 格式的压缩图像数据可用时触发的回调函数；参数 postview 为当拍照后的图片发生缩放操作时触发的回调函数，并非所有的硬件都支持。

3. release 方法

release 方法用于释放照相机对象，当一个应用程序完成拍照操作且不再使用时，需要调用该方法释放其占用的系统资源。其语法格式为：

```
public final void release()
```

4. PictureCallback 方法

PictureCallback 方法为一个重要的照相机回调函数，用于在照相机拍照产生图片时触发。用户可以在这个回调函数中执行保存图片文件的操作，也可以将拍照图片显示在界面上。在使用时，往往需要重载其中的 onPictureTaken 方法。

通过以下代码来演示怎样实现一个最简单的照相机。

```java
public class MainActivity extends Activity
{
    private SurfaceView surfaceView;
    private SurfaceHolder surfaceHolder;
    private Camera camera;
    /*第一次调用活动*/
    @Override
    public void onCreate(Bundle savedInstanceState)
    {
        super.onCreate(savedInstanceState);
        setContentView(R.layout.main);
        Button btn1=(Button)findViewById(R.id.button1);
        Button btn2=(Button)findViewById(R.id.button2);
        Button btn3=(Button)findViewById(R.id.button3);
        final ImageView imageView = (ImageView) findViewById(R.id.imageView1);
        final PictureCallback jpeg=new Camera.PictureCallback()
        {
            @Override
            public void onPictureTaken(byte[] data, Camera camera)
            {
                // TODO：自动存根法
                camera.startPreview();
                Bitmap bm = BitmapFactory.decodeByteArray(data, 0, data.length);
                imageView.setImageBitmap(bm);//将图片显示到下方的 ImageView 中
            }
        };
        btn1.setOnClickListener(new View.OnClickListener()
        {
            @Override
            public void onClick(View v)
            {
                camera=Camera.open();
            }
```

```
        });
        btn2.setOnClickListener(new View.OnClickListener()
        {
            @Override
            public void onClick(View v)
            {
                camera.takePicture(null, null, jpeg);
            }
        });
        btn3.setOnClickListener(new View.OnClickListener()
        {
            @Override
            public void onClick(View v)
            {
                camera.release();    //翻译系统资源
                camera=null;
            }
        });
    }
}
```

运行程序，效果如图 9-10 所示。

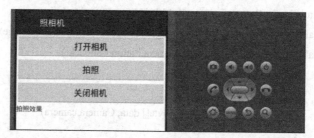

图 9-10　照相效果

5. ShutterCallback 方法

ShutterCallback 方法是一个重要的照相机回调函数，当图像被捕获时触发。可在此播放一个快门声音以提醒用户。在使用时，往往需要重载其中的 onShutter 方法。

6. startPreview 方法

startPreview 方法用于开始捕获图像，并将前一帧图像显示在屏幕上，即实现了预览的效果。只有 setPreviewDisplay 方法调用后，才会启用实际的预览效果。其语法格式为：

```
public final void startPreview()
```

7. stopPreview 方法

stopPreview 方法用于停止捕获图像，并将前一帧图像显示在屏幕上，同时复位照相机以等

待下一次调用 startPreview 方法。使用 stopPreview 方法结束预览。其语法格式为:

```
public final void stopPreview()
```

8．setPreviewDisplay 方法

setPreviewDisplay 方法用于设置预览所使用的 Sureface，预览效果将在 SurfaceView 对象中显示。需要注意的是，必须先调用 setPreviewDisplay 方法，才能够调用 startPreview 方法进行预览。其语法格式为:

```
public final void setPreviewDisplay(SurfaceHolder holder)
```

其中，参数 holder 为放置预览的 Surface，可以通过 SurfaceView 的 getHolder 方法获取。

通过以下代码来演示怎样实现照相机预览的效果。

```java
public class MainActivity extends Activity
{
    private SurfaceView surfaceView;
    private SurfaceHolder surfaceHolder;
    private Camera camera;
    /* 第一次调用活动 */
    @Override
    public void onCreate(Bundle savedInstanceState)
    {
        super.onCreate(savedInstanceState);
        setContentView(R.layout.main);
        Button btn1=(Button)findViewById(R.id.button1);
        final ImageView imageView = (ImageView) findViewById(R.id.imageView1);
        surfaceView=(SurfaceView)findViewById(R.id.surfaceView1);//获取 SurfaceView 对象
        surfaceHolder=surfaceView.getHolder();          //获取 SurfaceHolder
        SurfaceHolder.Callback surfaceCallback=new SurfaceHolder.Callback()
        {
            // SurfaceHolder 回调函数
            @Override
            public void surfaceDestroyed(SurfaceHolder holder)
            {
                //SurfaceView 销毁
                // TODO：自动存根法
                camera.stopPreview();
                camera.release();
                camera=null;
            }
            @Override
            public void surfaceCreated(SurfaceHolder holder)
            {
                //SurfaceView 创建
```

```java
                    camera=Camera.open();
                    try
                    {
                            camera.setPreviewDisplay(holder);
                    } catch (IOException e)
                    {
                            e.printStackTrace();
                    }
                    camera.startPreview();
            }
            @Override
        public void surfaceChanged(SurfaceHolder holder, int format, int width,int height)
            {
                    //SurfaceView 改变
                    // TODO：自动存根法
            }
    };
    surfaceHolder.addCallback(surfaceCallback);                    //添加回调函数
    surfaceHolder.setType(SurfaceHolder.SURFACE_TYPE_PUSH_BUFFERS);
    final PictureCallback jpeg=new Camera.PictureCallback()
    {
            @Override
        public void onPictureTaken(byte[] data, Camera camera)
            {
                    // TODO：自动存根法
                    camera.startPreview();
                    Bitmap bm = BitmapFactory.decodeByteArray(data, 0, data.length);
                    imageView.setImageBitmap(bm);//将图片显示到下方的 ImageView 中
            }
    };
    final Camera.ShutterCallback shutter=new Camera.ShutterCallback()
    {
            @Override
        public void onShutter()
            {
                    Toast.makeText(getApplicationContext(), "成功拍照！ ", Toast.LENGTH_LONG).show();
            }
    };
    btn1.setOnClickListener(new View.OnClickListener()
    {
            @Override
```

```
            public void onClick(View v)
            {
                    camera.takePicture(shutter, null, jpeg);

            }
        });
    }
}
```

9. setParameters 方法

setParameters 方法用于设置照相机的参数。相机的参数保存在 Camera.Parameters 类型的变量中，其中涵盖了照相机的各项参数，如照片格式、尺寸、预览尺寸等。其语法格式为：

```
public void setParameters(Camera.Parameters params)
```

10. getParameters 方法

getParameters 方法用于获取照片机的参数，返回值为 Camera.Parameters 类型，其中包含了照相机的各项参数，如照片格式、尺寸、预览尺寸等。该方法往往结合 setParameters 方法一起使用来重新设置照相机的参数。其语法格式为：

```
public Camera.Parameters getParameters()
```

通过以下代码来演示怎样获取和修改相机参数。

```
public class firstActivity extends Activity
{
    private SurfaceView surfaceView;
    private SurfaceHolder surfaceHolder;
    private Camera camera;
    /** 第一次调用活动 */
    @Override
    public void onCreate(Bundle savedInstanceState)
    {
        super.onCreate(savedInstanceState);
        setContentView(R.layout.main);
        Button btn1=(Button)findViewById(R.id.button1);
        final ImageView imageView = (ImageView) findViewById(R.id.imageView1);
        surfaceView=(SurfaceView)findViewById(R.id.surfaceView1);    //获取 SurfaceView 对象
        surfaceHolder=surfaceView.getHolder();                        //获取 SurfaceHolder
        surfaceHolder.Callback surfaceCallback=new SurfaceHolder.Callback()
        {
            // SurfaceHolder 回调函数
            @Override
            public void surfaceDestroyed(SurfaceHolder holder)
            {
```

```java
                    //SurfaceView 销毁
                    camera.stopPreview();
                    camera.release();
                    camera=null;
                }
                @Override
                public void surfaceCreated(SurfaceHolder holder)
                {
                    //SurfaceView 创建
                    camera=Camera.open();
                    try {
                        camera.setPreviewDisplay(holder);
                    } catch (IOException e)
                    {
                        e.printStackTrace();
                    }
                    camera.startPreview();
                }
                @Override
                public void surfaceChanged(SurfaceHolder holder, int format, int width,int height)
                {
                    //SurfaceView 改变
                    // TODO Auto-generated method stub
                    Camera.Parameters parameters=camera.getParameters();
                    parameters.setPreviewSize(320, 240);
                    parameters.setPictureFormat(PixelFormat.JPEG);
                    camera.setParameters(parameters);
                    camera.startPreview();
                }
            };
            surfaceHolder.addCallback(surfaceCallback);              //添加回调函数
            surfaceHolder.setType(SurfaceHolder.SURFACE_TYPE_PUSH_BUFFERS);
            final PictureCallback jpeg=new Camera.PictureCallback()
            {
                @Override
                public void onPictureTaken(byte[] data, Camera camera)
                {
                    camera.startPreview();
                    Bitmap bm = BitmapFactory.decodeByteArray(data, 0, data.length);
                    imageView.setImageBitmap(bm);//将图片显示到下方的 ImageView 中
                }
```

```
            };
            final Camera.ShutterCallback shutter=new Camera.ShutterCallback()
            {
                    @Override
                    public void onShutter()
                    {
                            Toast.makeText(getApplicationContext(),"成功拍照！",Toast.LENGTH_LONG).show();
                    }
            };
            btn1.setOnClickListener(new View.OnClickListener()
            {
                    @Override
                    public void onClick(View v)
                    {
                            camera.takePicture(shutter, null, jpeg);
                    }
            });
    }
}
```

11．autoFocus 方法

autoFocus 方法用于设置启动照相机的自动对焦功能，注册一个对焦完成回调函数 AutoFocusCallback，并需要重载其中的 onAutoFocus 方法。当相机对焦成功时即执行该回调函数中的 onAutoFocus 方法。autoFocus 方法必须在 startPreview 方法和 stopPreview 方法中直接调用。其语法格式为：

```
public final void autofocus(Camera.AutoFocusCallback cb)
```

其中，参数 cb 为自动对焦成功时触发的回调函数。

12．cancelAutoFocus 方法

cancelAutoFocus 方法用于在当前进程中取消照相机的自动对焦功能。无论当前进程是否使用自动对焦，该方法都将返回一个默认的对焦位置。其语法格式为：

```
public final void cancelAutoFocus()
```

13．unlock 方法

unlock 方法用于解锁相机，使其能够在另外的进程中被使用。一般来说，一个照相机被锁定在一个进程中，直到 release 方法被调用。使用该方法可以临时释放照相机，使之能够被其他进程使用，也就是能够使相机在进程之间快速切换。当其他进程结束时，可以调用 reconnect 方法来重新使用照相机。其语法格式为：

```
public final void unlock()
```

14. reconnect 方法

reconnect 方法用于当其他进程结束使用照相机后，重新连接相机。reconnect 方法和 unlock 方法经常一起使用，使得另外的进程可以临时使用照相机。其语法格式为：

```
public    final void reconnect()
```

15. strtSmoothZoom 方法

strtSmoothZoom 方法用于平滑地变焦到指定的数值。在该方法的执行过程中，将会触发变焦监听器 OnZoomChangeListener。该方法的使用需要实际硬件支持，如果手机上的照相机不支持变焦功能，将导致调用失效。其语法格式为：

```
public final void startSmoothZoom(int value)
```

其中，参数 value 为指定的变焦值，该值的取值范围依赖于硬件，一般为 0～getMaxZoom()。

9.3.2 照相机实例分析

下面通过两个例子来介绍手机照相机。

1. 照相机拍照功能

下面通过使用 Cammera 类提供的方法实现手机的拍照功能。

【实现步骤】

（1）在 Eclipse 环境下，建立一个名为 E9_3_2_1 的工程。

（2）打开 res/layout 目录下的 main.xml 文件，代码修改为：

```xml
<?xml version="1.0" encoding="utf-8"?>
<LinearLayout
    xmlns:android="http://schemas.android.com/apk/res/android"
    android:orientation="vertical"
    android:layout_width="fill_parent"
    android:layout_height="fill_parent">
<SurfaceView
    android:id="@+id/sView"
    android:layout_width="fill_parent"
    android:layout_height="fill_parent"/>
</LinearLayout>
```

（3）在 res/layout 目录下新建一个名为 save.xml 的文件，代码为：

```xml
<?xml version="1.0" encoding="utf-8"?>
<LinearLayout xmlns:android="http://schemas.android.com/apk/res/android"
    android:orientation="vertical"
    android:layout_width="fill_parent"
    android:layout_height="fill_parent">
<LinearLayout
    android:orientation="horizontal"
```

```xml
        android:layout_width="fill_parent"
        android:layout_height="wrap_content">
    <TextView
        android:layout_width="wrap_content"
        android:layout_height="wrap_content"
        android:layout_marginRight="8dip"
        android:text="@string/photo_name"/>
    <!-- 定义一个文本框来让用户输入照片名 -->
    <EditText
        android:id="@+id/phone_name"
        android:layout_width="fill_parent"
        android:layout_height="wrap_content"/>
</LinearLayout>
<!-- 定义一个图片框来显示照片 -->
<ImageView
    android:id="@+id/show"
    android:layout_width="240px"
    android:layout_height="315px"
    android:scaleType="fitCenter"
    android:layout_marginTop="10dp"/>
</LinearLayout>
```

（4）打开 src/com.example.e9_3_2_1 目录下的 MainActivity.java 文件，代码修改为：

```java
public class MainActivity extends Activity
{
    SurfaceView sView;
    SurfaceHolder surfaceHolder;
    int screenWidth, screenHeight;
    // 定义系统所用的照相机
    Camera camera;
    //是否在浏览中
    boolean isPreview = false;
    @Override
    public void onCreate(Bundle savedInstanceState)
    {
        super.onCreate(savedInstanceState);
        // 设置全屏
        requestWindowFeature(Window.FEATURE_NO_TITLE);
        getWindow().setFlags(WindowManager.LayoutParams.FLAG_FULLSCREEN,
            WindowManager.LayoutParams.FLAG_FULLSCREEN);
        setContentView(R.layout.main);
        WindowManager wm = (WindowManager) getSystemService(
```

```java
                    Context.WINDOW_SERVICE);
            Display display = wm.getDefaultDisplay();
            // 获取屏幕的宽和高
            screenWidth = display.getWidth();
            screenHeight = display.getHeight();
            // 获取界面中 SurfaceView 组件
            sView = (SurfaceView) findViewById(R.id.sView);
            // 获得 SurfaceView 的 SurfaceHolder
            surfaceHolder = sView.getHolder();
            // 为 SurfaceHolder 添加一个回调监听器
            surfaceHolder.addCallback(new Callback()
            {
                @Override
                public void surfaceChanged(SurfaceHolder holder, int format, int width,
                    int height)
                {
                }
                @Override
                public void surfaceCreated(SurfaceHolder holder)
                {
                    // 打开摄像头
                    initCamera();
                }
                @Override
                public void surfaceDestroyed(SurfaceHolder holder)
                {
                    // 如果 camera 不为 null ,释放摄像头
                    if (camera != null)
                    {
                        if (isPreview)
                            camera.stopPreview();
                        camera.release();
                        camera = null;
                    }
                }
            });
            // 设置该 SurfaceView 自己不维护缓冲
            surfaceHolder.setType(SurfaceHolder.SURFACE_TYPE_PUSH_BUFFERS);
        }
        private void initCamera()
        {
```

```java
        if (!isPreview)
        {
            camera = Camera.open();
        }
        if (camera != null && !isPreview)
        {
            try
            {
                Camera.Parameters parameters = camera.getParameters();
                // 设置预览照片的大小
                parameters.setPreviewSize(screenWidth, screenHeight);
                // 每秒显示 4 帧
                parameters.setPreviewFrameRate(4);
                // 设置图片格式
                parameters.setPictureFormat(PixelFormat.JPEG);
                // 设置 JPG 照片的质量
                parameters.set("jpeg-quality", 85);
                //设置照片的大小
                parameters.setPictureSize(screenWidth, screenHeight);
                camera.setParameters(parameters);
                //通过 SurfaceView 显示取景画面
                camera.setPreviewDisplay(surfaceHolder);
                // 开始预览
                camera.startPreview();
                // 自动对焦
                camera.autoFocus(null);
            }
            catch (Exception e)
            {
                e.printStackTrace();
            }
            isPreview = true;
        }
    }
    @Override
    public boolean onKeyDown(int keyCode, KeyEvent event)
    {
        switch (keyCode)
        {
            // 当用户按照相键、中央键时执行拍照
            case KeyEvent.KEYCODE_DPAD_CENTER:
```

```java
                    case KeyEvent.KEYCODE_CAMERA:
                        if (camera != null && event.getRepeatCount() == 0)
                        {
                            // 拍照
                            camera.takePicture(null, null , myjpegCallback);
                            return true;
                        }
                        break;
            }
            return super.onKeyDown(keyCode, event);
    }
    PictureCallback myjpegCallback = new PictureCallback()
    {
        @Override
        public void onPictureTaken(byte[] data, Camera camera)
        {
            // 根据拍照所得的数据创建位图
            final Bitmap bm = BitmapFactory.decodeByteArray(data
                , 0, data.length);
            // 加载/layout/save.xml 文件对应的布局资源
            View saveDialog = getLayoutInflater().inflate(
                    R.layout.save, null);
            final EditText photoName = (EditText) saveDialog
                    .findViewById(R.id.phone_name);
            // 获取 saveDialog 对话框上的 ImageView 组件
            ImageView show = (ImageView) saveDialog.findViewById(R.id.show);
            // 显示刚刚拍得的照片
            show.setImageBitmap(bm);
            //使用对话框显示 saveDialog 组件
            new AlertDialog.Builder(MainActivity.this)
                    .setView(saveDialog)
                    .setPositiveButton("保存", new OnClickListener()
                    {
                        @Override
                        public void onClick(DialogInterface dialog,
                            int which)
                        {
                            // 创建一个位于 SD 卡上的文件
                            File file = new File(Environment.getExternalStorageDirectory()
                                , photoName.getText().toString() + ".jpg");
                            FileOutputStream outStream = null;
```

```java
                            try
                            {
                                // 打开指定文件对应的输出流
                                outStream = new FileOutputStream(file);
                                // 把位图输出到指定文件中
                                bm.compress(CompressFormat.JPEG, 100, outStream);
                                outStream.close();
                            }
                            catch (IOException e)
                            {
                                e.printStackTrace();
                            }
                        }
                    })
                    .setNegativeButton("取消", null)
                    .show();
            //重新浏览
            camera.stopPreview();
            camera.startPreview();
            isPreview = true;
        }
    };
}
```

2. 照相机视频录制功能

MediaRecorder 除了可用于录制音频外，还可用于录制视频。使用 MediaRecorder 录制视频与录制音频相似，只是录制视频时不仅需要采集声音，还需要采集图像。为了让 MediaRecorder 录制时采集图像，应该在调用 setAutioSource(int audio_source)方法中调用 setVideoSource(int video_source)方法来设置图像来源。

除此之外，还需要在调用 setOutputFormat()设置输出文件格式之后按如下步骤操作。

① 调用 MediaRecorder 对象的 setVideoEncoder()、setVideoEncodingBitRate(int bitRate)、setVideoFrameRate 设置所录制的视频的编码格式、编码位率、每秒多少帧等，这些参数将可以控制所录制的视频品质、文件的大小。一般来说，视频品质越好，视频文件越大。

② 调用 MediaRecorder 的 setPreviewDisplay(Surface sv)方法设置使用哪个 SurfaceView 来显示视频预览。

本实例为利用 MediaRecorder 实现视频录制功能。界面中提供了两个按钮，分别用于控制开始、结束录制，还提供了一个 SurfaceView 用于显示视频预览。

【实现步骤】

（1）在 Eclipse 环境下，建立一个名为 E9_3_2_2 的工程。

（2）打开 res/layout 目录下的 main.xml 文件，代码修改为：

```xml
<?xml version="1.0" encoding="utf-8"?>
```

```xml
<LinearLayout
    xmlns:android="http://schemas.android.com/apk/res/android"
    android:orientation="vertical"
    android:layout_width="fill_parent"
    android:layout_height="fill_parent"
    android:gravity="center_horizontal">
    <LinearLayout
        android:orientation="horizontal"
        android:layout_width="wrap_content"
        android:layout_height="wrap_content"
        android:gravity="center_horizontal">
        <ImageButton
            android:id="@+id/record"
            android:layout_width="wrap_content"
            android:layout_height="wrap_content"
            android:src="@drawable/record"/>
        <ImageButton
            android:id="@+id/stop"
            android:layout_width="wrap_content"
            android:layout_height="wrap_content"
            android:src="@drawable/stop"/>
    </LinearLayout>
    <!-- 显示视频预览的 SurfaceView -->
    <SurfaceView
        android:id="@+id/sView"
        android:layout_width="fill_parent"
        android:layout_height="fill_parent"/>
</LinearLayout>
```

（3）打开 src/com.example.e9_3_2_2 目录下的 MainActivity.java 文件，代码为：

```java
public class MainActivity extends Activity implements OnClickListener
{
    // 程序中的两个按钮
    ImageButton record , stop;
    // 系统的视频文件
    File videoFile ;
    MediaRecorder mRecorder;
    // 显示视频预览的 SurfaceView
    SurfaceView sView;
    // 记录是否正在进行录制
    private boolean isRecording = false;
```

```java
@Override
public void onCreate(Bundle savedInstanceState)
{
    super.onCreate(savedInstanceState);
    setContentView(R.layout.main);
    // 获取程序界面中的两个按钮
    record = (ImageButton) findViewById(R.id.record);
    stop = (ImageButton) findViewById(R.id.stop);
    // 让 stop 按钮不可用
    stop.setEnabled(false);
    // 为两个按钮的单击事件绑定监听器
    record.setOnClickListener(this);
    stop.setOnClickListener(this);
    // 获取程序界面中的 SurfaceView
    sView = (SurfaceView) this.findViewById(R.id.sView);
    // 下面设置 Surface 不需要自己维护缓冲区
    sView.getHolder().setType(SurfaceHolder.SURFACE_TYPE_PUSH_BUFFERS);
    //设置分辨率
    sView.getHolder().setFixedSize(320, 280);
    // 设置该组件让屏幕不会自动关闭
    sView.getHolder().setKeepScreenOn(true);
}
@Override
public void onClick(View source)
{
    switch (source.getId())
    {
        // 单击录制按钮
        case R.id.record:
            if (!Environment.getExternalStorageState().equals(
                android.os.Environment.MEDIA_MOUNTED))
            {
                Toast.makeText(MainActivity.this
                    , "SD 卡不存在,请插入 SD 卡!"
                    , 5000)
                    .show();
                return;
            }
            try
            {
                // 创建保存录制视频的视频文件
```

```java
                    videoFile = new File(Environment
                            .getExternalStorageDirectory()
                            .getCanonicalFile() + "/myvideo.mp4");
                    // 创建 MediaPlayer 对象
                    mRecorder = new MediaRecorder();
                    mRecorder.reset();
                    // 设置从麦克风采集声音
                    mRecorder.setAudioSource(MediaRecorder.AudioSource.MIC);
                    // 设置从摄像头采集图像
                    mRecorder.setVideoSource(MediaRecorder.VideoSource.CAMERA);
                    // 设置视频文件的输出格式（必须在设置声音编码格式、
                    // 图像编码格式之前设置）
                    mRecorder.setOutputFormat(MediaRecorder
                            .OutputFormat.MPEG_4);
                    // 设置声音编码的格式
                    mRecorder.setAudioEncoder(MediaRecorder
                            .AudioEncoder.DEFAULT);
                    // 设置图像编码的格式
                    mRecorder.setVideoEncoder(MediaRecorder
                            .VideoEncoder.MPEG_4_SP);
                    mRecorder.setVideoSize(320, 280);
                    //每秒 4 帧
                    mRecorder.setVideoFrameRate(4);
                    mRecorder.setOutputFile(videoFile.getAbsolutePath());
                    // 指定使用 SurfaceView 来预览视频
                    mRecorder.setPreviewDisplay(sView.getHolder().getSurface());
                    mRecorder.prepare();
                    // 开始录制
                    mRecorder.start();
                    System.out.println("====recording====");
                    // 让 record 按钮不可用
                    record.setEnabled(false);
                    // 让 stop 按钮可用
                    stop.setEnabled(true);
                    isRecording = true;
                }
                catch (Exception e)
                {
                    e.printStackTrace();
                }
                break;
```

```java
            // 单击停止按钮
            case R.id.stop:
                // 如果正在进行录制
                if (isRecording)
                {
                    // 停止录制
                    mRecorder.stop();
                    // 释放资源
                    mRecorder.release();
                    mRecorder = null;
                    // 让 record 按钮可用
                    record.setEnabled(true);
                    // 让 stop 按钮不可用
                    stop.setEnabled(false);
                }
                break;
        }
    }
}
```

（4）授权录制视频权限。打开 AndroidManifest.xml 文件，添加如下代码授权权限：

```xml
</application>
    <!-- 授予该程序录制声音的权限 -->
    <uses-permission android:name="android.permission.RECORD_AUDIO"/>
    <!-- 授予该程序使用摄像头的权限 -->
    <uses-permission android:name="android.permission.CAMERA"/>
    <uses-permission android:name="android.permission.MOUNT_UNMOUNT_FILESYSTEMS"/>
        <!-- 授予使用外部存储器的权限 -->
    <uses-permission android:name="android.permission.WRITE_EXTERNAL_STORAGE"/>
</manifest>
```

运行程序，效果如图 9-11 所示。

图 9-11　录制视频界面图

第 10 章 Android 辅助工具

10.1 Map 地图

Map 地图对大家来说应该不算陌生,它让人们体会到了高科技的奥妙。

10.1.1 位置服务实例

Android 支持 GPS 和网络地图,通常将各种不同的定位技术称为 LBS。LBS 是基于位置的服务(Location Based Service)的简称,它是通过电信移动运营商的无线电通信网络(如 GSM 网、CDMA 网)或外部定位方式(如 GPS)获取移动终端用户的位置信息(地理坐标或大地坐标),在地理信息系统(Geographic Information System,简称 GIS)平台的支持下,为用户提供相应服务的一种增值业务。

1. Android.location 功能类

Android 支持地理定位服务的 API。该地理定位服务可以用来获取当前设备的地理位置。应用程序可以定时请求更新设备当前的地理定位信息。应用程序也可以借助一个 Intent 接收器来实现如下功能。以经纬度和半径划定的一个区域,当设备出入该区域时,可发现提醒信息。

1) Android Location API

以下是几个 Android 关于定位功能的比较重要的类。

- LocationManager:该类提供访问定位服务的功能,也提供获取最佳定位提供者的功能。另外,临近警报功能也可以借助该类实现。
- LocationProvider:该类是定位提供者的抽象类。定位提供者具备周期性报告设备地理位置的功能。
- LocationListener:提供定位信息发生改变时的回调功能。必须事先在定位管理器中注册监听器对象。
- Criteria:该类使得应用能够通过在 LocationProvider 中设置的属性来选择合适的定位提供者。

2) Map API

Android 也提供了一组访问 Map 的 API,借助 Map 及定位 API,用户就能在地图上显示当前的地理位置。在 Android 中定义了一个名为 com.google.android.maps 的包,其中包含了一系列用于在 Map 上显示、控制和层叠信息的功能类,以下是该包中最重要的几个类。

- MapActivity:这个类是用于显示 Map 的 Activity 类,它需要连接底层网络。

- MapView：用于显示地图的 View 组件，它必须和 MapActivity 配合使用。
- MapController：用于控制地图的移动。
- Overlay：这是一个可显示于地图之上的可绘制的对象。
- GeoPoint：一个包含经纬度位置的对象。

2．Android 定位的基本流程

了解了 Android 中和定位处理相关的类后，下面将简要介绍在 Android 中实现定位处理的基本流程。

1）准备 Activity 类

目标是使用 Map API 来显示地图，然后使用定位 API 来获取设备的当前定位信息以在 Map 上设置设备的当前位置，用户定位会随着用户的位置移动而发生改变。

先需要一个继承了 MapActivity 的 Activity 类，代码如下：

```
class MyGPSAactivity extends MapActivity
{
}
```

要成功引用 Map API，还必须先在 AndroidManifest.xml 中定义如下信息：

```
<uses-library android:name="com.google.android.map">
```

2）使用 Map View

要让地图显示的话，需要将 MapView 加入到应用中来。例如，在布局文件 main.xml 中加入如下代码：

```
<com.google.android.maps.MapView
        android:id="@+id/MyGMap"
        android:layout_width="fill_parent"
        android:layout_height="fill_parent"
        android:enabled="true"
        android.clickable="true"
        android:apiKey="API_Key_String"/>
```

另外，要使用 MAP 服务的话，还需要一个 API key，可以通过如下方式获取 API key。

① 打开 USER_HOME/Lcoal Settings/Application Data/Android 目录下的 debug.keystore 文件。

② 使用 keytool 工具来生成认证信息（MD5），使用如下命令：

```
Keytool –list –alias androiddebugkey –keystore <path_to_debug_keystore>.keystore –storepass
Android –keypass android
```

③ 打开 "Sign Up for the Android Maps API" 页面，输入之前生成的认证信息（MD5）后将获得 API key。

④ 替换上面 AndroidManifest.xml 配置文件中 "API_Key_String" 为刚才获取的 API key。

接下来继续补全 MyGPSActivity 类的代码，在此以使用 MapView，代码如下：

```
class myGPSActivity extends MapActivity
{
    @Override
    public void onCreate(Bundle savedInstanceState)
```

```
        {
                    //创建并初始化地图
                    gMapView=(MapView)findViewById(R.id.myGMap);
                    GeoPoint p=new GeoPoint((int)(lat*1000000),(int)(long*1000000));
                    gMapView.setSatellite(true);
                    mc=gMapView.getController();
                    mc.setCenter(p);
                    mc.setZoom(14);
        }
}
```

另外，如果要使用定位信息的话，必须设置一些权限，在 AndroidManifest.xml 中的具体配置如下：

```
<uses-permission android:name="android.permission.INTERNET"></uses-permission>
<uses-permission android:name="android.permission.ACCESS_COARSE_LOCATION"></uses-permission>
<uses-permission android:name="android.permission.access_fine_location"></uses-permission>
```

3）使用定位管理器

可以通过使用 Context.getSystemService 方法，并传入 Context.LOCATION_SERVICE 参数获取定位管理器的实例。代码如下：

```
locationManager lm=(LocationManager)getSystemService(Context.LOCATION_SERVICE);
```

之后，需要将原先的 MyGPSActivity 做一些修改，让其实现一个 LocationListener 接口，使其能够监听定位信息的改变。

```
class myGPSActivity extends MapActivity implements LocationListener
{
    ...
    ublic void onLocationChanged(Location location){}
    public void onProviderDisabled(String provider){}
    public void onProviderEnabled(String provider){}
    public void onStatusChanged(String provider,int status,Bundle extras){}
    protected boolean isRouteDisplayed()
    {
        return false;
    }
}
```

下面添加一些代码，对 LocationManager 进行一些初始化工作，并在它的 onCreate()方法中注册定位监听器。

```
@Override
public void onCreate(Bundle savedInstanceState)
{
    LocationManager lm=(LocationManager)getSystemService(Context.LOCATION_SERVICE);
    lm.requestLocationUpdates(LocationManager.GPS_PROVIDER,1000L,500.0f,this);
}
```

此时代码中的 onLocationChanged 方法就会在用户的位置发生 500m 距离的改变之后进行调用。这里默认使用的 LocationProvider 为 "gps"（GSP_PROVIDER），但是可以根据需要，使用特定的 Criteria 对象调用 LocationManager 类的 getBestProvider 方法获取其他的 LocationProvider。以下代码为 onLocationChanged 方法的参考实现。

```
public void onLocationChanged(Location location)
{
    if(location!=null)
    {
        double lat=location.getLatitude();
        double lng=location.getLongitude();
        p=new GeoPoint((int)lat*1000000,(int)lng*1000000);
        mc.animateTo(p);
    }
}
```

通过以上代码，获取了当前的新位置并更新地图上的位置显示，还可为应用程序添加一些诸如缩放效果、地图标注、文本等功能。

4）添加缩放控件

实现缩放控件的添加代码如下：

```
//将缩放控件添加到地图上
ZoomControls zoomControls=(ZoomControls)gMapView.getZoomControl();
zoomControls.setLayoutParams(new ViewGroup.LayoutParams(LayoutParams.WRAP_CONTENT,
LayoutParams.WRAP_CONTENT));
gMapView.addView(zoomControls);
gMapView.displayZoomControls(true);
```

5）添加 Map Overlay

最后一步，添加 Map Overlay。通过以下代码可定义一个 overlay。

```
class MyLocationOverlay extends com.google.android.maps.Overlay
{
    @Override
    public boolean draw(Canvas canvas,MapView mapView,boolean shadow,long when)
    {
        super.draw(canvas,mapView,shadow);
        Paint paint=new Paint();
        //将经纬度转换成实际屏幕坐标
        Point myScreenCoords=new Point();
        mapView.getProjection().toPixels(p,myScreenCoords);
        paint.setStrokeWidth(1);
        paint.setARGB(255,255,255,255);
        paint.setStyle(Paint.Style.STROKE);
        Bitmap bmp=BitmapFactory.decodeResource(getResources(),R.drawable.marker);
```

```
            canvas.drawBitmap(bmp,myScreenCoords.x,myScreenCoords.y,paint);
            canvas.drawText("Hello Android....",myScreenCoords.x,myScreenCoords.y,paint);
            return true;
        }
    }
```

上面的 overlay 会在地图上显示一个"Hello Android"的文本，然后把这个 overlay 添加到地图上去。

```
    MyLocationOverlay myLocationOverlay=new MyLocationOverlay();
    List<Overlay>list=gMapView.getOverlays();
    list.add(myLocationOverlay);
```

3. 获取当前位置的坐标

本实例使用 GPS 定位技术获取当前位置的坐标信息。

【实现步骤】

（1）在 Eclipse 环境下新建一名为 E10_1_1_1 的工程。

（2）打开 res/layout 目录下的 main.xml 文件，代码修改为：

```xml
<?xml version="1.0" encoding="utf-8"?>
<LinearLayout xmlns:android="http://schemas.android.com/apk/res/android"
    android:orientation="vertical"
    android:layout_width="fill_parent"
    android:layout_height="fill_parent" >
<TextView
    android:id="@+id/myLocationText"
    android:layout_width="fill_parent"
    android:layout_height="wrap_content"
    android:text="@string/hello_world"/>
</LinearLayout>
```

（3）打开 src/com.example.e10_1_1_1 目录下的 MainActivity.java 文件，代码修改为：

```java
public class MainActivity extends Activity
{
    /** 第一次调用活动 */
    @Override
    //使用 LocationManager 周期获得当前设备的一个类。调用 Context.getSystemService 并传
    //入服务名 LOCATION_SERVICE("location")获取 LocationManager 实例
    public void onCreate(Bundle savedInstanceState)
    {
        super.onCreate(savedInstanceState);
        setContentView(R.layout.main);
        LocationManager locationManager;
        String serviceName = Context.LOCATION_SERVICE;
        locationManager = (LocationManager)getSystemService(serviceName);
```

```java
        Criteria criteria = new Criteria();
        criteria.setAccuracy(Criteria.ACCURACY_FINE);
        criteria.setAltitudeRequired(false);
        criteria.setBearingRequired(false);
        criteria.setCostAllowed(true);
        criteria.setPowerRequirement(Criteria.POWER_LOW);
        String provider = locationManager.getBestProvider(criteria, true);
        Location location = locationManager.getLastKnownLocation(provider);
        updateWithNewLocation(location);
        locationManager.requestLocationUpdates(provider, 2000, 10,locationListener);
    }
    //定义 LocationListener 对象来监听坐标改变时的事件
    private final LocationListener locationListener = new LocationListener()
    {
    public void onLocationChanged(Location location)
        {
            updateWithNewLocation(location);
        }
        public void onProviderDisabled(String provider)
        {
            updateWithNewLocation(null);
        }
        public void onProviderEnabled(String provider)
        { }
        public void onStatusChanged(String provider, int status,
        Bundle extras){ }
    };
    //定义 updateWidthNewLocation(Location location)来更新显示用户界面
    private void updateWithNewLocation(Location location)
    {
        String latLongString;
        TextView myLocationText;
        myLocationText = (TextView)findViewById(R.id.myLocationText);
        if (location != null) {
        double lat = location.getLatitude();
        double lng = location.getLongitude();
        latLongString = "纬度:" + lat + "\n 经度:" + lng;
        } else {
        latLongString = "获取地理信息失败";
        }
        myLocationText.setText("当前坐标位置为:\n" +
        latLongString);
    }
}
```

（4）授权权限。打开 AndroidManifest.xml 文件，添加如下代码：

```
</application>
<uses-permission android:name="android.permission.ACCESS_FINE_LOCATION"/>
</manifest>
```

因为模拟器上没有 GPS 设备，所以需要在 Eclipse 的 DDMS 工具中提供模拟的 GPS 数据，即依次单击 Android 工具栏中的"窗口"菜单下的"打开透视图"子菜单下的"DDMS"选项下的"Emulator Control"选项，在弹出的对话框中找到"Location Control"选项，并输入坐标，最后，单击"Send"按钮即可完成，效果如图 10-1 所示。

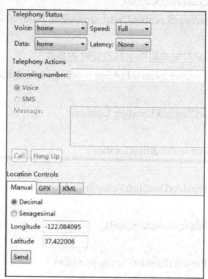

图 10-1　设置坐标

因为用到了 Google API，所以要在项目中引入 Google API，单击项目选择"窗口"菜单下的"首选项"，在弹出对话框中选择"Google API"版本，如图 10-2 所示。

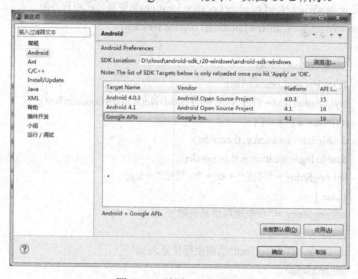

图 10-2　引用 Google API

10.1.2 定位实例

在本实例中，通过 Google MapView 和 GeoPoint 实现地图经纬度应用的流程。当在表单中输入一个经度和维度值后，单击"查询"按钮会在地图中显示此位置。

【实现步骤】

（1）在 Eclipse 环境下，建立一个名为 E10_1_2 工程。

（2）打开 res/layout 目录下的 main.xml 文件，代码修改为：

```xml
<?xml version="1.0" encoding="utf-8"?>
<AbsoluteLayout
    android:id="@+id/layout1"
    android:layout_width="fill_parent"
    android:layout_height="fill_parent"
    xmlns:android="http://schemas.android.com/apk/res/android">
    <EditText
        android:id="@+id/myEdit1"
        android:layout_width="105px"
        android:layout_height="40px"
        android:textSize="16sp"
        android:layout_x="130px"
        android:layout_y="12px"
        android:numeric="decimal"></EditText>
    <EditText
        android:id="@+id/myEdit2"
        android:layout_width="105px"
        android:layout_height="40px"
        android:textSize="16sp"
        android:layout_x="130px"
        android:layout_y="52px"
        android:numeric="decimal"></EditText>
    <TextView
        android:id="@+id/myText1"
        android:layout_width="wrap_content"
        android:layout_height="wrap_content"
        android:text="@string/str_longitude"
        android:layout_x="10px"
        android:layout_y="22px">
    </TextView>
    <TextView
        android:id="@+id/myText2"
```

```xml
            android:layout_width="wrap_content"
            android:layout_height="wrap_content"
            android:text="@string/str_latitude"
            android:layout_x="10px"
            android:layout_y="62px">
    </TextView>
    <Button
        android:id="@+id/myButton1"
        android:layout_width="77px"
        android:layout_height="45px"
        android:text="@string/str_button1"
        android:layout_x="240px"
        android:layout_y="12px"></Button>
    <Button
        android:id="@+id/myButton2"
        android:layout_width="40px"
        android:layout_height="40px"
        android:text="@string/str_button2"
        android:textSize="20sp"
        android:layout_x="277px"
        android:layout_y="55px"></Button>
    <Button
        android:id="@+id/myButton3"
        android:layout_width="40px"
        android:layout_height="40px"
        android:text="@string/str_button3"
        android:textSize="20sp"
        android:layout_x="240px"
        android:layout_y="55px"></Button>
    <!-- Google MapView Widget -->
    <com.google.android.maps.MapView
        android:id="@+id/myMapView1"
        android:layout_width="fill_parent"
        android:layout_height="fill_parent"
        android:layout_x="0px"
        android:layout_y="102px"
        android:enabled="true"
        android:clickable="true"
        android:apiKey="0by7ffx8jX0A_LWXeKCMTWAh8CqHAlqvzetFqjQ">
    </com.google.android.maps.MapView>
</AbsoluteLayout>
```

(3)在 EditText 输入框中输入坐标的经度和纬度,将坐标转换为 GeoPoint 对象后,再利用 MapController 的 animateTo()方法将地图的中心点移到 GeoPoint 坐标上。打开 src/com.example. e10_1_2 目录下的 MainActivity.java 文件,代码修改为:

```java
public class MainActivity extends MapActivity
{
    private MapController mMapController01;
    private MapView mMapView01;
    private Button mButton01,mButton02,mButton03;
    private EditText mEditText01;
    private EditText mEditText02;
    private int intZoomLevel=0;
    /* Map 启动时的默认坐标*/
    private double dLat=120.391177;
    private double dLng=39.9067452;
    @Override
    protected void onCreate(Bundle icicle)
    {
        super.onCreate(icicle);
        setContentView(R.layout.main);
        /* 创建 MapView 对象 */
        mMapView01 = (MapView)findViewById(R.id.myMapView1);
        mMapController01 = mMapView01.getController();
        /* 设置 MapView 的显示选项(卫星、街道) */
        mMapView01.setSatellite(false);
        mMapView01.setStreetView(true);
        /* 默认放大的层级 */
        intZoomLevel = 17;
        mMapController01.setZoom(intZoomLevel);
        /* 设置 Map 的中点为默认经纬度 */
        refreshMapView();
        mEditText01 = (EditText)findViewById(R.id.myEdit1);
        mEditText02 = (EditText)findViewById(R.id.myEdit2);
        /* 送出查询的 Button */
        mButton01 = (Button)findViewById(R.id.myButton1);
        mButton01.setOnClickListener(new Button.OnClickListener()
        {
            @Override
            public void onClick(View v)
            {
                /* 经纬度空白检查 */
                if(mEditText01.getText().toString().equals("")||
```

```java
                        mEditText02.getText().toString().equals(""))
        {
            showDialog("经度或纬度填写不正确!");
        }
        else
        {
            /* 取得输入的经纬度 */
            dLng=Double.parseDouble(mEditText01.getText().toString());
            dLat=Double.parseDouble(mEditText02.getText().toString());
            /* 依输入的经纬度重整 Map */
            refreshMapView();
        }
    }
});
/* 放大 Map 的 Button */
mButton02 = (Button)findViewById(R.id.myButton2);
mButton02.setOnClickListener(new Button.OnClickListener()
{
    @Override
    public void onClick(View v)
    {
        intZoomLevel++;
        if(intZoomLevel>mMapView01.getMaxZoomLevel())
        {
            intZoomLevel = mMapView01.getMaxZoomLevel();
        }
        mMapController01.setZoom(intZoomLevel);
    }
});
/* 缩小 Map 的 Button */
mButton03 = (Button)findViewById(R.id.myButton3);
mButton03.setOnClickListener(new Button.OnClickListener()
{
    @Override
    public void onClick(View v)
    {
        intZoomLevel--;
        if(intZoomLevel<1)
        {
            intZoomLevel = 1;
        }
```

```java
            mMapController01.setZoom(intZoomLevel);
        }
    });
}
/* 重整 Map 的 method */
public void refreshMapView()
{
    GeoPoint p = new GeoPoint((int)(dLat* 1E6), (int)(dLng* 1E6));
    mMapView01.displayZoomControls(true);
    /* 将 Map 的中点移至 GeoPoint */
    mMapController01.animateTo(p);
    mMapController01.setZoom(intZoomLevel);
}
@Override
protected boolean isRouteDisplayed()
{
    return false;
}

/* 显示 Dialog 的 method */
private void showDialog(String mess)
{
    new AlertDialog.Builder(MainActivity.this).setTitle("Message")
    .setMessage(mess)
    .setNegativeButton("确定", new DialogInterface.OnClickListener()
    {
        public void onClick(DialogInterface dialog, int which)
        {
        }
    });
    .show();
}
}
```

10.1.3 地址查询实例

在 Google 中提供了一个 Geocoder 服务，在非商用情况下，使用 Geocoder 可以反查 Address 地址对象服务。在本实例中，通过使用 Geocoder 实现地址反查功能。

【实现步骤】

（1）在 Eclipse 环境下，建立一个名为 E10_1_3 的工程。

（2）打开 res/layout 目录下，打开 main.xml 文件，代码修改为：

```xml
<?xml version="1.0" encoding="utf-8"?>
<LinearLayout
  xmlns:android="http://schemas.android.com/apk/res/android"
  android:orientation="vertical"
  android:layout_width="fill_parent"
  android:layout_height="fill_parent">
  <TextView
    android:id="@+id/myTextView1"
    android:layout_width="fill_parent"
    android:layout_height="wrap_content"
    android:text="@string/hello_world"/>
  <EditText
    android:id="@+id/myEditText1"
    android:layout_width="wrap_content"
    android:layout_height="wrap_content"
    android:text=""/>
  <LinearLayout
    android:orientation="horizontal"
    android:layout_width="wrap_content"
    android:layout_height="wrap_content">
  <Button
    android:id="@+id/myButton1"
    android:layout_width="wrap_content"
    android:layout_height="wrap_content"
    android:text="@string/str_button1"/>
  <Button
    android:id="@+id/myButton2"
    android:layout_width="wrap_content"
    android:layout_height="wrap_content"
    android:text="@string/str_button2"/>
  <Button
    android:id="@+id/myButton3"
    android:layout_width="wrap_content"
    android:layout_height="wrap_content"
    android:text="@string/str_button3"/>
  </LinearLayout>
  <com.google.android.maps.MapView
    android:id="@+id/myMapView1"
    android:layout_width="fill_parent"
    android:layout_height="fill_parent"
```

```
            android:enabled="true"
            android:clickable="true"
            android:apiKey="0by7ffx8jX0A_LWXeKCMTWAh8CqHAlqvzetFqjQ" />
</LinearLayout>
```

（3）打开 src/com.example.e10_1_3 目录下的 MainActivity.java 文件，修改代码为：

```
public class MainActivity extends MapActivity
{
    private MapController mMapController01;
    private MapView mMapView01;
    private Button mButton01,mButton02,mButton03;
    private EditText mEditText01;
    private int intZoomLevel=15;
    private String TAG = "HIPPO_GEO_DEBUG";
    @Override
    protected void onCreate(Bundle icicle)
    {
        // TODO：自动存根法
        super.onCreate(icicle);
        setContentView(R.layout.main);
        mEditText01 = (EditText)findViewById(R.id.myEditText1);
        mEditText01.setText
        (
            getResources().getText(R.string.str_default_address).toString()
        );
        /* 创建 MapView 对象 */
        mMapView01 = (MapView)findViewById(R.id.myMapView1);
        mMapController01 = mMapView01.getController();
        // 设置 MapView 的显示选项（卫星、街道）
        mMapView01.setSatellite(true);
        mMapView01.setStreetView(true);
        mButton01 = (Button)findViewById(R.id.myButton1);
        mButton01.setOnClickListener(new Button.OnClickListener()
        {
            @Override
            public void onClick(View v)
            {
                // TODO Auto-generated method stub
                if(mEditText01.getText().toString()!="")
                {
                    refreshMapViewByGeoPoint
                    (
```

```java
                    getGeoByAddress
                    (
                        mEditText01.getText().toString()
                    ),mMapView01,intZoomLevel,true
                );
            }
        }
    });
    /* 放大 */
    mButton02 = (Button)findViewById(R.id.myButton2);
    mButton02.setOnClickListener(new Button.OnClickListener()
    {
        @Override
        public void onClick(View v)
        {
            // TODO Auto-generated method stub
            intZoomLevel++;
            if(intZoomLevel>mMapView01.getMaxZoomLevel())
            {
                intZoomLevel = mMapView01.getMaxZoomLevel();
            }
            mMapController01.setZoom(intZoomLevel);
        }
    });
    /* 缩小 */
    mButton03 = (Button)findViewById(R.id.myButton3);
    mButton03.setOnClickListener(new Button.OnClickListener()
    {
        @Override
        public void onClick(View v)
        {
            // TODO Auto-generated method stub
            intZoomLevel--;
            if(intZoomLevel<1)
            {
                intZoomLevel = 1;
            }
            mMapController01.setZoom(intZoomLevel);
        }
    });
    /* 初次查询地点 */
```

```java
            refreshMapViewByGeoPoint
            (
               getGeoByAddress
               (
                    getResources().getText(R.string.str_default_address).toString()
               ),mMapView01,intZoomLevel,true
            );
        }
        private GeoPoint getGeoByAddress(String strSearchAddress)
        {
           GeoPoint gp = null;
           try
           {
              if(strSearchAddress!="")
              {
                 Geocoder mGeocoder01 = new Geocoder(dicha.this, Locale.getDefault());
                 List<Address> lstAddress = mGeocoder01.getFromLocationName(strSearchAddress, 1);
                 if (!lstAddress.isEmpty())
                 {
                    Address adsLocation = lstAddress.get(0);
                    double geoLatitude = adsLocation.getLatitude()*1E6;
                    double geoLongitude = adsLocation.getLongitude()*1E6;
                    gp = new GeoPoint((int) geoLatitude, (int) geoLongitude);
                 }
                 else
                 {
                    Log.i(TAG, "Address GeoPoint NOT Found.");
                 }
              }
           }
           catch (Exception e)
           {
              e.printStackTrace();
           }
           return gp;
        }
        public static void refreshMapViewByGeoPoint(GeoPoint gp, MapView mv, int zoomLevel, boolean bIfSatellite)
        {
           try
           {
```

```
            mv.displayZoomControls(true);
            /* 取得 MapView 的 MapController */
            MapController mc = mv.getController();
            /* 移至该地理坐标地址 */
            mc.animateTo(gp);

            /* 放大地图层级 */
            mc.setZoom(zoomLevel);

            /* 设置 MapView 的显示选项（卫星、街道）*/
            if(bIfSatellite)
            {
                mv.setSatellite(true);
                mv.setStreetView(true);
            }
            else
            {
                mv.setSatellite(false);
            }
        }
        catch(Exception e)
        {
            e.printStackTrace();
        }
    }
    @Override
    protected boolean isRouteDisplayed()
    {
        // TODO：自动存根法
        return false;
    }
}
```

10.1.4 导航实例

在 Android SDK 中，可使用手机内置地图程序传递导航坐标的方式来规划路径。在实例中，通过 Directions Route 实现了路径导航功能。

在实现中，先调用 getLocationProvider()来获取当前 Location，以取得当前所在位置的地理坐标，并通过提供的 EditText Widget 来让用户输入将要前往的地址，通过地址来反复取得目的地的地理坐标。通过 2 个 GeoPoint 对象，并通过 Intent 方式来调用内置的地图程序。

【实现步骤】
(1) 在 Eclipse 环境下，建立一个名为 E10_1_4 的工程。
(2) 打开 res/layout 目录下的 main.xml 文件，代码修改为：

```xml
<?xml version="1.0" encoding="utf-8"?>
<LinearLayout
    xmlns:android="http://schemas.android.com/apk/res/android"
    android:orientation="vertical"
    android:layout_width="fill_parent"
    android:layout_height="fill_parent">
    <!-- 建立一个 TextView -->
    <TextView
        android:id="@+id/myTextView1"
        android:layout_width="fill_parent"
        android:layout_height="wrap_content"
        android:text="@string/hello_world"/>
    <!-- 建立一个 EditText -->
    <EditText
        android:id="@+id/myEditText1"
        android:layout_width="wrap_content"
        android:layout_height="wrap_content"
        android:text="" />
    <LinearLayout
        android:orientation="horizontal"
        android:layout_width="wrap_content"
        android:layout_height="wrap_content">
        <!-- 建立第一个 Button -->
        <Button
            android:id="@+id/myButton1"
            android:layout_width="wrap_content"
            android:layout_height="wrap_content"
            android:text="@string/str_button1"/>
        <!-- 建立第二个 Button -->
        <Button
            android:id="@+id/myButton2"
            android:layout_width="wrap_content"
            android:layout_height="wrap_content"
            android:text="@string/str_button2"/>
        <!-- 建立第三个 Button -->
        <Button
            android:id="@+id/myButton3"
            android:layout_width="wrap_content"
```

```xml
        android:layout_height="wrap_content"
        android:text="@string/str_button3"/>
</LinearLayout>
<!-- 设定 Google map 的 API Key -->
<com.google.android.maps.MapView
    android:id="@+id/myMapView1"
    android:layout_width="fill_parent"
    android:layout_height="fill_parent"
    android:enabled="true"
    android:clickable="true"
    android:apiKey="0by7ffx8jX0A_LWXeKCMTWAh8CqHAlqvzetFqjQ"/>
</LinearLayout>
```

（3）打开 src/com.example.e10_1_4 目录下的 MainActivity.java 文件，代码修改为：

```java
public class MainActivity extends MapActivity
{
    private TextView mTextView01;
    private LocationManager mLocationManager01;
    private String strLocationProvider = "";
    private Location mLocation01;
    private MapController mMapController01;
    private MapView mMapView01;
    private Button mButton01,mButton02,mButton03;
    private EditText mEditText01;
    private int intZoomLevel=0;
    private GeoPoint fromGeoPoint, toGeoPoint;
    @Override
    protected void onCreate(Bundle icicle)
    {
        // TODO: 自动存根法
        super.onCreate(icicle);
        setContentView(R.layout.main);
        mTextView01 = (TextView)findViewById(R.id.myTextView1);
        mEditText01 = (EditText)findViewById(R.id.myEditText1);
        mEditText01.setText
        (
            getResources().getText
            (R.string.str_default_address).toString()
        );
        /* 创建 MapView 对象 */
        mMapView01 = (MapView)findViewById(R.id.myMapView1);
        mMapController01 = mMapView01.getController();
```

```java
// 设置 MapView 的显示选项（卫星、街道）
mMapView01.setSatellite(true);
mMapView01.setStreetView(true);
// 放大的层级
intZoomLevel = 15;
mMapController01.setZoom(intZoomLevel);
/* 创建 LocationManager 对象取得系统 Location 服务 */
mLocationManager01 =
(LocationManager)getSystemService(Context.LOCATION_SERVICE);
/*
 * 自定义函数，访问 Location Provider,
 * 并将之存储在 strLocationProvider 当中
 */
getLocationProvider();
/* 传入 Location 对象，显示于 MapView */
fromGeoPoint = getGeoByLocation(mLocation01);
refreshMapViewByGeoPoint(fromGeoPoint,mMapView01, intZoomLevel);
/* 创建 LocationManager 对象，监听
 * Location 更改时的事件，更新 MapView*/
mLocationManager01.requestLocationUpdates
(strLocationProvider, 2000, 10, mLocationListener01);
mButton01 = (Button)findViewById(R.id.myButton1);
mButton01.setOnClickListener(new Button.OnClickListener()
{
  @Override
  public void onClick(View v)
  {
    // TODO：自动存根法
    if(mEditText01.getText().toString()!="")
    {
      /* 取得 User 要前往地址的 GeoPoint 对象 */
      toGeoPoint=getGeoByAddress(mEditText01.getText().toString());
      /* 路径规划 Intent */
      Intent intent = new Intent();
      intent.setAction(android.content.Intent.ACTION_VIEW);
      /* 传入路径规划所需要的目标地址 */
      intent.setData
      (
        Uri.parse("http://maps.google.com/maps?f=d&saddr="+
        GeoPointToString(fromGeoPoint)+
        "&daddr="+GeoPointToString(toGeoPoint)+
```

```
              "&hl=cn" +
              "")
          );
          startActivity(intent);
        }
      }
    });
    /* 放大地图 */
    mButton02 = (Button)findViewById(R.id.myButton2);
    mButton02.setOnClickListener(new Button.OnClickListener()
    {
      @Override
      public void onClick(View v)
      {
        // TODO Auto-generated method stub
        intZoomLevel++;
        if(intZoomLevel>mMapView01.getMaxZoomLevel())
        {
          intZoomLevel = mMapView01.getMaxZoomLevel();
        }
        mMapController01.setZoom(intZoomLevel);
      }
    });
    /* 缩小地图 */
    mButton03 = (Button)findViewById(R.id.myButton3);
    mButton03.setOnClickListener(new Button.OnClickListener()
    {
      @Override
      public void onClick(View v)
      {
        // TODO Auto-generated method stub
        intZoomLevel--;
        if(intZoomLevel<1)
        {
          intZoomLevel = 1;
        }
        mMapController01.setZoom(intZoomLevel);
      }
    });
}
/* 捕捉当手机 GPS 坐标更新时的事件 */
```

```java
public final LocationListener mLocationListener01 =
new LocationListener()
{
  @Override
  public void onLocationChanged(Location location)
   {
      // TODO Auto-generated method stub

      /* 当手机收到位置更改时，将 Location 传入 getMyLocation */
      mLocation01 = location;
      fromGeoPoint = getGeoByLocation(location);
      refreshMapViewByGeoPoint(fromGeoPoint,
            mMapView01, intZoomLevel);
   }
  @Override
  public void onProviderDisabled(String provider)
   {
      // TODO: 自动存根法
      mLocation01 = null;
   }
  @Override
  public void onProviderEnabled(String provider)
   {
      // TODO: 自动存根法
   }
  @Override
  public void onStatusChanged(String provider,
            int status, Bundle extras)
   {
      // TODO: 自动存根法
   }
};
/* 传入 Location 对象，取回其 GeoPoint 对象 */
private GeoPoint getGeoByLocation(Location location)
{
  GeoPoint gp = null;
  try
   {
      /* 当 Location 存在 */
      if (location != null)
       {
```

```java
            double geoLatitude = location.getLatitude()*1E6;
            double geoLongitude = location.getLongitude()*1E6;
            gp = new GeoPoint((int) geoLatitude, (int) geoLongitude);
        }
    }
    catch(Exception e)
    {
        e.printStackTrace();
    }
    return gp;
}
/* 输入地址,取得其 GeoPoint 对象 */
private GeoPoint getGeoByAddress(String strSearchAddress)
{
    GeoPoint gp = null;
    try
    {
        if(strSearchAddress!="")
        {
            Geocoder mGeocoder01 = new Geocoder
            (MainActivity.this, Locale.getDefault());

            List<Address> lstAddress = mGeocoder01.getFromLocationName
                            (strSearchAddress, 1);
            if (!lstAddress.isEmpty())
            {
                Address adsLocation = lstAddress.get(0);
                double geoLatitude = adsLocation.getLatitude()*1E6;
                double geoLongitude = adsLocation.getLongitude()*1E6;
                gp = new GeoPoint((int) geoLatitude, (int) geoLongitude);
            }
        }
    }
    catch (Exception e)
    {
        e.printStackTrace();
    }
    return gp;
}
/* 传入 geoPoint 更新 MapView 里的 Google Map */
public static void refreshMapViewByGeoPoint
```

```
    (GeoPoint gp, MapView mapview, int zoomLevel)
    {
      try
      {
        mapview.displayZoomControls(true);
        MapController myMC = mapview.getController();
        myMC.animateTo(gp);
        myMC.setZoom(zoomLevel);
        mapview.setSatellite(false);
      }
      catch(Exception e)
      {
        e.printStackTrace();
      }
    }
    /* 传入经纬度更新 MapView 里的 Google Map */
    public static void refreshMapViewByCode
    (double latitude, double longitude, MapView mapview, int zoomLevel)
    {
      try
      {
        GeoPoint p = new GeoPoint((int) latitude, (int) longitude);
        mapview.displayZoomControls(true);
        MapController myMC = mapview.getController();
        myMC.animateTo(p);
        myMC.setZoom(zoomLevel);
        mapview.setSatellite(false);
      }
      catch(Exception e)
      {
        e.printStackTrace();
      }
    }
    /* 将 GeoPoint 里的经纬度以 String 返回 */
    private String GeoPointToString(GeoPoint gp)
    {
      String strReturn="";
      try
      {
        /* 当 Location 存在 */
        if (gp != null)
```

```java
            {
                double geoLatitude = (int)gp.getLatitudeE6()/1E6;
                double geoLongitude = (int)gp.getLongitudeE6()/1E6;
                strReturn = String.valueOf(geoLatitude)+","+
                    String.valueOf(geoLongitude);
            }
        }
        catch(Exception e)
        {
            e.printStackTrace();
        }
        return strReturn;
    }
    /* 取得 LocationProvider */
    public void getLocationProvider()
    {
        try
        {
            Criteria mCriteria01 = new Criteria();
            mCriteria01.setAccuracy(Criteria.ACCURACY_FINE);
            mCriteria01.setAltitudeRequired(false);
            mCriteria01.setBearingRequired(false);
            mCriteria01.setCostAllowed(true);
            mCriteria01.setPowerRequirement(Criteria.POWER_LOW);
            strLocationProvider =
            mLocationManager01.getBestProvider(mCriteria01, true);

            mLocation01 = mLocationManager01.getLastKnownLocation
            (strLocationProvider);
        }
        catch(Exception e)
        {
            mTextView01.setText(e.toString());
            e.printStackTrace();
        }
    }
    @Override
    protected boolean isRouteDisplayed()
    {
        // TODO Auto-generated method stub
        return false;
```

        }
    }

10.2 蓝牙

蓝牙在实际应用开发中使用频率不是很高，当用户的应用需要在设备间进行数据传输时，可能会用到蓝牙，当然也可以考虑用其他无线技术，根据实际情况很容易就能做出选择。鉴于蓝牙的廉价与普遍性，应用也非常广泛。Android 平台支持蓝牙网络协议栈，其允许一台设备与其他无线蓝牙设备交换数据。

Android 应用程序框架提供了访问蓝牙功能的 API，这些 API 使应用程序能够点对点、多点连接到其他无线蓝牙设备，本节先简单介绍蓝牙开发中需要用的主要类。

使用蓝牙的 API，Android 的应用程序可以实现以下功能：
- 扫描远端的蓝牙设备。
- 与远端蓝牙设备进行认证配对。
- 通过 SDP 服务发现、建立 RFCOMM 通道。
- 连接到远端设备。
- 得到输入/输出流，与远端设备交换数据。

在学习 Android 蓝牙前先了解两个基本概念。

10.2.1 RFCOMM 协议

RFCOMM 是一个基于欧洲电信标准协会 ETSI07.10 规程的串行线性仿真协议。此协议提供 RS232 控制和状态信号，如基带上的损坏、CTS 以及数据信号等，为上层业务（如传统的串行线缆应用）提供传送能力。

RFCOMM 是一个简单的传输协议，其目的是针对如何在两个不同设备上的应用之间保证一条完整的通信路径，并在它们之间保持一通信段。

RFCOMM 是为了兼容传统的串口应用，同时取代有线的通信方式，蓝牙协议栈需要提供与有线串口一致的通信接口而开发出的协议。RFCOMM 协议提供对基于 L2CAP 协议的串口仿真，基于 ETSI07.10 可支持在两个 BT 设备之间同时保持高达 60 路的通信连接。

RFCOMM 只针对直接互连设备之间的连接，或者是设备与网络接入设备之间的互连。通信设备两端必须兼容于 RFCOMM 协议，有两类设备：DTE（Data Terminal Endpoint，通信终端，如 PC，PRINTER）和 DCE（Data Circuit Endpoint，通信段的一部分，如 ModemI）。此两类设备不进行区分。

10.2.2 MAC 硬件地址

MAC（Medium/MediaAccess Control，介质访问控制）地址是烧录在 NetworkInterfaceCard（网卡，NIC）里的。MAC 地址也叫硬件地址，是由 48 比特长 16 进制的数字组成的。0～23 位

叫做组织唯一标志符（organizationally unique），是识别 LAN（局域网）节点的标识。24～47 位是由厂家自己分配的，其中第 40 位是组播地址标志位。网卡的物理地址通常是由网卡生产厂家烧入网卡的 EPROM（一种闪存芯片，通常可以通过程序擦写）中，它存储的是传输数据时真正赖以标识发出数据的主机和接收数据的主机的地址。

Android 平台提供蓝牙 API 来实现蓝牙设备之间的通信，蓝牙设备之间的通信主要包括四个步骤：设置蓝牙设备、寻找局域网内可能或者匹配的设备、连接设备、设备之间的数据传输。以下是建立蓝牙连接的所需要的一些基本类。

- BluetoothAdapter 类：代表一个本地的蓝牙适配器。它是所有蓝牙交互的的入口点。利用它你可以发现其他蓝牙设备，查询绑定了的设备，使用已知的 MAC 地址实例化一个蓝牙设备和建立一个 BluetoothServerSocket（作为服务器端）来监听来自其他设备的连接。
- BluetoothDevice 类：代表一个远端的蓝牙设备，使用它请求远端蓝牙设备连接或者获取远端蓝牙设备的名称、地址、种类和绑定状态（其信息封装在 BluetoothSocket 中）。
- BluetoothSocket 类：代表一个蓝牙套接字的接口（类似于 TCP 中的套接字），它是应用程序通过输入、输出流与其他蓝牙设备通信的连接点。
- BluebothServerSocket 类：代表打开服务连接来监听可能到来的连接请求（属于 server 端），为了连接两个蓝牙设备必须有一个设备作为服务器打开一个服务套接字。当远端设备发起连接请求的时候，并且已经连接到的时候，BluebothServerSocket 类将会返回一个 BluetoothSocket。
- BluetoothClass 类：描述了一个蓝牙设备的一般特点和能力。它的只读属性集定义了设备的主、次设备类和一些相关服务。然而，它并没有准确地描述所有该设备所支持的蓝牙文件和服务，而是作为对设备种类来说的一个小小暗示。

要操作蓝牙，先要在 AndroidManifest.xml 里加入权限：

```
<uses-permissionandroid:name="android.permission.BLUETOOTH_ADMIN" />
<uses-permissionandroid:name="android.permission.BLUETOOTH" />
```

Android 所有关于蓝牙开发的类都在 android.bluetooth 包下，只有 8 个类，常用的有 4 个类。

1. BluetoothAdapter

BluetoothAdapter 为蓝牙适配器，直到建立 BluetoothSocket 连接之前，都要不断操作它。BluetoothAdapter 里的方法很多，常用的有以下几个。

- cancelDiscovery()：取消发现，也就是说当正在搜索设备的时候调用这个方法将不再继续搜索。
- disable()：关闭蓝牙。
- enable()：打开蓝牙，这个方法打开蓝牙不会弹出提示，更多时需要问下用户是否打开，以下两行代码同样是打开蓝牙，但会提示用户：

```
Intentenabler = new Intent(BluetoothAdapter.ACTION_REQUEST_ENABLE);
startActivityForResult(enabler,reCode);    //同 startActivity(enabler);
```

- getAddress()：获取本地蓝牙地址。
- getDefaultAdapter()：获取默认 BluetoothAdapter，实际上，也只有这一种方法获取 BluetoothAdapter。
- getName()：获取本地蓝牙名称。

- getRemoteDevice(String address)：根据蓝牙地址获取远程蓝牙设备。
- getState()：获取本地蓝牙适配器当前状态（感觉可能调试的时候更需要）。
- isDiscovering()：判断当前是否正在查找设备，是则返回 true。
- isEnabled()：判断蓝牙是否打开，已打开返回 true，否则返回 false。
- listenUsingRfcommWithServiceRecord(String name,UUID uuid)：根据名称，UUID 创建并返回 BluetoothServerSocket，这是创建 BluetoothSocket 服务器端的第一步。
- startDiscovery()：开始搜索，这是搜索的第一步。

2．BluetoothDevice

BluetoothDevice 为描述一个蓝牙设备，其一个常用方法为：

createRfcommSocketToServiceRecord(UUIDuuid)

根据 UUID 创建并返回一个 BluetoothSocket，这个方法也是获取 BluetoothDevice 的目的——创建 BluetoothSocket。

这个类的其他方法，如 getAddress()、getName()等，同 BluetoothAdapter。

3．BluetoothServerSocket

如果去除了 Bluetooth Server 相信大家一定再熟悉不过了，既然是 Socket，方法就应该都差不多，这个类只有三个方法。

- 两个重载的 accept()，accept(int timeout)：两者的区别在于后面的方法指定了过时时间，需要注意的是，执行这两个方法的时候，直到接收到了客户端的请求（或是过期之后），都会阻塞线程，应该放在新线程里运行。还有一点值得注意的是，这两个方法都返回一个 BluetoothSocket，最后的连接也是服务器端与客户端的两个 BluetoothSocket 的连接。
- close()：关闭。

4．BluetoothSocket

BluetoothSocket 为客户端，与 BluetoothServerSocket 相对。其一共有 5 个方法，不出意外，都会用到。

- close()：关闭；
- connect()：连接；
- getInptuStream()：获取输入流；
- getOutputStream()：获取输出流；
- getRemoteDevice()：获取远程设备。

这里指的是获取 BluetoothSocket 指定连接的那个远程蓝牙设备。

10.2.3 编程实现蓝牙综合实例

下面介绍 Android 编程实现蓝牙的主要步骤。

1．启动蓝牙功能

首先通过调用静态方法 getDefaultAdapter()获取蓝牙适配器 BluetoothAdapter，以后就可以使用该对象了。如果返回为空，则无法继续执行了。例如：

```
BluetoothAdapter mBluetoothAdapter = BluetoothAdapter.getDefaultAdapter();
if (mBluetoothAdapter == null)
{
    // 设备不支持蓝牙
}
```

其次，调用 isEnabled()来查询当前蓝牙设备的状态，如果返回为 false，则表示蓝牙设备没有开启，接下来需要封装一个 ACTION_REQUEST_ENABLE 请求到 intent 里面，调用 startActivityForResult()方法使能蓝牙设备。例如：

```
if (!mBluetoothAdapter.isEnabled())
{
    Intent enableBtIntent = new Intent(BluetoothAdapter.ACTION_REQUEST_ENABLE);
    startActivityForResult(enableBtIntent, REQUEST_ENABLE_BT);
}
```

2．查找设备

使用 BluetoothAdapter 类里的方法，可以查找远端设备（大概 10m 以内）或者查询在你手机上已经匹配（或者说绑定）的其他手机。此时要确定对方蓝牙设备已经开启或者已经开启了"被发现使能"功能（对方设备是可以被发现的是你能够发起连接的前提条件）。如果该设备是可以被发现的，会反馈回来一些对方的设备信息，如名字、MAC 地址等，利用这些信息，你的设备就可以选择去向对方发送一个初始化连接。

如果你是第一次与该设备连接，那么一个配对的请求就会自动显示给用户。当设备配对好之后，它的一些基本信息（主要是名字和 MAC）被保存下来并可以使用蓝牙的 API 来读取。使用已知的 MAC 地址就可以对远端的蓝牙设备发起连接请求。

匹配好的设备和连接上的设备的不同点：匹配好只是说明对方设备发现了你的存在，并拥有一个共同的识别码，并且可以连接。连接上：表示当前设备共享一个 RFCOMM 信道并且两者之间可以交换数据。也就是说，蓝牙设备在建立 RFCOMM 信道之前，必须是已经配对好的。

3．查询匹配好的设备

在建立连接之前必须先查询配对好了的蓝牙设备集（周围的蓝牙设备可能不止一个），以便你选取哪一个设备进行通信，例如，可以查询所有配对的蓝牙设备，并使用一个数组适配器将其打印显示出来：

```
Set<BluetoothDevice> pairedDevices = mBluetoothAdapter.getBondedDevices();
// If there are paired devices
if (pairedDevices.size() > 0) {
//Loop through paired devices
for (BluetoothDevice device : pairedDevices) {
// Add the name and address to an array adapter to show in a ListView
mArrayAdapter.add(device.getName() + "\n" + device.getAddress());
}
```

建立一个蓝牙连接只需要 MAC 地址就已经足够了。

4. 扫描设备

扫描设备，只需要简单地调用 startDiscovery()方法，这个扫描的过程大概持续是 12s，应用程序为了 ACTION_FOUND 动作需要注册一个 BroadcastReceiver 来接受设备扫描到的信息。对于每一个设备，系统都会广播 ACTION_FOUND 动作。例如：

```
private final BroadcastReceiver mReceiver = new BroadcastReceiver()
{
    public void onReceive(Context context, Intent intent)
    {
        String action = intent.getAction();
        // 发现设备
        if (BluetoothDevice.ACTION_FOUND.equals(action))
        {
            BluetoothDevice device = intent.getParcelableExtra(BluetoothDevice.EXTRA_DEVICE);
            // 为数组适配器添加名称和地址，并显示在 ListView 中
            mArrayAdapter.add(device.getName() + "\n" + device.getAddress());
        }
    }
};
// 注册 BroadcastReceiver
IntentFilter filter = new IntentFilter(BluetoothDevice.ACTION_FOUND);
registerReceiver(mReceiver, filter);
```

注意：扫描的过程是一个很耗费资源的过程，一旦你找到需要的设备之后，在发起连接请求之前，确保你的程序调用 cancelDiscovery()方法停止扫描。显然，如果你已经连接上一个设备，启动扫描会减少你的通信带宽。

5. 使能被发现（Enabling discoverability）

如果想使设备能够被其他设备发现，将 ACTION_REQUEST_DISCOVERABLE 动作封装在 Intent（意图）中并调用 startActivityForResult(Intent, int)方法就可以了。它将在不使应用程序退出的情况下使设备能够被发现。默认情况下的使能时间是 120s，当然可以通过添加 EXTRA_DISCOVERABLE_DURATION 字段来改变使能时间（最大不超过 300s，这是出于对你设备上的信息安全考虑）。例如：

```
Intent discoverableIntent = new Intent(BluetoothAdapter.ACTION_REQUEST_DISCOVERABLE);
discoverableIntent.putExtra(BluetoothAdapter.EXTRA_DISCOVERABLE_DURATION, 300);
startActivity(discoverableIntent);
```

运行该段代码之后，系统会弹出一个对话框来提示启动设备使能被发现（此过程中如果蓝牙功能没有开启，系统会帮助开启），并且如果准备对该远端设备发现一个连接，不需要开启使能设备被发现功能，因为该功能只是在应用程序作为服务器端的时候才需要。

6. 连接设备

在应用程序中，想建立两个蓝牙设备之间的连接，必须实现客户端和服务器端的代码（因

为任何一个设备都必须可以作为服务端或者客户端)。一个开启服务来监听,一个发起连接请求(使用服务器端设备的 MAC 地址)。当它们都拥有一个蓝牙套接字在同一 RFCOMM 信道上的时候,可以认为它们之间已经连接上了。服务端和客户端通过不同的方式或其他的蓝牙套接字。当一个连接被监听到的时候,服务端获取到蓝牙套接字。当客户可打开一个 RFCOMM 信道给服务器端的时候,客户端即可获得蓝牙套接字。

注意:在此过程中,如果两个蓝牙设备还没有配对好,Android 系统会通过一个通知或者对话框的形式来通知用户。RFCOMM 连接请求会在用户选择之前阻塞。

7.服务端的连接

当想要连接两台设备时,一个必须作为服务端(通过持有一个打开的 BluetoothServerSocket),目的是监听外来连接请求,当监听到以后提供一个连接上的 BluetoothSocket 给客户端,当客户端从 BluetoothServerSocket 得到 BluetoothSocket 以后就可以销毁 BluetoothServerSocket,除非你还想监听更多的连接请求。

建立服务套接字和监听连接的基本步骤:

首先通过调用 listenUsingRfcommWithServiceRecord(String,UUID)方法来获取 BluetoothServerSocket 对象,参数 String 代表了该服务的名称,UUID 代表了和客户端连接的一个标识(128位格式的字符串 ID,相当于 PIN 码),UUID 必须双方匹配才可以建立连接。其次调用 accept()方法来监听可能到来的连接请求,当监听到以后,返回一个连接上的蓝牙套接字 BluetoothSocket。最后,在监听到一个连接以后,需要调用 close()方法来关闭监听程序。(一般蓝牙设备之间是点对点的传输)

注意:accept()方法不应该放在主 Acitvity 里面,因为它是一种阻塞调用(在没有监听到连接请求之前程序就一直停在那里)。解决方法是新建一个线程来管理。例如:

```
private class AcceptThread extends Thread
{
    private final BluetoothServerSocket mmServerSocket;
    public AcceptThread()
    {
        BluetoothServerSocket tmp = null;
        try
        {
            tmp = mAdapter.listenUsingRfcommWithServiceRecord(NAME, MY_UUID);
        } catch (IOException e) {}
        mmServerSocket = tmp;
    }
    public void run()
    {
        BluetoothSocket socket = null;
        while (true)
        {
            try {
```

```
                socket = mmServerSocket.accept();
            } catch (IOException e)
            {
                break;
            }
            // 假如连接被接受
            if (socket != null)
            {
                manageConnectedSocket(socket);
                mmServerSocket.close();
                break;
            }
        }
    }
    /*将取消监听套接字,并导致线程完成 */
    public void cancel()
    {
        try {
            mmServerSocket.close();
        } catch (IOException e) { }
    }
}
```

8. 客户端的连接

为了初始化一个与远端设备的连接,需要先获取代表该设备的一个 BluetoothDevice 对象。通过 BluetoothDevice 对象来获取 BluetoothSocket 并初始化连接。

具体步骤使用 BluetoothDevice 对象里的方法 createRfcommSocketToServiceRecord(UUID)来获取 BluetoothSocket。UUID 就是匹配码。然后,调用 connect()方法来实现 UUID 连接。如果远端设备接收了该连接,它们将在通信过程中共享 RFCOMM 信道,并且用 connect()方法返回。例如:

```
private class ConnectThread extends Thread
{
    private final BluetoothSocket mmSocket;
    private final BluetoothDevice mmDevice;
    public ConnectThread(BluetoothDevice device)
    {
        BluetoothSocket tmp = null;
        mmDevice = device;
        // 将一个 BluetoothSocket 与 BluetoothDevice 进行连接
        try {
            tmp = device.createRfcommSocketToServiceRecord(MY_UUID);
        } catch (IOException e) { }
```

```
            mmSocket = tmp;
        }
        public void run()
        {
            // 取消发现，因为它会减慢连接
            mAdapter.cancelDiscovery();
            try {
                mmSocket.connect();
            } catch (IOException connectException)
            {
                try {
                    mmSocket.close();
                } catch (IOException closeException) { }
                return;
            }
            manageConnectedSocket(mmSocket);
        }
```

注意：conncet()方法也是阻塞调用，一般建立一个独立的线程中来调用该方法。在设备 discover 过程中不应该发起连接 connect()，这样会明显减慢速度以至于连接失败。且数据传输完成只有调用 close()方法来关闭连接，这样可以节省系统内部资源。

9. 管理连接（主要涉及数据的传输）

当设备连接上以后，每个设备都拥有各自的 BluetoothSocket，就可以实现设备之间数据的共享了。

（1）首先通过调用 getInputStream()和 getOutputStream()方法来获取输入/输出流，然后通过调用 read(byte[]) 和 write(byte[])方法来读取或者写数据。

（2）实现细节：以为读取和写操作都是阻塞调用，需要建立一个专用文件实现管理。

（3）实现程序：

```
private class ConnectedThread extends Thread {
    private final BluetoothSocket mmSocket;
    private final InputStream mmInStream;
    private final OutputStream mmOutStream;
    public ConnectedThread(BluetoothSocket socket) {
        mmSocket = socket;
        InputStream tmpIn = null;
        OutputStream tmpOut = null;
        // Get the input and output streams, using temp objects because
        // member streams are final
        try {
            tmpIn = socket.getInputStream();
            tmpOut = socket.getOutputStream();
```

```java
} catch (IOException e) { }
mmInStream = tmpIn;
mmOutStream = tmpOut;
}
public void run()
{
    byte[] buffer = new byte[1024];        //缓冲存储的流
    int bytes;                             // 返回read()方法中的字节
    // 监听 InputStream，直到发生异常
    while (true)
    {
        try {
            // 从 InputStream 中读取
            bytes = mmInStream.read(buffer);
            // 将所获得的字节发送到 UI 活动
            mHandler.obtainMessage(MESSAGE_READ, bytes, -1, buffer).sendToTarget();
        } catch (IOException e)
        {
            break;
        }
    }
}
/* 调用主要活动将数据发送到远程设备 */
public void write(byte[] bytes)
{
    try {
        mmOutStream.write(bytes);
    } catch (IOException e) { }
}
/* 调用主要活动，关闭连接 */
public void cancel()
{
    try {
        mmSocket.close();
    } catch (IOException e) { }
}
}
```

10.3 中国象棋

象棋又称为中国象棋，象棋在中国有着悠久的历史，属于二人对抗性游戏的一种，由于用具简单，趣味性强，成为流行极为广泛的棋艺活动。暗棋阁也是其中一种传统玩法，中国象棋是我国正式开展的 78 个体育运动项目之一，为促进该项目在世界范围内的普及和推广，在中国古代，象棋被列为士大夫们的修身之艺，现在则被视为怡神益智的一种有益的活动。在棋战中，人们可以从攻与防、虚与实、整体与局部等复杂关系的变化中悟出某种哲理。

本实例实现了一个中国象棋游戏。

【实现步骤】

（1）在 Eclipse 环境下，建立一个名为 xiangqi 的工程。

（2）打开 res/layout 目录下的 main.xml 文件，代码修改为：

```xml
<?xml version="1.0" encoding="utf-8"?>
<LinearLayout xmlns:android="http://schemas.android.com/apk/res/android"
    android:orientation="vertical"
    android:layout_width="fill_parent"
    android:layout_height="fill_parent" >
<TextView
    android:layout_width="fill_parent"
    android:layout_height="wrap_content"
    android:text="@string/hello_world"/>
</LinearLayout>
```

（3）打开 src/com.example.xiangqi 目录下的 MainActivity.java 文件，在文件中定义了类 MainActivity，这是本实例游戏控制器类，功能是在合适时初始化相应的用户界面，根据其他界面的要求切换到需要的界面中。代码如下：

```java
public class Help extends SurfaceView implements SurfaceHolder.Callback
{
    MainActivity activity;                                      //Activity 的引用
    private TutorialThread thread;                              //刷帧的线程
    Bitmap back;                                                //返回按钮
    Bitmap helpBackground;                                      //背景图片
    public Help(Context context,MainActivity activity)
    {
        //构造器
        super(context);
        this.activity = activity;                               //得到 Activity 引用
        getHolder().addCallback(this);
        this.thread = new TutorialThread(getHolder(), this);    //初始化重绘线程
        initBitmap();                                           //初始化图片资源
    }
```

```java
public void initBitmap()
{
    //初始化所用到的图片
    back = BitmapFactory.decodeResource(getResources(), R.drawable.back);//返回按钮
    helpBackground = BitmapFactory.decodeResource(
                        getResources(),
                        R.drawable.helpbackground);            //初始化背景图片
}
public void onDraw(Canvas canvas){                              //自己写的绘制方法
    canvas.drawBitmap(helpBackground, 0, 90, new Paint());      //绘制背景图片
    canvas.drawBitmap(back, 200, 370, new Paint());             //绘制按钮
}
public void surfaceChanged(SurfaceHolder holder, int format, int width, int height) {}
public void surfaceCreated(SurfaceHolder holder)
{
    //被创建时启动刷帧线程
    this.thread.setFlag(true);                                  //设置循环标志位
    this.thread.start();                                        //启动刷帧线程
}
public void surfaceDestroyed(SurfaceHolder holder)
{
    //被摧毁时停止刷帧线程
    boolean retry = true;                                       //循环标志位
    thread.setFlag(false);                                      //设置循环标志位
    while (retry)
    {
        try {
            thread.join();                                      //等待线程结束
            retry = false;                                      //停止循环
        }catch (InterruptedException e){}    //不断地循环,直到刷帧线程结束
    }
}
public boolean onTouchEvent(MotionEvent event) {                //屏幕监听
    if(event.getAction() == MotionEvent.ACTION_DOWN){
        if(event.getX()>200 && event.getX()<200+back.getWidth()
                && event.getY()>370 && event.getY()<370+back.getHeight())
        {
            //单击了返回按钮
            activity.myHandler.sendEmptyMessage(1);//发送 Handler 消息
        }
```

```java
            return super.onTouchEvent(event);
    }
    class TutorialThread extends Thread{                    //刷帧线程
        private int span = 1000;                            //睡眠的毫秒数
        private SurfaceHolder surfaceHolder;                //SurfaceHolder 的引用
        private Help helpView;                              //父类的引用
        private boolean flag = false;                       //循环标记位
        public TutorialThread(SurfaceHolder surfaceHolder, Help helpView)
        {
            //构造器
            this.surfaceHolder = surfaceHolder;             //得到 SurfaceHolder 引用
            this.helpView = helpView;                       //得到 helpView 引用
        }
        public void setFlag(boolean flag) {                 //设置循环标记位
            this.flag = flag;
        }
        public void run()
        {
            //重写的 run 方法
            Canvas c;                                       //画布
            while (this.flag) {                             //循环
                c = null;
                try {
                    c = this.surfaceHolder.lockCanvas(null);
                    synchronized (this.surfaceHolder) {     //同步
                        helpView.onDraw(c);                 //调用绘制方法
                    }
                } finally {                                 //用 finally 语句保证下面的代码一定会被执行
                    if (c != null) {                        //更新屏幕显示内容
                        this.surfaceHolder.unlockCanvasAndPost(c);
                    }
                }
                try{
                    Thread.sleep(span);                     //睡眠指定毫秒数
                }catch(Exception e){                        //捕获异常
                    e.printStackTrace();                    //打印异常堆栈信息
                }
            }
        }
    }
}
```

（4）在 src/com.example.xiangqi 目录下新建一个名为 WelcomeView.java 的文件，在文件中定义了类 WelcomeView，该类为一个辅助界面类，是进入游戏系统后显示的欢迎界面框架，其实现代码如下：

```java
public class WelcomeView extends SurfaceView implements SurfaceHolder.Callback
{
    MainActivity activity;
    private TutorialThread thread;                          //刷帧的线程
    private WelcomeThread moveThread;                       //物件移动的线程
    Bitmap welcomebackage;                                  //大背景
    Bitmap logo;
    Bitmap boy;                                             //小孩的图片
    Bitmap oldboy;                                          //老头的图片
    Bitmap bordbackground;                                  //文字背景
    Bitmap logo2;
    Bitmap menu;                                            //菜单按钮
    int logoX = -120;                                       //初始化需要移动的图片的相应坐标
    int boyX = -100;
    int oldboyX = -120;
    int logo2X = 320;
    int bordbackgroundY = -100;                             //背景框的 Y 坐标
    int menuY = 520;                                        //菜单的 Y 坐标
    public WelcomeView(Context context,MainActivity activity)
    {
        //构造器
        super(context);
        this.activity = activity;                           //得到 activity 引用
        getHolder().addCallback(this);
        this.thread = new TutorialThread(getHolder(), this); //初始化刷帧线程
        this.moveThread = new WelcomeThread(this);           //初始化图片移动线程
        initBitmap();                                        //初始化所有图片
    }
    public void initBitmap(){                               //初始化所有图片
        welcomebackage = BitmapFactory.decodeResource(getResources(), R.drawable.welcomebackage);
        logo = BitmapFactory.decodeResource(getResources(), R.drawable.logo);
        boy = BitmapFactory.decodeResource(getResources(), R.drawable.boy);
        oldboy = BitmapFactory.decodeResource(getResources(), R.drawable.oldboy);
        bordbackground = BitmapFactory.decodeResource(getResources(), R.drawable.bordbackground);
        logo2 = BitmapFactory.decodeResource(getResources(), R.drawable.logo2);
        menu = BitmapFactory.decodeResource(getResources(), R.drawable.menu);
    }
    public void onDraw(Canvas canvas){                      //自己写的绘制方法,并非重写的
```

```java
            //后画的会覆盖前面画的
            canvas.drawColor(Color.BLACK);//清屏
            canvas.drawBitmap(welcomebackage, 0, 100, null);              //绘制欢迎界面
            canvas.drawBitmap(logo, logoX, 110, null);                    //绘制 logo
            canvas.drawBitmap(boy, boyX, 210, null);                      //绘制小孩的图片
            canvas.drawBitmap(oldboy, oldboyX, 270, null);                //绘制老头的图片
canvas.drawBitmap(bordbackground, 150, bordbackgroundY, null);            //绘制文字背景
            canvas.drawBitmap(logo2, logo2X, 100, null);                  //绘制 logo2
            canvas.drawBitmap(menu, 200, menuY, null);                    //绘制菜单
        }
        public void surfaceChanged(SurfaceHolder holder, int format, int width, int height)
        {}
        public void surfaceCreated(SurfaceHolder holder)
        {
            //创建时启动相应进程
            this.thread.setFlag(true);                                    //设置循环标志位
            this.thread.start();                                          //启动线程
            this.moveThread.setFlag(true);                                //设置循环标志位
            this.moveThread.start();                                      //启动线程
        }
        public void surfaceDestroyed(SurfaceHolder holder)
        {
            //摧毁时释放相应进程
            boolean retry = true;
            thread.setFlag(false);                                        //设置循环标志位
            moveThread.setFlag(false);
            while (retry)
            {
                //循环
                try {
                    thread.join();                                        //等待线程结束
                    moveThread.join();
                    retry = false;                                        //停止循环
                }
                catch (InterruptedException e) {                          //不断地循环，直到刷帧线程结束
                }
            }
        }
        @Override
        public boolean onTouchEvent(MotionEvent event)
        {
```

```java
            //屏幕监听
            if(event.getAction() == MotionEvent.ACTION_DOWN)
            {
                if(event.getX()>200 && event.getX()<200+menu.getWidth()
                        && event.getY()>355 && event.getY()<355+menu.getHeight())
                {
                    //单击菜单按钮
                    activity.myHandler.sendEmptyMessage(1);
                }
            }
            return super.onTouchEvent(event);
        }
        class TutorialThread extends Thread{              //刷帧线程
            private int span = 100;                        //睡眠的毫秒数
            private SurfaceHolder surfaceHolder;           //SurfaceHolder 引用
            private WelcomeView welcomeView;               //WelcomeView 引用
            private boolean flag = false;
            public TutorialThread(SurfaceHolder surfaceHolder, WelcomeView welcomeView)
            {
                //构造器
                this.surfaceHolder = surfaceHolder;        //得到 SurfaceHolder 引用
                this.welcomeView = welcomeView;            //得到 WelcomeView 引用
            }
            public void setFlag(boolean flag)
            {
                //设置循环标记位
                this.flag = flag;
            }
            @Override
            public void run() {//重写的 run 方法
                Canvas c;//画布
                while (this.flag)
                {
                    //循环
                    c = null;
                    try {
                        // 锁定整个画布，在内存要求比较高的情况下，建议参数不要为 null
                        c = this.surfaceHolder.lockCanvas(null);
                        synchronized (this.surfaceHolder) {    //同步
                            welcomeView.onDraw(c);              //绘制
                        }
```

```
                    } finally {//使用 finally 语句保证下面的代码一定会被执行
                        if (c != null)
                        {
                            //更新屏幕显示内容
                            this.surfaceHolder.unlockCanvasAndPost(c);
                        }
                    }
                    try{
                        Thread.sleep(span);             //睡眠指定毫秒数
                    }
                    catch(Exception e){                 //捕获异常
                        e.printStackTrace();            //打印堆栈信息
                    }
                }
            }
        }
    }
```

（5）在 src/com.example.xiangqi 目录下新建一个名为 WelcomeThread.java 的文件，在文件中定义了类 WelcomeThread，此类也为一个辅助界面类，用于生成欢迎界面的动画效果。实现代码如下：

```
    public class WelcomeThread extends Thread
    {
        private boolean flag = true;                    //循环标志位
        WelcomeView welcomeView;                        //WelcomeView 的引用
        public WelcomeThread(WelcomeView welcomeView)
        {
            //构造器
            this.welcomeView = welcomeView;             //得到 WelcomeView 的引用
        }
        public void setFlag(boolean flag){              //设置循环标志位
            this.flag = flag;
        }
        public void run(){                              //重写的 run 方法
            try{
                Thread.sleep(300);                      //睡眠 300ms，保证界面已经显示
            }
            catch(Exception e){                         //捕获异常
                e.printStackTrace();                    //打印异常信息
            }
            while(flag)
            {
                welcomeView.logoX += 10;                //移动欢迎界面的 logo
```

```
            if(welcomeView.logoX>0){              //到位后停止移动
                welcomeView.logoX = 0;
            }
            welcomeView.boyX += 20;               //移动小男孩图片
            if(welcomeView.boyX>70){              //到位置后停止移动
                welcomeView.boyX = 70;
            }
            welcomeView.oldboyX += 15;            //移动小老头
            if(welcomeView.oldboyX>0){            //到位后停止移动
                welcomeView.oldboyX = 0;
            }
            welcomeView.bordbackgroundY += 50;    //移动文字背景
            if(welcomeView.bordbackgroundY>240){
                welcomeView.bordbackgroundY = 240;
            }
            welcomeView.logo2X -= 30;             //更改图片的坐标
            if(welcomeView.logo2X<150)
            {
                welcomeView.logo2X = 150;         //停止移动
            }
            if(welcomeView.logo2X == 150){        //当logo2到位后按钮才移动出现
                welcomeView.menuY -= 30;
                if(welcomeView.menuY<355)
                {
                    welcomeView.menuY = 355;
                }
            }
        try{
            Thread.sleep(100);                    //睡眠指定毫秒数
        }catch(Exception e){                      //捕获异常
            e.printStackTrace();                  //打印异常信息
        }
    }
  }
}
```

（6）在 src/com.example.xiangqi 目录下新建一个名为 CAIMenuView.java 的文件，在文件中定义了 CAIMenuView 类，该类功是在欢迎界面中单击"菜单"按钮时进入菜单界面，实现代码如下：

```
public class CAIMenuView extends SurfaceView implements SurfaceHolder.Callback {
    MainActivity activity;                        //总 Activity 的引用
    private TutorialThread thread;                //刷帧的线程
```

```java
    Bitmap startGame;                                               //开始游戏图片
    Bitmap openSound;                                               //打开声音图片
    Bitmap closeSound;                                              //关闭声音的图片
    Bitmap help;                                                    //帮助的图片
    Bitmap exit;                                                    //退出游戏的图片
    public CAIMenuView(Context context,MainActivity activity) {     //构造器
        super(context);
        this.activity = activity;                                   //得到 Activity 引用
        getHolder().addCallback(this);
        this.thread = new TutorialThread(getHolder(), this);        //启动刷帧线程
        initBitmap();                                               //初始化图片资源
    }
    public void initBitmap(){                                       //初始化图片资源图片
        startGame = BitmapFactory.decodeResource(getResources(), R.drawable.startgame);
                                                                    //开始游戏按钮
        openSound = BitmapFactory.decodeResource(getResources(), R.drawable.opensound);
                                                                    //开始声音按钮
        closeSound = BitmapFactory.decodeResource(getResources(), R.drawable.closesound);
                                                                    //关闭声音按钮
    help = BitmapFactory.decodeResource(getResources(), R.drawable.help);//帮助按钮
    exit = BitmapFactory.decodeResource(getResources(), R.drawable.exit);//退出按钮
    }
    public void onDraw(Canvas canvas){                              //自己写的绘制方法
        canvas.drawColor(Color.BLACK);                              //清屏
        canvas.drawBitmap(startGame, 50, 50, null);                 //绘制图片
        if(activity.isSound){                                       //放声音时,绘制关闭声音图片
            canvas.drawBitmap(closeSound, 50, 150, null);           //绘制关闭声音
        }else{                                                      //没有放声音时绘制打开声音图片
            canvas.drawBitmap(openSound, 50, 150, null);            //绘制开始声音
        }
        canvas.drawBitmap(help, 50, 250, null);                     //绘制帮助按钮
        canvas.drawBitmap(exit, 50, 350, null);                     //绘制退出按钮
    }
    public void surfaceChanged(SurfaceHolder holder, int format, int width, int height) {
    }
    public void surfaceCreated(SurfaceHolder holder) {              //创建时启动刷帧
        this.thread.setFlag(true);                                  //设置循环标志位
        this.thread.start();                                        //启动线程
    }
    public void surfaceDestroyed(SurfaceHolder holder) {            //摧毁时释放刷帧线程
        boolean retry = true;                                       //循环标志位
```

```java
            thread.setFlag(false);                              //设置循环标志位
            while (retry) {                                     //循环
                try {
                    thread.join();                              //等待线程结束
                    retry = false;                              //停止循环
                }catch (InterruptedException e){}               //不断地循环,直到刷帧线程结束
            }
        }
        public boolean onTouchEvent(MotionEvent event) {        //屏幕监听
            if(event.getAction() == MotionEvent.ACTION_DOWN){
                if(event.getX()>105 && event.getX()<220
                        &&event.getY()>60 && event.getY()<95){  //单击的是开始游戏按钮
                    activity.myHandler.sendEmptyMessage(2);
                }else if(event.getX()>105 && event.getX()<220
                        &&event.getY()>160 && event.getY()<195){ //单击的是声音按钮
                    activity.isSound = !activity.isSound;       //将声音开关取反
                    if(!activity.isSound){                      //当没有放声音时
                        if(activity.gamesound != null){         //检查当前是否已经有声音正在播放
                            if(activity.gamesound.isPlaying()){ //当游戏声音正在播放时,
                                activity.gamesound.pause();     //停止声音的播放
                            }
                        }
                    }else{                                      //当需要播放声音时
                        if(activity.gamesound != null){         //当 gamesound 不为空时
                            if(!activity.gamesound.isPlaying()){ //且当前声音没有在播放
                                activity.gamesound.start();     //则播放声音
                            }
                        }
                    }
                }else if(event.getX()>105 && event.getX()<220
                        &&event.getY()>260 && event.getY()<295){ //单击的是帮助按钮
                    //向 Activity 发送 Hander 消息通知切换 View
                    activity.myHandler.sendEmptyMessage(3);
                }else if(event.getX()>105 && event.getX()<220
                        &&event.getY()>360 && event.getY()<395){ //单击的是退出游戏按钮
                    System.exit(0);//直接退出游戏
                }
            }
            return super.onTouchEvent(event);
        }
        class TutorialThread extends Thread{                    //刷帧线程
```

```java
        private int span = 500;                                    //睡眠的毫秒数
        private SurfaceHolder surfaceHolder;                       //SurfaceHolder 的引用
        private CAIMenuView menuView;                              //MenuView 的引用
        private boolean flag = false;                              //循环标记位
        public TutorialThread(SurfaceHolder surfaceHolder, CAIMenuView menuView)
    {//构造器
            this.surfaceHolder = surfaceHolder;                    //得到 SurfaceHolder 引用
            this.menuView = menuView;                              //得到 menuView 引用
        }
        public void setFlag(boolean flag) {                        //设置循环标记位
            this.flag = flag;
        }
        public void run() {                                        //重写的 run 方法
            Canvas c;                                              //画布
            while (this.flag) {                                    //循环
                c = null;
                try {
                    // 锁定整个画布,在内存要求比较高的情况下,建议参数不要为 null
                    c = this.surfaceHolder.lockCanvas(null);
                    synchronized (this.surfaceHolder) {            //同步锁
                        menuView.onDraw(c);                        //调用绘制方法
                    }
                } finally {                                        //使用 finally 保证下面代码一定被执行
                    if (c != null) {
                        //更新屏幕显示内容
                        this.surfaceHolder.unlockCanvasAndPost(c);
                    }
                }
                try{
                    Thread.sleep(span);                            //睡眠指定毫秒数
                }catch(Exception e){                               //捕获异常
                    e.printStackTrace();                           //有异常时打印异常堆栈信息
                }
            }
        }
    }
```

(7) 在 src/com.example.xiangqi 目录下新建一个名为 Help.java 的文件,在文件中定义了类 Help,该类也为一个辅助界面类,功能为显示游戏系统的使用方法。实现代码如下:

```java
public class Help extends SurfaceView implements SurfaceHolder.Callback
{
```

```java
        MainActivity activity;                                              //Activity 的引用
        private TutorialThread thread;                                      //刷帧的线程
        Bitmap back;                                                        //返回按钮
        Bitmap helpBackground;                                              //背景图片
        public Help(Context context,MainActivity activity)
        {
            //构造器
            super(context);
            this.activity = activity;                                       //得到 Activity 引用
            getHolder().addCallback(this);
            this.thread = new TutorialThread(getHolder(), this);            //初始化重绘线程
            initBitmap();                                                   //初始化图片资源
        }
        public void initBitmap()
        {
            //初始化所用到的图片
back = BitmapFactory.decodeResource(getResources(), R.drawable.back);//返回按钮
            helpBackground = BitmapFactory.decodeResource(
                        getResources(),
                        R.drawable.helpbackground);                         //初始化背景图片
        }
        public void onDraw(Canvas canvas){                                  //自己写的绘制方法
            canvas.drawBitmap(helpBackground, 0, 90, new Paint());          //绘制背景图片
            canvas.drawBitmap(back, 200, 370, new Paint());                 //绘制按钮
        }
        public void surfaceChanged(SurfaceHolder holder, int format, int width, int height) {}
        public void surfaceCreated(SurfaceHolder holder)
        {
            //被创建时启动刷帧线程
            this.thread.setFlag(true);                                      //设置循环标志位
            this.thread.start();                                            //启动刷帧线程
        }
        public void surfaceDestroyed(SurfaceHolder holder)
        {
            //被摧毁时停止刷帧线程
            boolean retry = true;                                           //循环标志位
            thread.setFlag(false);                                          //设置循环标志位
            while (retry)
            {
                try {
                    thread.join();                                          //等待线程结束
```

```java
                    retry = false;                                      //停止循环
                }catch (InterruptedException e){}                       //不断地循环，直到刷帧线程结束
            }
        }
        public boolean onTouchEvent(MotionEvent event) {                //屏幕监听
            if(event.getAction() == MotionEvent.ACTION_DOWN){
                if(event.getX()>200 && event.getX()<200+back.getWidth()
                        && event.getY()>370 && event.getY()<370+back.getHeight())
                {
                    //单击了返回按钮
                    activity.myHandler.sendEmptyMessage(1);             //发送 Handler 消息
                }
            }
            return super.onTouchEvent(event);
        }
        class TutorialThread extends Thread{                            //刷帧线程
            private int span = 1000;                                    //睡眠的毫秒数
            private SurfaceHolder surfaceHolder;                        //SurfaceHolder 的引用
            private Help helpView;                                      //父类的引用
            private boolean flag = false;                               //循环标记位
            public TutorialThread(SurfaceHolder surfaceHolder, Help helpView)
            {
                //构造器
                this.surfaceHolder = surfaceHolder;                     //得到 SurfaceHolder 引用
                this.helpView = helpView;                               //得到 helpView 引用
            }
            public void setFlag(boolean flag)                           //设置循环标记位
            {
                this.flag = flag;
            }
            public void run()
            {
                //重写的 run 方法
                Canvas c;                                               //画布
                while (this.flag) {                                     //循环
                    c = null;
                    try {
                        c = this.surfaceHolder.lockCanvas(null);
                        synchronized (this.surfaceHolder) {              //同步
                            helpView.onDraw(c);                          //调用绘制方法
                        }
                    } finally {//用 finally 语句保证下面的代码一定会被执行
```

```
                    if (c != null) {                          //更新屏幕显示内容
                        this.surfaceHolder.unlockCanvasAndPost(c);
                    }
                }
                try{
                    Thread.sleep(span);                        //睡眠指定毫秒数
                }catch(Exception e){                           //捕获异常
                    e.printStackTrace();                       //打印异常堆栈信息
                }
            }
        }
    }
}
```

(8)在src/com.example.xiangqi目录下新建一个名为Games.java的文件,该文件定义的Games类为一个核心类,功能用于实现游戏界面的框架。其实现代码如下:

```
//定义继承于 SurfaceView 的类 Games,并定义了类中所需要的成员变量
public class Games extends SurfaceView implements SurfaceHolder.Callback
{
    private TutorialThread thread;          //刷帧的线程
    TimeThread timeThread ;
    MainActivity activity;                  //声明 Activity 的引用
    Bitmap qiPan;                           //棋盘
    Bitmap qizibackground;                  //棋子的背景图片
    Bitmap win;                             //胜利的图片
    Bitmap lost;                            //失败的图片
    Bitmap ok;                              //确定按钮
    Bitmap vs;                              //黑方红方 VS 的图片
    Bitmap right;                           //向右的指针
    Bitmap left;                            //向左的指针
    Bitmap current;                         // "当前" 文字
    Bitmap exit2;                           //退出按钮图片
    Bitmap sound2;                          //声音按钮图片
    Bitmap sound3;                          //当前是否播放了声音
    Bitmap time;                            //冒号
    Bitmap redtime;                         //红色冒号
    Bitmap background;                      //背景图片
    MediaPlayer go;                         //下棋声音
    Paint paint;                            //画笔
    boolean caiPan = true;                  //是否为玩家走棋
    boolean focus = false;                  //当前是否有选中的棋子
    int selectqizi = 0;                     //当然选中的棋子
```

```java
        int startI, startJ;                                  //记录当前棋子的开始位置
        int endI, endJ;                                      //记录当前棋子的目标位置
        Bitmap[] heiZi = new Bitmap[7];                      //黑子的图片数组
        Bitmap[] hongZi = new Bitmap[7];                     //红子的图片数组
        Bitmap[] number = new Bitmap[10];                    //数字的图片数组，用于显示时间
        Bitmap[] redNumber = new Bitmap[10];                 //红色数字的图片，用于显示时间
        GuiZe guiZe;                                         //规则类
        int status = 0;                                      //游戏状态。0 游戏中，1 胜利，2 失败
        int heiTime = 0;                                     //黑方总共思考时间
        int hongTime = 0;                                    //红方总共思考时间
        int[][] qizi = new int[][]{                          //棋盘
            {2,3,6,5,1,5,6,3,2},
            {0,0,0,0,0,0,0,0,0},
            {0,4,0,0,0,0,0,4,0},
            {7,0,7,0,7,0,7,0,7},
            {0,0,0,0,0,0,0,0,0},

            {0,0,0,0,0,0,0,0,0},
            {14,0,14,0,14,0,14,0,14},
            {0,11,0,0,0,0,0,11,0},
            {0,0,0,0,0,0,0,0,0},
            {9,10,13,12,8,12,13,10,9},
        };
        //分别定义系统中的构造器和对应的构造方法
        public Games(Context context,MainActivity activity) {   //构造器
            super(context);
            this.activity = activity;                        //得到 Activity 的引用
            getHolder().addCallback(this);
            go = MediaPlayer.create(this.getContext(), R.raw.go);//加载下棋的声音
            this.thread = new TutorialThread(getHolder(), this); //初始化刷帧线程
            this.timeThread = new TimeThread(this);          //初始化思考时间的线程
            init();                                          //初始化所需资源
            guiZe = new GuiZe();                             //初始化规则类
        }
        public void init(){                                  //初始化方法
            paint = new Paint();                             //初始化画笔
            qiPan = BitmapFactory.decodeResource(getResources(), R.drawable.qipan);
                                                             //棋盘图片
            qizibackgroud = BitmapFactory.decodeResource(getResources(), R.drawable.qizi);
                                                             //棋子的背景
            win = BitmapFactory.decodeResource(getResources(), R.drawable.win);    //胜利的图片
```

```java
lost = BitmapFactory.decodeResource(getResources(), R.drawable.lost);          //失败的图片
ok = BitmapFactory.decodeResource(getResources(), R.drawable.ok);              //确定按钮图片
vs = BitmapFactory.decodeResource(getResources(), R.drawable.vs);              //VS 字样的图片
right = BitmapFactory.decodeResource(getResources(), R.drawable.right);        //向右的指针
left = BitmapFactory.decodeResource(getResources(), R.drawable.left);          //向左的指针
current = BitmapFactory.decodeResource(getResources(), R.drawable.current);    //文字"当前"
exit2 = BitmapFactory.decodeResource(getResources(), R.drawable.exit2);        //退出按钮图片
sound2 = BitmapFactory.decodeResource(getResources(), R.drawable.sound2);      //声音按钮图片
time = BitmapFactory.decodeResource(getResources(), R.drawable.time);          //黑色冒号
redtime = BitmapFactory.decodeResource(getResources(), R.drawable.redtime);    //红色冒号
sound3 = BitmapFactory.decodeResource(getResources(), R.drawable.sound3);

heiZi[0] = BitmapFactory.decodeResource(getResources(), R.drawable.heishuai);  //黑帅
heiZi[1] = BitmapFactory.decodeResource(getResources(), R.drawable.heiju);     //黑车
heiZi[2] = BitmapFactory.decodeResource(getResources(), R.drawable.heima);     //黑马
heiZi[3] = BitmapFactory.decodeResource(getResources(), R.drawable.heipao);    //黑炮
heiZi[4] = BitmapFactory.decodeResource(getResources(), R.drawable.heishi);    //黑士
heiZi[5] = BitmapFactory.decodeResource(getResources(), R.drawable.heixiang);  //黑象
heiZi[6] = BitmapFactory.decodeResource(getResources(), R.drawable.heibing);   //黑兵

hongZi[0] = BitmapFactory.decodeResource(getResources(), R.drawable.hongjiang);  //红将
hongZi[1] = BitmapFactory.decodeResource(getResources(), R.drawable.hongju);     //红车
hongZi[2] = BitmapFactory.decodeResource(getResources(), R.drawable.hongma);     //红马
hongZi[3] = BitmapFactory.decodeResource(getResources(), R.drawable.hongpao);    //红炮
hongZi[4] = BitmapFactory.decodeResource(getResources(), R.drawable.hongshi);    //红仕
hongZi[5] = BitmapFactory.decodeResource(getResources(), R.drawable.hongxiang);  //红相
hongZi[6] = BitmapFactory.decodeResource(getResources(), R.drawable.hongzu);     //红卒

number[0] = BitmapFactory.decodeResource(getResources(), R.drawable.number0);//黑色数字 0
number[1] = BitmapFactory.decodeResource(getResources(), R.drawable.number1);//黑色数字 1
number[2] = BitmapFactory.decodeResource(getResources(), R.drawable.number2);//黑色数字 2
number[3] = BitmapFactory.decodeResource(getResources(), R.drawable.number3);//黑色数字 3
number[4] = BitmapFactory.decodeResource(getResources(), R.drawable.number4);//黑色数字 4
number[5] = BitmapFactory.decodeResource(getResources(), R.drawable.number5);//黑色数字 5
number[6] = BitmapFactory.decodeResource(getResources(), R.drawable.number6);//黑色数字 6
number[7] = BitmapFactory.decodeResource(getResources(), R.drawable.number7);//黑色数字 7
number[8] = BitmapFactory.decodeResource(getResources(), R.drawable.number8);//黑色数字 8
number[9] = BitmapFactory.decodeResource(getResources(), R.drawable.number9);//黑色数字 9

redNumber[0] = BitmapFactory.decodeResource(getResources(), R.drawable.rednumber0);
                                                                               //红色数字 0
```

```java
        redNumber[1] = BitmapFactory.decodeResource(getResources(), R.drawable.rednumber1);
                                                                    //红色数字 1
        redNumber[2] = BitmapFactory.decodeResource(getResources(), R.drawable.rednumber2);
                                                                    //红色数字 2
        redNumber[3] = BitmapFactory.decodeResource(getResources(), R.drawable.rednumber3);
                                                                    //红色数字 3
        redNumber[4] = BitmapFactory.decodeResource(getResources(), R.drawable.rednumber4);
                                                                    //红色数字 4
        redNumber[5] = BitmapFactory.decodeResource(getResources(), R.drawable.rednumber5);
                                                                    //红色数字 5
        redNumber[6] = BitmapFactory.decodeResource(getResources(), R.drawable.rednumber6);
                                                                    //红色数字 6
        redNumber[7] = BitmapFactory.decodeResource(getResources(), R.drawable.rednumber7);
                                                                    //红色数字 7
        redNumber[8] = BitmapFactory.decodeResource(getResources(), R.drawable.rednumber8);
                                                                    //红色数字 8
        redNumber[9] = BitmapFactory.decodeResource(getResources(), R.drawable.rednumber9);
                                                                    //红色数字 9

        background = BitmapFactory.decodeResource(getResources(), R.drawable.bacnground);
    }
    /**
     * 该方法是自己定义的,并非重写的
     * 该方法是死的,只根据数据绘制屏幕
     */
    public void onDraw(Canvas canvas){                  //自己写的绘制方法
        canvas.drawColor(Color.WHITE);
        canvas.drawBitmap(background, 0,0, null);       //清背景
        canvas.drawBitmap(qiPan, 10, 10, null);         //绘制棋盘
        for(int i=0; i<qizi.length; i++){
            for(int j=0; j<qizi[i].length; j++){        //绘制棋子
                if(qizi[i][j] != 0){
                    canvas.drawBitmap(qizibackground, 9+j*34, 10+i*35, null);
                                                        //绘制棋子的背景
                    if(qizi[i][j] == 1){                //为黑帅时
                        canvas.drawBitmap(heiZi[0], 12+j*34, 13+i*35, paint);
                    }
                    else if(qizi[i][j] == 2){           //为黑车时
                        canvas.drawBitmap(heiZi[1], 12+j*34, 13+i*35, paint);
                    }
```

```
        else if(qizi[i][j] == 3){                    //为黑马时
            canvas.drawBitmap(heiZi[2], 12+j*34, 13+i*35, paint);
        }
        else if(qizi[i][j] == 4){                    //为黑炮时
            canvas.drawBitmap(heiZi[3], 12+j*34, 13+i*35, paint);
        }
        else if(qizi[i][j] == 5){                    //为黑士时
            canvas.drawBitmap(heiZi[4], 12+j*34, 13+i*35, paint);
        }
        else if(qizi[i][j] == 6){                    //为黑象时
            canvas.drawBitmap(heiZi[5], 12+j*34, 13+i*35, paint);
        }
        else if(qizi[i][j] == 7){                    //为黑兵时
            canvas.drawBitmap(heiZi[6], 12+j*34, 13+i*35, paint);
        }
        else if(qizi[i][j] == 8){                    //为红将时
            canvas.drawBitmap(hongZi[0], 12+j*34, 13+i*35, paint);
        }
        else if(qizi[i][j] == 9){                    //为红车时
            canvas.drawBitmap(hongZi[1], 12+j*34, 13+i*35, paint);
        }
        else if(qizi[i][j] == 10){                   //为红马时
            canvas.drawBitmap(hongZi[2], 12+j*34, 13+i*35, paint);
        }
        else if(qizi[i][j] == 11){                   //为红炮时
            canvas.drawBitmap(hongZi[3], 12+j*34, 13+i*35, paint);
        }
        else if(qizi[i][j] == 12){                   //为红仕时
            canvas.drawBitmap(hongZi[4], 12+j*34, 13+i*35, paint);
        }
        else if(qizi[i][j] == 13){                   //为红相时
            canvas.drawBitmap(hongZi[5], 12+j*34, 13+i*35, paint);
        }
        else if(qizi[i][j] == 14){                   //为红卒时
            canvas.drawBitmap(hongZi[6], 12+j*34, 13+i*35, paint);
        }
    }
  }
}
canvas.drawBitmap(vs, 10, 360, paint);                //绘制VS背景图
//绘制黑方的时间
```

```java
            canvas.drawBitmap(time, 81, 411, paint);                //绘制冒号
            int temp = this.heiTime/60;                             //换算时间
            String timeStr = temp+"";                               //转换成字符串
            if(timeStr.length()<2){                                 //当不足两位时前面填 0
                timeStr = "0" + timeStr;
            }
            for(int i=0;i<2;i++){                                   //循环绘制时间
                int tempScore=timeStr.charAt(i)-'0';
                canvas.drawBitmap(number[tempScore], 65+i*7, 412, paint);
            }
            //画分钟
            temp = this.heiTime%60;
            timeStr = temp+"";                                      //转换成字符串
            if(timeStr.length()<2){
                timeStr = "0" + timeStr;                            //当长度小于 2 时在前面添加一个 0
            }
            for(int i=0;i<2;i++){                                   //循环
                int tempScore=timeStr.charAt(i)-'0';
                canvas.drawBitmap(number[tempScore], 85+i*7, 412, paint);
                                                                    //绘制
            }
            //开始绘制红方时间
            canvas.drawBitmap(this.redtime, 262, 410, paint);       //红方的冒号
            int temp2 = this.hongTime/60;                           //换算时间
            String timeStr2 = temp2+"";                             //转换成字符串
            if(timeStr2.length()<2){                                //当不足两位时前面填 0
                timeStr2 = "0" + timeStr2;
            }
            for(int i=0;i<2;i++){                                   //循环绘制时间
                int tempScore=timeStr2.charAt(i)-'0';
                canvas.drawBitmap(redNumber[tempScore], 247+i*7, 411, paint);//绘制
            }
            //画分钟
            temp2 = this.hongTime%60;                               //求出当前的秒数
            timeStr2 = temp2+"";                                    //转换成字符串
            if(timeStr2.length()<2){
                timeStr2 = "0" + timeStr2;                          //不足两位时前面用 0 补
            }
            for(int i=0;i<2;i++){                                   //循环绘制
                int tempScore=timeStr2.charAt(i)-'0';
                canvas.drawBitmap(redNumber[tempScore], 267+i*7, 411, paint);
```

```java
                                                          //绘制时间数字
        }
        if(caiPan == true){                               //当该玩家走棋时，即红方走棋
            canvas.drawBitmap(right, 155, 420, paint);    //绘制向右的指针
        }
        else{                                             //黑方走棋，即计算机走棋时
            canvas.drawBitmap(left, 120, 420, paint);     //绘制向左的指针
        }
        canvas.drawBitmap(current, 138, 445, paint);      //绘制当前文字
        canvas.drawBitmap(sound2, 10, 440, paint);        //绘制声音
        if(activity.isSound){                             //如果正在播放声音
            canvas.drawBitmap(sound3, 80, 452, paint);    //绘制
        }
        canvas.drawBitmap(exit2, 250, 440, paint);        //绘制退出按钮
        if(status == 1){                                  //当胜利时
            canvas.drawBitmap(win, 85, 150, paint);       //绘制胜利图片
            canvas.drawBitmap(ok, 113, 240, paint);
        }
        if(status == 2){                                  //失败后
            canvas.drawBitmap(lost, 85, 150, paint);      //绘制失败界面
            canvas.drawBitmap(ok, 113, 236, paint);
        }
    }
    /**
     * 该方法是游戏主要逻辑接口
     * 接受玩家输入
     * 根据单击的位置和当前的游戏状态做出相应的处理
     * 而当需要切换 View 时，通过给 Activity 发送 Handler 消息来处理
     * 注意的是只取屏幕被按下的事件
     */
    @Override
    public boolean onTouchEvent(MotionEvent event) {                  //重写的屏幕监听
        if(event.getAction() == MotionEvent.ACTION_DOWN){             //只取鼠标按下的事件
            if(event.getX()>10&&event.getX()<10+sound2.getWidth()
                    && event.getY()>440 && event.getY()<440+sound2.getHeight()){
                                                                      //按下了声音按钮
                activity.isSound = !activity.isSound;                 //声音取反
                if(activity.isSound){                                 //当需要放声音时
                    if(activity.gamesound != null){                   //gamesound 不为空时
                        if(!activity.gamesound.isPlaying()){          //当前没有音乐时
                            activity.gamesound.start();               //播放音乐
```

```java
                    }
                }
            }
            else{
                if(activity.gamesound != null){              //gamesound 不为空时
                    if(activity.gamesound.isPlaying()){      //当前有音乐时
                        activity.gamesound.pause();          //停止音乐
                    }
                }
            }
        }//end 按下了声音按钮
        if(event.getX()>250&&event.getX()<250+exit2.getWidth()
                && event.getY()>440 && event.getY()<440+exit2.getHeight()){
                                                             //按下了退出按钮
            activity.myHandler.sendEmptyMessage(1);
                                                             //发送消息，切换到 MenuView
        }
        if(status == 1){                                     //胜利后
            if(event.getX()>135&&event.getX()<190
                    && event.getY()>249 && event.getY()<269){
                                                             //单击了确定按钮
                activity.myHandler.sendEmptyMessage(1);//发送消息，切换到 MenuView
            }
        }
        else if(status == 2){                                //失败后
            if(event.getX()>135&&event.getX()<190
                    && event.getY()>245 && event.getY()<265){
                                                             //单击了确定按钮
                activity.myHandler.sendEmptyMessage(1);//发送消息，切换到 MenuView
            }
        }
        /**
         * 游戏过程中的逻辑处理
         * 当单击棋盘时，先判断当前是否为玩家走棋，
         * 然后再判断当然玩家是否已经有选中的棋子，如果没有则选中
         * 如果之前有选中的棋子，再判断单击的位置是空地、对方棋子还是自己的棋子
         * 是空地判断是否可走
         * 是对方棋子同样判断是否可以走，能走自然吃子
         * 是自己的棋子则选中该棋子
         */
        else if(status == 0){                                //游戏中时
```

```java
                    if(event.getX()>10&&event.getX()<310
                        && event.getY()>10 && event.getY()<360){
                                                //单击的位置在棋盘内时
                        if(caiPan == true){                 //如果是该玩家走棋
                            int i = -1, j = -1;
                            int[] pos = getPos(event);      //根据坐标换算成所在的行和列
                            i = pos[0];
                            j = pos[1];
                            if(focus == false){             //之前没有选中的棋子
                                if(qizi[i][j] != 0){        //单击的位置有棋子
                                    if(qizi[i][j] > 7){
                                                    //单击的是自己的棋子,
                                                    //即下面的黑色棋子
                                        selectqizi = qizi[i][j];
                                                    //将该棋子设为选中的棋子
                                        focus = true;//标记当前有选中的棋子
                                        startI = i;
                                        startJ = j;
                                    }
                                }
                            }
                            else{                           //之前选中过棋子
                                if(qizi[i][j] != 0){        //单击的位置有棋子
                                    if(qizi[i][j] > 7){ //如果是自己的棋子
                                        selectqizi = qizi[i][j];
                                                    //将该棋子设为选中的棋子
                                        startI = i;
                                        startJ = j;
                                    }
                                    else{               //如果是对方的棋子
                                        endI = i;
                                        endJ = j;       //保存该点
                                        boolean canMove = guiZe.canMove(qizi, startI,
startJ, endI, endJ);
                                        if(canMove){
                                                    //如果可以移动过去
                                            caiPan = false;
                                                    //不让玩家走了
                                            if(qizi[endI][endJ] == 1 || qizi[endI][endJ]
== 8){
                                                    //如果是"帅"或"将"
                                                this.success();
```

```java
                                            //胜利了
                                        }
                                        else{
                                            if(activity.isSound){
                                                go.start();
                                            //播放下棋声音
                                            }
                                            qizi[endI][endJ] = qizi[startI][startJ];
                                        //移动棋子
                                            qizi[startI][startJ] = 0;
                                        //将原来处设空
                                            startI = -1;
                                            startJ = -1;
                                            endI = -1;
                                            endJ = -1;
                                        //还原保存点
                                            focus = false;
                                        //标记当前没有选中棋子

                                            Move cm = guiZe.searchAGoodMove(qizi);
                                        //根据当前局势查询一个最好的走法
                                            if(activity.isSound){
                                                go.start();
                                        //播放下棋声音
                                            }
                                            qizi[cm.toX][cm.toY] = qizi[cm.fromX][cm.fromY];
                                        //移动棋子
                                            qizi[cm.fromX][cm.fromY] = 0;
                                            caiPan = true;
                                        //恢复玩家响应
                                        }
                                    }
                                }                    //end 单击的位置有棋子
                                else{                //如果单击的位置没有棋子
                                    endI = i;
                                    endJ = j;
                                    boolean canMove = guiZe.canMove(qizi, startI, startJ, endI, endJ);
                                                     //查看是否可走
                                    if(canMove){     //如果可以移动
```

```java
                                            caiPan = false;      //不让玩家走了
                                            if(activity.isSound){
                                                    go.start();  //播放下棋声音
                                            }
                                            qizi[endI][endJ] = qizi[startI][startJ];
                                                                 //移动棋子
                                            qizi[startI][startJ] = 0; //将原来处置空
                                            startI = -1;
                                            startJ = -1;
                                            endI = -1;
                                            endJ = -1;           //还原保存点
                                            focus = false;       //标志位设 false

                                            Move cm = guiZe.searchAGoodMove(qizi);
                                                                 //得到一步走法
                                            if(qizi[cm.toX][cm.toY] == 8){
                                                                 //计算机吃了您的将
                                                    status = 2;  //切换游戏状态为失败
                                            }
                                            if(activity.isSound){//需要播放声音时
                                                    go.start();  //播放下棋声音
                                            }
                                            qizi[cm.toX][cm.toY] = qizi[cm.fromX][cm.fromY];
                                                                 //移动棋子
                                            qizi[cm.fromX][cm.fromY] = 0;
                                            caiPan = true;       //恢复玩家响应
                                    }
                                }
                            }                                    //end 之前选中过棋子
                        }                                        //end 单击的位置在棋盘内时
                    }                                            //end 游戏中时
                }
                return super.onTouchEvent(event);
            }

            public int[] getPos(MotionEvent e){                  //将坐标换算成数组的维数
                int[] pos = new int[2];
                double x = e.getX();                             //得到单击位置的 X 坐标
                double y = e.getY();                             //得到单击位置的 Y 坐标
                if(x>10 && y>10 && x<10+qiPan.getWidth() && y<10+qiPan.getHeight()){
```

```java
                                                        //单击的是棋盘时
            pos[0] = Math.round((float)((y-21)/36));    //取得所在的行
            pos[1] = Math.round((float)((x-21)/35));    //取得所在的列
        }
        else{                                           //单击的位置不是棋盘时
            pos[0] = -1;                                //将位置设为不可用
            pos[1] = -1;
        }
        return pos;                                     //将坐标数组返回
    }

    public void success(){                              //胜利了
        status = 1;                                     //切换到胜利状态
    }

    public void surfaceChanged(SurfaceHolder holder, int format, int width,
            int height) {
    }

    public void surfaceCreated(SurfaceHolder holder) {  //重写的
        this.thread.setFlag(true);
        this.thread.start();                            //启动刷帧线程
        timeThread.setFlag(true);
        timeThread.start();                             //启动思考时间的线程
    }

    public void surfaceDestroyed(SurfaceHolder holder) {//view 被释放时调用的
        boolean retry = true;
        thread.setFlag(false);                          //停止刷帧线程
        timeThread.setFlag(false);                      //停止思考时间线程
        while (retry) {
            try {
                thread.join();
                timeThread.join();                      //等待线程结束
                retry = false;                          //设置循环标志位为 false
            }
            catch (InterruptedException e) {            //不断地循环,直到等待的线程结束
            }
        }
    }
    class TutorialThread extends Thread{                //刷帧线程
```

```java
            private int span = 300;                                //睡眠的毫秒数
            private SurfaceHolder surfaceHolder;                   //SurfaceHolder 的引用
            private Games gameView;                                //gameView 的引用
            private boolean flag = false;                          //循环标志位
            public TutorialThread(SurfaceHolder surfaceHolder, Games gameView) {//构造器
                this.surfaceHolder = surfaceHolder;                //得到 SurfaceHolder 引用
                this.gameView = gameView;                          //得到 GameView 的引用
            }
            public void setFlag(boolean flag) {                    //设置循环标记
                this.flag = flag;
            }
            public void run() {                                    //重写的方法
                Canvas c;                                          //画布
                while (this.flag) {                                //循环绘制
                    c = null;
                    try {
                        c = this.surfaceHolder.lockCanvas(null);
                        synchronized (this.surfaceHolder) {
                            gameView.onDraw(c);                    //调用绘制方法
                        }
                    } finally {                                    //用 finally 保证下面代码一定被执行
                        if (c != null) {
                            //更新屏幕显示内容
                            this.surfaceHolder.unlockCanvasAndPost(c);
                        }
                    }
                    try{
                        Thread.sleep(span);                        //睡眠 span（时间）毫秒
                    }catch(Exception e){                           //抛出异常信息
                        e.printStackTrace();                       //打印异常堆栈信息
                    }
                }
            }
        }
```

（9）在 src/com.example.xiangqi 目录下新建一个名为 Move.java 的文件，在文件中定义了象棋的走法类 Move，在走法中包含了什么棋子、起始点的位置、目标点的位置以及传感器时所用到的 score。实现代码为：

```java
        package com.example.xiangqi;
        public class Move
        {
```

```java
        int ChessID;                                    //表明是什么棋子
        int fromX;                                      //起始的坐标
        int fromY;
        int toX;                                        //目的地的坐标
        int toY;
        int score;                                      //值,估值时会用到
        public Move(int ChessID, int fromX,int fromY,int toX,int toY,int score){//构造器
            this.ChessID = ChessID;                     //棋子的类型
            this.fromX = fromX;                         //棋子的起始坐标
            this.fromY = fromY;
            this.toX = toX;                             //棋子的目标点 X 坐标
            this.toY = toY;                             //棋子的目标点 Y 坐标
            this.score = score;
        }
    }
```

（10）在 src/com.example.xiangqi 目录下新建一个名为 TimeThread.java 的文件,在此文件中定义了类 TimeThread,此类能根据是哪一方该先走,并增加这一方的思考时间。实现代码如下:

```java
            package com.example.xiangqi;
            public class TimeThread extends Thread
            {
                private boolean flag = true;            //循环标志
                Games gameView;
                public TimeThread(Games gameView)
                {
                    //构造器
                    this.gameView = gameView;           //得到 GameView 引用
                }
                public void setFlag(boolean flag)
                {
                    //设置循环标记位
                    this.flag = flag;
                }
                @Override
                public void run()
                {
                    //重写的 run 方法
                    while(flag){                        //循环
                        if(gameView.caiPan == false)
                        {
                            //当前为黑方走棋、思考
                            gameView.heiTime++;         //黑方时间自加
```

```
                }
                else if(gameView.caiPan == true)
                {
                    //当前为红方走棋、思考
                    gameView.hongTime++;              //红方时间自加
                }
                try{
                    Thread.sleep(1000);               //睡眠 1s
                }
                catch(Exception e){                   //捕获异常
                    e.printStackTrace();              //打印异常信息
                }
            }
        }
    }
```

（11）在 src/com.example.xiangqi 目录下新建一个名为 GuiZe.java 的文件，在文件中定义了象棋规则类 GuiZe。其实现代码如下：

```
public class GuiZe {
    boolean isRedGo = false;                         //是不是红方走棋
    public boolean canMove(int[][] qizi, int fromY, int fromX, int toY, int toX)
    {
        int i = 0;
        int j = 0;
        int moveChessID;                             //起始位置是什么棋子
        int targetID;                                //目的地是什么棋子或空地
        if(toX<0){                                   //当左边出界时
            return false;
        }
        if(toX>8){                                   //当右边出界时
            return false;
        }
        if(toY<0){                                   //当上边出界时
            return false;
        }
        if(toY>9){                                   //当下边出界时
            return false;
        }
        if(fromX==toX && fromY==toY){                //目的地与出发点相同
            return false;
        }
        moveChessID = qizi[fromY][fromX];            //得到起始棋子
```

```java
            targetID = qizi[toY][toX];                      //得到终点棋子
            if(isSameSide(moveChessID,targetID)){           //如果是同一阵营的
                return false;
            }
            switch(moveChessID)
            {
                case 1:                                     //黑帅
                    if(toY>2||toX<3||toX>5){                //出了九宫格
                        return false;
                    }
                    if((Math.abs(fromY-toY)+Math.abs(toX-fromX))>1)
                    {                                       //只能走一步
                        return false;
                    }
                    break;
                case 5:                                     //黑士
                    if(toY>2||toX<3||toX>5){                //出了九宫格
                        return false;
                    }
                    if(Math.abs(fromY-toY) != 1 || Math.abs(toX-fromX) != 1){//走斜线
                        return false;
                    }
                    break;
                case 6:                                     //黑象
                    if(toY>4){                              //不能过河
                        return false;
                    }
                    if(Math.abs(fromX-toX) != 2 || Math.abs(fromY-toY) != 2){    //相走"田"字
                        return false;
                    }
                    if(qizi[(fromY+toY)/2][(fromX+toX)/2] != 0){
                        return false;                       //相眼处有棋子
                    }
                    break;
                case 7:                                     //黑兵
                    if(toY < fromY){                        //不能回头
                        return false;
                    }
                    if(fromY<5 && fromY == toY){            //过河前只能直走
                        return false;
                    }
```

```
            if(toY - fromY + Math.abs(toX-fromX) > 1){    //只能走一步,并且是直线
                return false;
            }
            break;
        case 8:                                            //红将
            if(toY<7||toX>5||toX<3){                      //出了九宫格
                return false;
            }
            if((Math.abs(fromY-toY)+Math.abs(toX-fromX))>1){//只能走一步
                return false;
            }
            break;
        case 2:                                            //黑车
        case 9:                                            //红车
            if(fromY != toY && fromX != toX){             //只能走直线
                return false;
            }
            if(fromY == toY){                              //走横线
                if(fromX < toX){                           //向右走
                    for(i = fromX + 1; i < toX; i++){      //循环
                        if(qizi[fromY][i] != 0){
                            return false;                  //返回 false
                        }
                    }
                }
                else{                                      //向左走
                    for(i = toX + 1; i < fromX; i++){      //循环
                        if(qizi[fromY][i] != 0){
                            return false;                  //返回 false
                        }
                    }
                }
            }
            else{                                          //走的是竖线
                if(fromY < toY){                           //向右走
                    for(j = fromY + 1; j < toY; j++){
                        if(qizi[j][fromX] != 0)
                            return false;                  //返回 false
                    }
                }
                else{                                      //向左走
```

```java
                    for(j= toY + 1; j < fromY; j++){
                        if(qizi[j][fromX] != 0)
                            return false;                    //返回 false
                    }
                }
            }
            break;
        case 10:                                              //红马
        case 3:                                               //黑马
            if(!((Math.abs(toX-fromX)==1 && Math.abs(toY-fromY)==2)
                    || (Math.abs(toX-fromX)==2 && Math.abs(toY-fromY)==1))){
                return false;                                 //马走的不是日字时
            }
            if(toX-fromX==2){                                 //向右走
                i=fromX+1;                                    //移动
                j=fromY;
            }
            else if(fromX-toX==2){                            //向左走
                i=fromX-1;                                    //移动
                j=fromY;
            }
            else if(toY-fromY==2){                            //向下走
                i=fromX;                                      //移动
                j=fromY+1;
            }
            else if(fromY-toY==2){                            //向上走
                i=fromX;                                      //移动
                j=fromY-1;
            }
            if(qizi[j][i] != 0)
                return false;                                 //绊马腿
            break;
        case 11:                                              //红炮
        case 4:                                               //黑炮
            if(fromY!=toY && fromX!=toX){                     //炮走直线
                return false;                                 //返回 false
            }
            if(qizi[toY][toX] == 0){                          //不吃子时
                if(fromY == toY){                             //横线
                    if(fromX < toX){                          //向右走
                        for(i = fromX + 1; i < toX; i++){
```

```java
                        if(qizi[fromY][i] != 0){
                            return false;                //返回false
                        }
                    }
                }
                else{                                    //向左走
                    for(i = toX + 1; i < fromX; i++){
                        if(qizi[fromY][i]!=0){
                            return false;                //返回false
                        }
                    }
                }
            }
            else{                                        //竖线
                if(fromY < toY){                         //向下走
                    for(j = fromY + 1; j < toY; j++){
                        if(qizi[j][fromX] != 0){
                            return false;                //返回false
                        }
                    }
                }
                else{                                    //向上走
                    for(j = toY + 1; j < fromY; j++){
                        if(qizi[j][fromX] != 0){
                            return false;                //返回false
                        }
                    }
                }
            }
        }
        else{                                            //吃子时
            int count=0;
            if(fromY == toY){                            //走的是横线
                if(fromX < toX){                         //向右走
                    for(i=fromX+1;i<toX;i++){
                        if(qizi[fromY][i]!=0){
                            count++;
                        }
                    }
                    if(count != 1){
                        return false;                    //返回false
```

```java
                    }
                }
                else{                                    //向左走
                    for(i=toX+1;i<fromX;i++){
                        if(qizi[fromY][i] != 0){
                            count++;
                        }
                    }
                    if(count!=1){
                        return false;                    //返回 false
                    }
                }
            }
            else{                                        //走的是竖线
                if(fromY<toY){                           //向下走
                    for(j=fromY+1;j<toY;j++){
                        if(qizi[j][fromX]!=0){
                            count++;                     //返回 false
                        }
                    }
                    if(count!=1){
                        return false;                    //返回 false
                    }
                }
                else{                                    //向上走
                    for(j=toY+1;j<fromY;j++){
                        if(qizi[j][fromX] != 0){
                            count++;                     //返回 false
                        }
                    }
                    if(count!=1){
                        return false;                    //返回 false
                    }
                }
            }
        }
        break;
    case 12:                                             //红仕
        if(toY<7||toX>5||toX<3){                         //出了九宫格
            return false;
        }
```

```java
            if(Math.abs(fromY-toY) != 1 || Math.abs(toX-fromX) != 1){//走斜线
                return false;
            }
            break;
        case 13:                                                //红相
            if(toY<5){                                          //不能过河
                return false;                                   //返回 false
            }
            if(Math.abs(fromX-toX) != 2 || Math.abs(fromY-toY) != 2){//相走"田"字
                return false;                                   //返回 false
            }
            if(qizi[(fromY+toY)/2][(fromX+toX)/2] != 0){
                return false;                                   //相眼处有棋子
            }
            break;
        case 14:                                                //红卒
            if(toY > fromY){                                    //不能回头
                return false;
            }
            if(fromY > 4 && fromY == toY){
                return false;                                   //不让走
            }
            if(fromY - toY + Math.abs(toX - fromX) > 1){        //只能走一步,并且是直线
                return false;                                   //返回 false 不让走
            }
            break;
        default:
            return false;
    }
    return true;
}
/**
 *
 * 计算机通过该方法得到当前棋局最好的走法
 */
public Move searchAGoodMove(int[][] qizi){                      //查询一个好的走法
    List<Move> ret = allPossibleMoves(qizi);                    //产生所有走法
    try {
        Thread.sleep(4000);                                     //睡眠 4s,以便调试
    } catch (InterruptedException e) {                          //捕获异常
        e.printStackTrace();                                    //打印堆栈信息
```

```java
            }
            return ret.get((int)(Math.random()*ret.size()));
        }
        public List<Move> allPossibleMoves(int qizi[][]){          //产生所有可能的走法
            List<Move> ret = new ArrayList<Move>();                //用来装所有可能的走法
            for (int x = 0; x < 10; x++){
                for (int y = 0; y < 9; y++){                       //循环所有的棋牌位置
                    int chessman = qizi[x][y];
                    if (chessman != 0){                            //当此位置不为空时,即有棋子时
                        if(chessman > 7){                          //是红方,即是玩家棋子时跳过
                            continue;
                        }
                        switch (chessman){
                            case 1:                                //黑帅
                                if(canMove(qizi, x, y, x, y+1)){//向下走一格
                                    ret.add(new Move(chessman, x, y, x, y+1, 0));
                                }
                                if(canMove(qizi, x, y, x, y-1)){//向上走一格
                                    ret.add(new Move(chessman, x, y, x, y-1, 0));
                                }
                                if(canMove(qizi, x, y, x+1, y)){   //向左走一格
                                    ret.add(new Move(chessman, x, y, x+1, y, 0));
                                }
                                if(canMove(qizi, x, y, x-1, y)){   //向右走一格
                                    ret.add(new Move(chessman, x, y, x-1, y, 0));
                                }
                                break;
                            case 5:                                //黑士
                            case 12:                               //红仕
                                if(canMove(qizi, x, y, x-1, y+1)){ //左下走
                                    ret.add(new Move(chessman, x, y, x-1, y+1, 1));
                                }
                                if(canMove(qizi, x, y, x-1, y-1)){ //左上走
                                    ret.add(new Move(chessman, x, y, x-1, y-1, 1));
                                }
                                if(canMove(qizi, x, y, x+1, y+1)){ //右下走
                                    ret.add(new Move(chessman, x, y, x+1, y+1, 1));
                                }
                                if(canMove(qizi, x, y, x+1, y-1)){ //右上走
                                    ret.add(new Move(chessman, x, y, x+1, y-1, 1));
                                }
```

```java
                    break;
            case 6:                                          //黑象
            case 13:                                         //红相
                if(canMove(qizi, x, y, x-2, y+2)){           //左上走
                    ret.add(new Move(chessman, x, y, x-2, y+2, 1));
                }
                if(canMove(qizi, x, y, x-2, y-2)){           //左下走
                    ret.add(new Move(chessman, x, y, x-2, y-2, 1));
                }
                if(canMove(qizi, x, y, x+2, y+2)){           //右下走
                    ret.add(new Move(chessman, x, y, x+2, y+2, 1));
                }
                if(canMove(qizi, x, y, x+2, y-2)){           //右上走
                    ret.add(new Move(chessman, x, y, x+2, y-2, 1));
                }
                break;
            case 7:                                          //黑兵
                if(canMove(qizi, x, y, x, y+1)){             //直走
                    ret.add(new Move(chessman, x, y, x, y+1, 2));
                }
                if(y >= 5){                                  //过河了
                    if (canMove(qizi, x, y, x - 1, y)) {     //过河后向左走
                        ret.add(new Move(chessman, x, y, x - 1, y, 2));
                    }
                    if (canMove(qizi, x, y, x + 1, y)) {     //过河走向右走
                        ret.add(new Move(chessman, x, y, x + 1, y, 2));
                    }
                }
                break;
            case 14:                                         //红兵
                if(canMove(qizi, x, y, x, y-1)){             //向前走
                    ret.add(new Move(chessman, x, y, x, y-1, 2));
                }
                if(y <=4 ){                                  //过河了
                    if (canMove(qizi, x, y, x - 1, y)) {     //过河后向左走
                        ret.add(new Move(chessman, x, y, x - 1, y, 2));
                    }
                    if (canMove(qizi, x, y, x + 1, y)) {     //过河走向右走
                        ret.add(new Move(chessman, x, y, x + 1, y, 2));
                    }
                }
```

```
                    break;
            case 8:                                            //红将
                if(canMove(qizi, x, y, x, y+1)){               //向下走一格
                    ret.add(new Move(chessman, x, y, x, y+1, 0));
                }
                if(canMove(qizi, x, y, x, y-1)){               //向上走一格
                    ret.add(new Move(chessman, x, y, x, y-1, 0));
                }
                if(canMove(qizi, x, y, x+1, y)){               //向右走一格
                    ret.add(new Move(chessman, x, y, x+1, y, 0));
                }
                if(canMove(qizi, x, y, x-1, y)){               //向左走一格
                    ret.add(new Move(chessman, x, y, x-1, y, 0));
                }
                break;
            case 2:                                            //黑车
            case 9:                                            //红车
                for(int i=y+1; i<10; i++){                     //向下走
                    if(canMove(qizi, x, y, x, i)){             //可以走时
                        ret.add(new Move(chessman, x, y, x, i, 0));
                    }else{                                     //不可以走时直接 break
                        break;
                    }
                }
                for(int i=y-1; i>-1; i++){                     //向上走
                    if(canMove(qizi, x, y, x, i)){             //可以走时
                        ret.add(new Move(chessman, x, y, x, i, 0));
                    }else{                                     //不可以走时
                        break;
                    }
                }
                for(int j=x-1; j>-1; j++){                     //向左走
                    if(canMove(qizi, x, y, j, y)){             //可以走时
                        ret.add(new Move(chessman, x, y, j, y, 0));
                    }else{                                     //不可以走时
                        break;
                    }
                }
                for(int j=x+1; j<9; j++){                      //向右走
                    if(canMove(qizi, x, y, j, y)){             //可以走时
                        ret.add(new Move(chessman, x, y, j, y, 0));
                    }else{                                     //不可以走时
```

```java
                    break;
                }
            }
            break;
        case 10:                                            //红马
        case 3:                                             //黑马
            if(canMove(qizi, x, y, x-1, y-2)){              //向上左走"日"字
                ret.add(new Move(chessman, x, y, x-1, y-2, 0));
            }
            if(canMove(qizi, x, y, x-1, y+2)){              //向下左走"日"字
                ret.add(new Move(chessman, x, y, x-1, y+2, 0));
            }
            if(canMove(qizi, x, y, x+1, y-2)){              //向上右走"日"字
                ret.add(new Move(chessman, x, y, x+1, y-2, 0));
            }
            if(canMove(qizi, x, y, x+1, y+2)){              //向下右走"日"字
                ret.add(new Move(chessman, x, y, x+1, y+2, 0));
            }
            if(canMove(qizi, x, y, x-2, y-1)){              //向上左走"田"字
                ret.add(new Move(chessman, x, y, x-2, y-1, 0));
            }
            if(canMove(qizi, x, y, x-2, y+1)){              //向下左走"田"字
                ret.add(new Move(chessman, x, y, x-2, y+1, 0));
            }
            if(canMove(qizi, x, y, x+2, y-1)){              //向上右走"田"字
                ret.add(new Move(chessman, x, y, x+2, y-1, 0));
            }
            if(canMove(qizi, x, y, x+2, y+1)){              //向下右走"田"字
                ret.add(new Move(chessman, x, y, x+2, y+1, 0));
            }
            break;
        case 11:                                            //红炮
        case 4:                                             //黑炮
            for(int i=y+1; i<10; i++){                      //向下走时
                if(canMove(qizi, x, y, x, i)){              //当可以走时
                    ret.add(new Move(chessman, x, y, x, i, 0));
                }
            }
            for(int i=y-1; i>-1; i--){                      //向上走时
                if(canMove(qizi, x, y, x, i)){              //当可以走时
                    ret.add(new Move(chessman, x, y, x, i, 0));
                }
```

```
                                }
                                for(int j=x-1; j>-1; j--){              //向左走时
                                    if(canMove(qizi, x, y, j, y)){      //当可以走时
                                        ret.add(new Move(chessman, x, y, j, y, 0));
                                    }
                                }
                                for(int j=x+1; j<9; j++){               //向右走时
                                    if(canMove(qizi, x, y, j, y)){      //当可以走时
                                        ret.add(new Move(chessman, x, y, j, y, 0));
                                    }
                                }
                                break;
                            }
                        }
                    }
                }
            return ret.isEmpty() ? null : ret;              //当ret中没有走法时，返回空，有时返回ret
        }
        public boolean isSameSide(int moveChessID, int targetID){//判断两个子是否为同一阵营
            if(targetID == 0){                              // 当目标地为空地时
                return false;
            }
            if(moveChessID>7&&targetID>7){                  //当都为红色棋子时
                return true;
            }
            else if(moveChessID<=7&&targetID<=7){           //都为黑色棋子时
                return true;
            }
            else{                                           //其他情况
                return false;
            }
        }
```

运行程序，初始界面如图10-3所示。

图10-3 初始界面

参 考 文 献

[1] 扶松柏. Android 开发从入门到精通. 北京：北京希望电子出版社, 兵器工业出版社. 2012.

[2] Wei-Meng Lee. Android 编程入门经典. 何晨光, 李洪刚, 等译. 北京：清华大学出版社, 2012.

[3] 李佐彬. Android 开发入门与实战体验. 北京：机械工业出版社, 2011.

[4] 赵启明. Android 典型技术模块开发详解. 北京：中国铁道出版社, 2011.

[5] 杨明羽. Android 语法范例参考大全. 北京：电子工业出版社, 2012.

[6] 李刚. 疯狂 Android 讲义. 北京：电子工业出版社, 2011.